The Human Dimensions of Forest and Tree Health

Julie Urquhart · Mariella Marzano
Clive Potter
Editors

The Human Dimensions of Forest and Tree Health

Global Perspectives

Editors
Julie Urquhart
Countryside & Community Research Institute
University of Gloucestershire
Gloucester, UK

Clive Potter
Imperial College London
London, UK

Mariella Marzano
Forest Research
Edinburgh, UK

ISBN 978-3-319-76955-4   ISBN 978-3-319-76956-1 (eBook)
https://doi.org/10.1007/978-3-319-76956-1

Library of Congress Control Number: 2018934662

© The Editor(s) (if applicable) and The Author(s) 2018
Chapters 5 and 11 are licensed under the terms of the Creative Commons Attribution 4.0 International License (http://creativecommons.org/licenses/by/4.0/). For further details see license information in the chapters.

This work is subject to copyright. All rights are solely and exclusively licensed by the Publisher, whether the whole or part of the material is concerned, specifically the rights of translation, reprinting, reuse of illustrations, recitation, broadcasting, reproduction on microfilms or in any other physical way, and transmission or information storage and retrieval, electronic adaptation, computer software, or by similar or dissimilar methodology now known or hereafter developed.

The use of general descriptive names, registered names, trademarks, service marks, etc. in this publication does not imply, even in the absence of a specific statement, that such names are exempt from the relevant protective laws and regulations and therefore free for general use.

The publisher, the authors and the editors are safe to assume that the advice and information in this book are believed to be true and accurate at the date of publication. Neither the publisher nor the authors or the editors give a warranty, express or implied, with respect to the material contained herein or for any errors or omissions that may have been made. The publisher remains neutral with regard to jurisdictional claims in published maps and institutional affiliations.

Cover image: © Tommy Clark/Getty

Printed on acid-free paper

This Palgrave Macmillan imprint is published by the registered company Springer International Publishing AG part of Springer Nature
The registered company address is: Gewerbestrasse 11, 6330 Cham, Switzerland

Dr. Rosalind Hague died suddenly in November 2017, shortly after completing her valuable contribution to this book. She had held academic positions at both the University of Nottingham and Nottingham Trent University, and also had a long association with the University of Leicester where she studied both as an undergraduate and to complete her Ph.D. on feminism, autonomy and identity. Her co-authors in this volume feel privileged to have had the opportunity to work with her at the time they did and benefit from her commitment to intellectual rigour and her enthusiasm for novel analytic approaches, as well as her humour and compassion. She will be remembered by all for her passion for teaching and supporting students, her commitment to developing feminist ideas and environmentalism. Ros had much more to give and will be greatly missed.

Preface and Acknowledgements

There is growing scientific, policy maker and public concern surrounding the threats posed by the growing incidence of invasive pests and pathogens to tree health worldwide. The upsurge in new tree pest and disease outbreaks, many of them with the potential to radically reshape our native woodlands and forests, is closely linked to the significant growth in global trade and transportation in recent decades. Alongside this, alien pests and pathogens are able to establish at latitudes and altitudes that previously would have been unsuitable for them to flourish due to climate and environmental change. Growing evidence suggests that tree pests and diseases are likely to have profound consequences for the ecosystem services provided by trees and forests with resulting substantial impacts on human well-being. Dealing with such outbreaks, therefore, will often involve complex interactions between a wide range of actors including government agencies, tree growers, transporters, suppliers, consumers and the wider public in what we broadly define as the 'Human Dimensions of Forest and Tree Health'.

Human Dimensions of Forest and Tree Health-Global Perspectives were conceived in response to recognition of a need to better understand the diverse human dimensions of forest health. Addressing this requires approaches from a range of academic disciplines, such as economics, sociology, environmental psychology, cultural geography, environmental ethics, anthropology, health studies and history, alongside traditional technical risk assessment tools and natural science expertise. As a first step in this process, this book has been produced by researchers who are engaged with the International Union of Forest Research Organizations' (IUFRO) newly formed working party '7.03.15 - Social dimensions of forest health'. It also draws on the themes from a series of workshops hosted by the book's editors as part of the UK's Tree Health and Plant Biosecurity Initiative (THAPBI). As such, it provides a state-of-the-art collection of contributions from diverse social scientists and economists working across the globe and represents the first book-length synthesis of an important area of applied academic research. It brings together arguments, relevant theoretical frameworks and the latest empirical research findings to consider the specifically human dimensions of tree pests and diseases. A central theme of the book is to consider the contribution of the social sciences in better understanding the social, economic and environmental drivers and impacts of tree disease and pest outbreaks. Taken together, the chapters make theoretical, methodological and applied contributions to our understanding that will have relevance to a broad range of academic, policy and practitioner audiences.

From the outset, the editors wanted to provide a collection of work that represented different geographical, cultural and socio-political contexts. Alongside a core of contributions from UK researchers, chapters are included from scholars in New Zealand, the USA, Sweden, Romania and Turkey. Thus, its international scope allows for a comparative assessment of tree health social science research and hopefully highlights transferrable lessons for improving biosecurity in a range of socio-economic and spatial contexts. Given the relative infancy of social science attention to tree health issues, the number and geographical scope of researchers working in this field is currently limited. A clear gap in coverage is in developing countries in Africa, Asia and South America. We hope that this book will provide inspiration to social science scholars on these continents to engage with this important growing area of academic and applied interest. We firmly believe that the social sciences, and arts and humanities, have much to offer to improve our understanding of the complex interactions between humans and tree pests and diseases.

The editors would like to thank the contributors to this volume for their hard work in response to several rounds of revisions that were requested of them and for their timely response to other more technical matters, often at very short notice. Thanks also to the contributors for their role as peer reviewers, who graciously accepted requests to review chapters and provided constructive and useful feedback. This process, we feel, has strengthened the quality of the contributions immensely. It has been a pleasure to work with this team of very impressive academics to turn our idea for this book into a reality.

We also appreciate the support of Rachael Ballard, our publisher at Palgrave-Macmillan, for inviting us to work on this book project and for her assistance with the publication processes.

Gloucester, UK   Julie Urquhart
Edinburgh, UK   Mariella Marzano
London, UK   Clive Potter
January 2018

Contents

1 Introducing the Human Dimensions of Forest
   and Tree Health 1
   Julie Urquhart, Mariella Marzano and Clive Potter

2 English Tree Populations: Economics, Agency
   and the Problem of the "Natural" 21
   Tom Williamson, Gerry Barnes and Toby Pillatt

3 Local Knowledge on Tree Health in Forest
   Villages in Turkey 47
   Akile Gürsoy

4 Mountain Pine Beetles and Ecological Imaginaries:
   The Social Construction of Forest Insect Disturbance 77
   Elizabeth W. Prentice, Hua Qin and Courtney G. Flint

5 Indigenous Biosecurity: Māori Responses to Kauri
   Dieback and Myrtle Rust in Aotearoa New Zealand 109
   *Simon Lambert, Nick Waipara, Amanda Black,
   Melanie Mark-Shadbolt and Waitangi Wood*

6	**User-Generated Content: What Can the Forest Health Sector Learn?** *John Fellenor, Julie Barnett and Glyn Jones*	139
7	**The Social Amplification of Tree Health Risks: The Case of Ash Dieback Disease in the UK** *Julie Urquhart, Julie Barnett, John Fellenor, John Mumford, Clive Potter and Christopher P. Quine*	165
8	**Implementing Plant Health Regulations with Focus on Invasive Forest Pests and Pathogens: Examples from Swedish Forest Nurseries** *E. Carina H. Keskitalo, Caroline Strömberg, Maria Pettersson, Johanna Boberg, Maartje Klapwijk, Jonàs Oliva Palau and Jan Stenlid*	193
9	**The Economic Analysis of Plant Health and the Needs of Policy Makers** *Glyn Jones*	211
10	**Stated Willingness to Pay for Tree Health Protection: Perceptions and Realities** *Colin Price*	235
11	**The Use of Rubrics to Improve Integration and Engagement Between Biosecurity Agencies and Their Key Partners and Stakeholders: A Surveillance Example** *Will Allen, Andrea Grant, Lynsey Earl, Rory MacLellan, Nick Waipara, Melanie Mark-Shadbolt, Shaun Ogilvie, E. R. (Lisa) Langer and Mariella Marzano*	269

12	Enhancing Socio-technological Innovation for Tree Health Through Stakeholder Participation in Biosecurity Science *Mariella Marzano, Rehema M. White and Glyn Jones*	299
13	Gaming with Deadwood: How to Better Teach Forest Protection When Bugs Are Lurking Everywhere *Marian Drăgoi*	331
14	The Effects of Mountain Pine Beetle on Drinking Water Quality: Assessing Communication Strategies and Knowledge Levels in the Rocky Mountain Region *Katherine M. Mattor, Stuart P. Cottrell, Michael R. Czaja, John D. Stednick and Eric R. V. Dickenson*	355
15	Forest Collaborative Groups Engaged in Forest Health Issues in Eastern Oregon *Emily Jane Davis, Eric M. White, Meagan L. Nuss and Donald R. Ulrich*	383
16	Environmental Ethics of Forest Health: Alternative Stories of Asian Longhorn Beetle Management in the UK *Norman Dandy, Emily Porth and Ros Hague*	419
17	Towards a More-Than-Human Approach to Tree Health *Alison Dyke, Hilary Geoghegan and Annemarieke de Bruin*	445
18	Towards an Agenda for Social Science Contributions on the Human Dimensions of Forest Health *Mariella Marzano and Julie Urquhart*	471
Index		489

Editors and Contributors

About the Editors

Julie Urquhart is a Senior Research Fellow at the Countryside & Community Research Institute, University of Gloucestershire. She is an environmental social scientist with research interests in plant biosecurity, human–environment relationships, public engagement and ecosystem services. She is an interdisciplinary researcher with a focus on research that has applied policy relevance and the public understanding of environmental challenges. Recent work involved exploring public risk perceptions in the context of tree pests and diseases and investigating the social and cultural value of small-scale fisheries.

Mariella Marzano is a senior social researcher at Forest Research with a background in sustainable natural resource management across agriculture, fisheries and forestry sectors in the UK and internationally. Mariella leads interdisciplinary research on topics such as tree and plant biosecurity, adaptive forest management, human dimensions of species management and risk communication.

Clive Potter is Professor of environmental social science in the Centre for Environmental Policy, Imperial College London. His research interests include plant biosecurity, ecosystem services and stakeholder engagement.

Contributors

Will Allen is an independent systems scientist, action researcher and evaluator. He has more than 25 years of experience in sustainable development and natural resource management. Through his work, he seeks to bridge local, indigenous and organizational perspectives and help diverse groups work together to develop a shared understanding around goals, actions and indicators. He also developed and managed the learning for sustainability (LfS) website—http://learningforsustainability.net—as an international clearinghouse for online resources around collaboration and innovation processes.

Gerry Barnes, M.B.E. is a Research Fellow at the University of East Anglia and is the co-author, with Tom Williamson, of *Hedgerow History: Ecology, History and Landscape Character* (2006); *Veteran Trees in the Landscape* (2011); and *Rethinking Ancient Woodland* (2015).

Julie Barnett is a social and health psychologist with particular interest and expertise in risk appreciation, new forms of data, public engagement processes and policy development and the maintenance and change of behaviour. Her most recent projects have explored the public understandings of tree and plant pests and diseases, policy engagement with new forms of data and the challenges faced by people with food allergies and intolerances. Julie has been part of a range of interdisciplinary projects and has attracted funding from a variety of organisations, including the EPSRC, ESRC, the Environment Agency and the European Union.

Amanda Black is a senior lecturer in bio-protection and has expertise in soil chemistry and biochemistry. Her areas of research include the influence of trace elements in nutrient cycling, including transcription and activity of specific enzymes, abiotic influences on soil pathogen spread and virulence, and the protection of culturally significant species from biosecurity risks and threats. She is also a proponent of Indigenous perspectives, particularly Māori perspective in mainstream science.

Johanna Boberg is a researcher in Forest Pathology, Department of Forest Mycology and Plant Pathology, Swedish University of Agricultural Sciences, Uppsala, Sweden.

Stuart P. Cottrell is an Associate Professor in the Department of Human Dimensions of Natural Resources at CSU. He teaches courses in ecotourism, sustainable tourism development and tourism research. His research focus includes sustainable tourism development—linking tourism, nature conservation and protected areas; collaborative conservation and protected area management; and perceptions of landscape disturbance—implications for human dimensions of natural resources, recreation and tourism.

Michael R. Czaja earned a Ph.D. in Human Dimensions of Natural Resources from Colorado State University in 2012. He is a programme manager within the Colorado State University Office of the Executive Vice President and an Affiliate Faculty member, Department of Human Dimensions of Natural Resources, Colorado State University. His doctoral work in wildland fire social science examined public perceptions of prescribed fire as a management tool. Mike's academic interests include ecosystem services, the relationship between natural resources and conflict, and wildland firefighter safety. He is qualified as a wildland Firefighter Type 2 and is a Public Information Officer trainee.

Norman Dandy is an interdisciplinary environmental researcher and project leader with expertise in forest governance, human–wildlife interactions and plant biosecurity. His work, currently with the Plunkett Foundation, focuses on management situations that require cooperative or collaborative responses.

Emily Jane Davis is an Assistant Professor and Extension Specialist in the Department of Forest Ecosystems and Society and the Extension Service at Oregon State University. She is a social scientist who analyses natural resource collaboration, public lands policy and rural community development. She also provides technical assistance to a variety of collaborative efforts in Oregon. Davis has previously worked in British Columbia on topics including community forestry and forest health.

Annemarieke de Bruin is a researcher at the Stockholm Environment Institute at the Environment Department, University of York, UK. Her research focusses on the relationship between people and the environment within the context of agricultural water management, rural livelihoods, water governance, social and cultural values of trees, and oak tree health management. She has studied these relationships through the use of participatory methodologies, including participatory GIS, cultural probes and approaches based on social learning. In her tree-related work, she is currently interested in the role non-humans play in tree health management.

Eric R. V. Dickenson is a Project Manager for the Water Quality Research and Development Division of the Southern Nevada Water Authority. He has his Ph.D. from the Colorado School of Mines. His research is directed to identify surrogates and indicators for chemical contaminant removal in indirect potable reuse systems.

Marian Drăgoi is currently an Associate Professor at the University of Suceava, Romania. He graduated the Faculty of Forestry and Forest Engineering of Brasov in 1984. The first 13 years he worked in forest planning and forest economics. Since 1997, he has been teaching information technology, forest management and forestry economics within the faculty of forestry and has been working in different research and consultancy projects focussed on forest policy and environmental economics. In 2000, he was a Fulbright scholar at North Carolina State University where he studied and developed market-oriented instruments for sustainable forest management.

Alison Dyke is a political ecologist based at the Stockholm Environment Institute at the University of York, UK. Alison's work uses participatory approaches to explore the interface between nature and society in a relation to management and equity across species boundaries in a variety of areas. Recently, Alison's work has focussed on trees, biosecurity, plant health, citizen science and wild harvests. Alison's work also maintains the theme of nature society relations in work on sustainability behaviours in relation to housing and community.

Lynsey Earl is a biosecurity surveillance advisor for the Ministry for Primary Industries, New Zealand, and has a background in veterinary epidemiology. Her work includes enhancing engagement with agricultural communities, scientists and the general public to encourage early detection of new and emerging pests and diseases. Her other areas of work include biosecurity response, spatial analysis and data management.

John Fellenor is a research associate currently working on a collaborative project exploring the public understandings of tree and plant pests and diseases. A theme running throughout John's work, and which captures his broader interests, concerns ontology as a necessary basis for epistemological and methodological design. Of particular interest is how ontological frames can be used to ground and explore policy design, issues of public concern and their intersection with new forms of media.

Courtney G. Flint is a Professor of Natural Resource Sociology at Utah State University. Her research focus is on community and regional response to environmental disturbance as well as the integration of social and environmental science to address natural resource-related opportunities and vulnerabilities. She serves on the Board of Scientific Counselors for the US Environmental Protection Agency.

Hilary Geoghegan is Associate Professor in Human Geography in the Department of Geography Environmental Science, University of Reading. Hilary researches at the interface of the social and natural sciences with expertise in the social and cultural dimensions of tree health, citizen science and climate change. Her interests in the relationship between people and the material world have led to research projects on trees, wetland birds, landscape, technology and modernity architecture. Most recently, Hilary has held an ESRC Future Research Leader Award to explore enthusiasm in tree health citizen science and currently leads a social science project on oak tree health.

Andrea Grant is an applied social scientist with Scion, New Zealand, specializing in risk and resilience. She has a research background spanning multiple disciplines including science and technology studies, natural resource management and rural risk sociology. Her current areas of research include urban pest eradication, general biosecurity surveillance, wildfire volunteering and resilience, and climate change adaptation.

Akile Gürsoy is an anthropologist presently a faculty member in the Sociology Department of Beykent University, Istanbul. Her publications are on health anthropology, migration and ideology in science. She has been the principal investigator of an anthropological research project looking at social change in forest villages in Turkey (2014–2017). Her career includes working for UNICEF, lecturing and holding administrative positions in a number of Turkish state and foundation universities. She was visiting professor in the USA, acted as advisory editor to Pergamon Press. In Venezuela, she was elected secretary general of the International Forum for Social Sciences and Health.

Ros Hague was a senior lecturer in political theory at Nottingham Trent University and had research interests in feminist and environmental theory. Her work specifically looked at theories of autonomy, questions of identity and the human relation to nature. Her publications include *Autonomy and Identity: The Politics of Who We Are* (Routledge).

Glyn Jones is an environmental economist at Fera Science and the University of Newcastle. His work has focussed mostly on invasive species and agri-environment policy. Current research includes the assessment of new technologies to improve agricultural performance and to detect plant pests in commercial and natural environments.

E. Carina H. Keskitalo is a Professor of Political Science, Department of Geography and Economic History, Umeå University.

Maartje Klapwijk is a researcher in Insect Ecology, Department of Ecology, Swedish University of Agricultural Sciences, Uppsala, Sweden.

Simon Lambert is an Associate Professor in Indigenous Studies at the University of Saskatchewan, Canada. His research is in the area of Indigenous environmental management and planning with particular interest in indigenous communities and disaster risk reduction.

E. R. (Lisa) Langer is a Senior Scientist and Research Leader at Scion (New Zealand's Forest Research Institute) and has led Scion's social research since 2003. Her principal focus has been transdisciplinary research on topics such as biowaste management for the Centre for Integrated Biowaste Research, community resilience to wildfire challenges, fire danger communication and urban pest eradication. She has extensive experience in working with communities, and in particular Māori communities, using action research and transdisciplinary approaches to enable community voices to shape research inquiry and drive suitable communication and management.

Rory MacLellan has been working in biosecurity for over 15 years managing surveillance programmes for fruit flies and gypsy moth, coordinating and contributing to several plant health surveillance national and international working groups. Rory often acts a conduit between researchers and producers ensuring the research that is occurring will be beneficial in the field. Originally from the east coast of Canada growing up on a farm and working in biosecurity within various commodities, Rory brings an international perspective from a variety of experience.

Melanie Mark-Shadbolt is an indigenous environmental social scientist based at the Bio-Protection Research Centre at Lincoln University, New Zealand. She is also the Māori Manager for New Zealand's Biological Heritage National Science Challenge and Kaiwhakahaere for Te Tira Whakamātaki, the Māori Biosecurity Network. Her areas of research include human dimensions of environmental health, specifically how indigenous people view biosecurity and how they participate in the management and protection of their environment and culturally significant species. With Te Tira Whakamātaki, she has won two biosecurity awards: the New Zealand Biosecurity Institute's inaugural Dave Galloway award for Innovation (2016) and the New Zealand Government's Māori Biosecurity Award (2017).

Katherine M. Mattor is a research scientist and instructor in the Forest and Rangeland Stewardship Department and an affiliate faculty member in the Department of Human Dimensions of Natural Resources at Colorado State University. Her research focuses on the evaluation of forest policies and programmes promoting resilient forests and communities, the communication and implementation of science, and the role of collaboration and community-based forestry programmes in sustainable forest management. She holds a Ph.D. and M.S. in forest science, policy from Colorado State University and a B.S. in environmental studies from the State University of NY, College of Environmental Science and Forestry.

John Mumford is an entomologist who has worked extensively on plant health risk assessment, regulation and policy in Europe. Much of his work on the implementation of field pest management has involved encouraging uptake of control measures by large numbers of individuals to achieve greater efficiency. He leads work packages on several EC projects which are currently addressing plant health risks that pose a threat to Europe from the great expansion in trade from China in the past decade.

Meagan L. Nuss has a diverse history across the field of forest management, with a background that spans researching the efficacy of collaborative efforts on national forests, providing educational resources to non-industrial private forestland owners, identifying ecosystem services on industrial agroforestry properties and documenting biomass stakeholders' willingness to develop harvesting guidelines. Most recently Meagan manages project development for a leading biomass energy design firm that specializes in community-scale heating, combined heat-and-power and district energy applications. She holds a master's degree in forest ecosystems and society from Oregon State University and a bachelor's degree in environmental studies from Lewis & Clark College.

Shaun Ogilvie is an indigenous New Zealander, with tribal affiliations to Te Arawa and Ngati Awa. He did his Ph.D. in marine biology and has been working in the area of environmental science for more than 20 years. Shaun has a professional and personal interest in research projects that are underpinned by indigenous worldviews and aspirations for sound environmental outcomes.

Jonàs Oliva Palau is a forest pathologist and Associate Professor at the Department of Forest Mycology and Plant Pathology, Swedish university of Agricultural Sciences, Uppsala, Sweden.

Maria Pettersson is an Associate Professor in Environmental and Natural Resources Law, Department of Business Administration, Technology and Social Sciences, Luleå University of Technology, Luleå, Sweden.

Toby Pillatt was a Research Fellow at the University of East Anglia and is now a researcher at the Department of Archaeology at the University of Leeds. He has particular interests in the archaeology of bee-keeping and, more generally, in human relationships with the natural world. He is co-author, with Tom Williamson and Gerry Barnes, of *Trees in England: Management and Disease since 1600* (2017).

Emily Porth is a cultural anthropologist and ecofeminist with expertise in stakeholder engagement, public outreach and social inclusion. She has long-standing interests in human–animal relationships, trees and storytelling for social change.

Elizabeth W. Prentice is a Ph.D. student in Rural Sociology and Sustainable Development in the Division of Applied Social Sciences at the University of Missouri, Columbia. Her research interests include political ecology, environmental sociology, environmental discourses and local environmental knowledge with special attention to the intersection of environmental and social justice issues. Her dissertation research focuses on the shifting economic context of the Mountain Pine Beetle outbreak in Northern Colorado.

Colin Price has M.A. and D.Phil. degrees from Oxford University. He has taught urban and regional economics in Oxford Brookes University, land use economics in Oxford University and environmental and forestry economics in Bangor University, where he was professor. He left to become a freelance academic. He has had three books published: *Landscape Economics* (1978), Macmillan; *The Theory and Application of Forest Economics* (1989), Blackwell; and *Time, Discounting and Value* (1993), Blackwell. He has authored over 200 journal papers and chapters. A second edition of *Landscape Economics* is recently published.

Hua Qin is an Assistant Professor of environmental and natural resource sociology at the University of Missouri, Columbia. His research interests involve five distinct but interrelated areas: population (migration) and the environment; vulnerability and adaptation to environmental change; community, natural resources and sustainability; applied research methods and data practices; and the sociology of environmental and resource sociology. He teaches undergraduate courses focused on the relationships among people, society and the natural world, as well as several foundational seminars on community, sustainable development and research methodology for the graduate programmes of the Division of Applied Social Sciences.

Christopher P. Quine is Head of Centre for Ecosystems, Society and Biosecurity and Head of Station at the Forest Research's Northern Research Station (NRS); he is a member of the Forest Research Executive Board, a Fellow of the Institute of Chartered Foresters, a trustee of the Scottish Forestry Trust, and a member of Scottish Natural Heritage's Expert Panel to the Scientific Advisory Committee.

John D. Stednick is a retired Professor of the Watershed Science Program, Department of Forest, Rangeland, and Watershed Stewardship at Colorado State University. Education includes a B.Sc. in Forest Sciences and Ph.D. in Forest Resources from University of Washington. Current interests include land use and water quality, forest hydrology, water quality hydrology, biogeochemistry, water chemistry, soil chemistry, hydrometry, watershed management, risk assessment, watershed analysis, environmental impact assessment and technology transfer.

Jan Stenlid is a Professor of forest pathology at the Swedish University of Agricultural Sciences, Uppsala, with expertise in plant protection and animal health.

Caroline Strömberg a researcher in the Department of Business Administration, Technology and Social Sciences, Luleå University of Technology, Luleå, Sweden.

Donald R. Ulrich began working with natural resource collaboratives as a Peace Corps Volunteer in the Philippines where he helped subsistence fishing communities organize and engage with the local government in sustainable planning. He then earned his M.S. from Oregon State's College of Forestry where he applied his Peace Corps experience, social science background and interest in natural resource management. After graduating, Donald pursued his passion for community development around natural resources in the forests of Oregon, ultimately looking at forest governance across the region with Dr. Emily Jane Davis.

Nick Waipara is a biosecurity scientist with a background specializing in plant pathology, conservation ecology and invasive species management. He also works in New Zealand's Biological Heritage National Science Challenge. His work includes developing partnerships with Tangata Whenua (Māori), agencies, industry and communities to improve understanding and engagement with the increasing range of social, cultural, environmental and economic issues arising in biosecurity-based programmes.

Eric M. White is a research social scientist with the Pacific Northwest Research Station in Olympia, WA. His current research is focused on the social and economic effects from a variety of uses on public and private lands. This includes recreation on public lands and the resulting economic effects on local communities. White earned his bachelor's and master's degree in forestry from Southern Illinois University, and a doctorate in forestry from Michigan State University.

Rehema M. White is a sustainability generalist developing scholarship and pursuing sustainable development practice from the University of St. Andrews, Scotland. She works on governance of natural resources (including multi-level collaboration, Global North–South links, roles of community); knowledge and sustainable development (including learning for sustainability and research modes); and sustainability in practice (including tree health).

Tom Williamson is Professor of Landscape History at the University of East Anglia and has written widely on landscape archaeology, agricultural history and the history of landscape design. His books include *Rabbits, Warrens and Archaeology* (2007) and *An Environmental History of Wildlife in England* (2013).

Waitangi Wood is the Tangata Whenua Roopu representative on Governance of the Kauri Dieback Programme, member of the Māori Biosecurity Focus Group for the Ministry for Primary Industries and is working with research groups to develop cultural safety agreements for collaborative research including protocols for Kauri and other germplasm collections.

List of Figures

Introducing the Human Dimensions of Forest and Tree Health

Fig. 1   The cumulative numbers of new tree pathogens (○) and insect pests (□) identified in the UK shown over time since 1900. The total accumulated number of pathogens and pests are also shown (▲) (Freer-Smith and Webber 2017)   2

Fig. 2   Subject area distribution for 25,663 journal articles identified as relating to tree pests and diseases. Search terms used [tree OR forest AND pest OR disease OR pathogen], medicine, physics and related subject areas were excluded (*Source* Elsevier Scopus, 29 December 2017)   4

Fig. 3   The top 8 journals for publications on tree and forest health (*Source* Elsevier Scopus)   4

Local Knowledge on Tree Health in Forest Villages in Turkey

Fig. 1   Map of Turkey with 12 NUTS districts   57
Fig. 2   Forest village in TR7 Central Anatolia (Erdoğan 2017)   58
Fig. 3   Forest village in TR9 Eastern Black Sea region (Türkeli 2014)   58
Fig. 4   *Thaumetopea pityocampa (Çam kese böceği)* found on all types of pine trees across Turkey (Gürsoy, TR8 Western Black Sea region, September 2015)   61

Fig. 5	Watch tents and sound-making bottles placed in fields to protect the crops and the fruit trees from bears and wild boar, Western Turkey, August (Kaplan 2015)	63
Fig. 6	Watch tents and sound-making bottles placed in fields to protect the crops and the fruit trees from bears and wild boar, Western Turkey, August (Kaplan 2015)	64
Fig. 7	Pear tree damaged by bear, TR4 Eastern Marmara region (Kaplan 2015)	65
Fig. 8	Branches of spruce trees believed to be infected due to humidity, TR9 Eastern Black Sea region (Türkeli 2014)	68

Indigenous Biosecurity: Māori Responses to Kauri Dieback and Myrtle Rust in Aotearoa New Zealand

Fig. 1	Dead mature kauri tree >500 years old. Commonly referred to as 'stag head' (*Source* Kauri Dieback Programme, www.kauridieback.co.nz)	114
Fig. 2	The distribution of Kauri Dieback disease in New Zealand 2017 (*Source* Kauri Dieback Programme, www.kauridieback.co.nz/more/where-has-it-been-detected/)	115
Fig. 3	Signage designed to raise public awareness of Kauri Dieback and the key hygiene measures in place to help reduce the spread of the pathogen (*Source* Kauri Dieback Programme, www.kauridieback.co.nz)	117
Fig. 4	Tāne Mahuta, Waipoua Forest, Northland, New Zealand (*Photograph source* Alastair Jamieson, Wild Earth Media, and Auckland Council)	119
Fig. 5	Governance and management structure for the Kauri Dieback Programme (*Source* Kauri Dieback Programme, www.kauridieback.co.nz)	120

The Social Amplification of Tree Health Risks: The Case of Ash Dieback Disease in the UK

Fig. 1	The social amplification of risk framework (from Kasperson 2012a)	169
Fig. 2	Conceptualisation of response to ash dieback outbreak using the SARF	172

Fig. 3	Methods adopted to explore the interactions between expert assessment, media attention and public concern about ash dieback in the UK	174
Fig. 4	Stated concerns about impacts of tree pests and diseases	183
Fig. 5	Narratives associated with beliefs about pathways of introduction for ash dieback	184
Fig. 6	Emails and calls to Defra and the Forestry Commission helpline during the ash dieback crisis in 2012	185

The Economic Analysis of Plant Health and the Needs of Policy Makers

Fig. 1	Expected impacts of invasive species over time	216
Fig. 2	Pest progression, willingness to respond and cost-effectiveness	226

Stated Willingness to Pay for Tree Health Protection: Perceptions and Realities

Fig. 1	Revealed WTP for landscape quality	260

Enhancing Socio-technological Innovation for Tree Health Through Stakeholder Participation in Biosecurity Science

Fig. 1	A schematic representation of the stakeholder map for the early detection system for tree health	309

Gaming with Deadwood: How to Better Teach Forest Protection When Bugs Are Lurking Everywhere

Fig. 1	Dynamics of the main types of fellings carried out in Romania since 1990 (*Source* Romanian national yearbooks)	336
Fig. 2	Location of Rarau-Giumalau natural reserve (*Source* http://natura2000.eea.europa.eu)	343
Fig. 3	Precise field-trip location in Rarau-Giumalau natural reserve (*Source* the management plan)	344
Fig. 4	Learning progress by working hours	345
Fig. 5	Penalties recorded by the two teams	345
Fig. 6	Net cumulative incomes per hour	346

The Effects of Mountain Pine Beetle on Drinking Water Quality: Assessing Communication Strategies and Knowledge Levels in the Rocky Mountain Region

Fig. 1	Map of mountain pine beetle outbreak in northern Colorado and southern Wyoming, USA	362

Forest Collaborative Groups Engaged in Forest Health Issues in Eastern Oregon

Fig. 1	Federal forest collaborative groups in eastern Oregon as of 2016	391
Fig. 2	Respondents' reported satisfaction with aspects of their collaborative groups; response for individual aspects ranged from 74 to 86 respondents (Overall survey n was 97)	407
Fig. 3	Respondents' reported satisfaction with aspects of their collaborative groups; response for individual aspects ranged from 76 to 82 respondents (Overall survey n was 97)	408

Towards a More-Than-Human Approach to Tree Health

Fig. 1	A turned bowl showing fungal spalting	460
Fig. 2	The PuRpOsE team inspects a tree with symptoms of AOD	461

List of Tables

Local Knowledge on Tree Health in Forest Villages in Turkey
Table 1 Forest area, demographic and economic variables of selected countries (*Source* Compiled from *2016 The Little Green Data Book*, World Bank Group, The World Bank, Washington, 2016)   50
Table 2 Locations and categories of trees   56

Mountain Pine Beetles and Ecological Imaginaries: The Social Construction of Forest Insect Disturbance
Table 1 Differences between newer- and longer-term residents in sociodemographic and perceptional variables   100

User-Generated Content: What Can the Forest Health Sector Learn?
Table 1 Organisation of categories into overarching areas   145

The Economic Analysis of Plant Health and the Needs of Policy Makers
Table 1 Approaches to identify cost-effective responses (adapted from Epanchin-Niell 2017)   219

The Use of Rubrics to Improve Integration and Engagement Between Biosecurity Agencies and Their Key Partners and Stakeholders: A Surveillance Example

Table 1	Seeing communication and engagement as a continuum (adapted from Morphy, n.d.)	274
Table 2	An abbreviated summary of the final rubric developed by workshop participants (July–August 2016) to evaluate a general surveillance programme, modified to be applicable to a wider range of stakeholders	281
Table 3	Assessment of the myrtle rust surveillance system in New Zealand (undertaken before its 2017 identification in the country)	286

Enhancing Socio-technological Innovation for Tree Health Through Stakeholder Participation in Biosecurity Science

Table 1	Data sources	307

The Effects of Mountain Pine Beetle on Drinking Water Quality: Assessing Communication Strategies and Knowledge Levels in the Rocky Mountain Region

Table 1	Survey objectives and question	364
Table 2	Topics identified by reported knowledge level	367
Table 3	Primary responses to survey questions by respondent type	369

Forest Collaborative Groups Engaged in Forest Health Issues in Eastern Oregon

Table 1	Collaborative group survey respondent affiliations	392
Table 2	Sectoral survey respondent affiliations	393
Table 3	Organizational characteristics of eastern Oregon's forest collaborative groups	398
Table 4	Overview of missions and forest management issues of focus	403

Towards a More-Than-Human Approach to Tree Health

Table 1	Outline of the 'In conversation with' workshop	457
Table 2	Principles of engagement (underlined text was added at the end of the event)	463

1

Introducing the Human Dimensions of Forest and Tree Health

Julie Urquhart, Mariella Marzano and Clive Potter

1 Introduction

Attending to plant health issues is, of course, not new—farmers and growers have had to deal with crop losses caused by diseases or insect pests for centuries (MacLeod et al. 2010). As Williamson et al. (Chapter 2 in this volume) assert, trees have had to contend with a whole range of 'pest' threats throughout history, including insects, bacteria, viruses and fungi (Boyd et al. 2013) and various mammal pests such as deer, squirrels, rabbits, beavers and elephants (Gill 1992; Chafota and Owen-Smith 2009;

J. Urquhart (✉)
Countryside & Community Research Institute,
University of Gloucestershire, Oxstalls Campus,
Gloucester, UK

M. Marzano
Forest Research, Edinburgh, UK

C. Potter
Centre for Environmental Policy, Imperial College London,
South Kensington Campus, London, UK

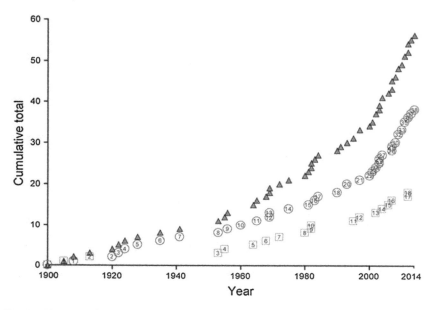

Fig. 1 The cumulative numbers of new tree pathogens (○) and insect pests (□) identified in the UK shown over time since 1900. The total accumulated number of pathogens and pests are also shown (▲) (Freer-Smith and Webber 2017)

Raum et al. under review). However, while dealing with tree pests may not be an entirely new phenomenon, there is growing evidence that the incidence of invasive tree and pathogen introductions is increasing (see Fig. 1), a trend closely linked to a significant upsurge in global trade and transportation in recent decades (Potter and Urquhart 2017; Brasier 2008). Climate change is also likely to provide new environments in which alien pests and pathogens are able to establish as well as potentially altering the behaviour of native pests.

The impacts of these invasive species will have profound consequences for the ecosystem services provided by trees and forests. The Millennium Ecosystem Assessment identified pests as one of the main drivers of biodiversity loss and ecosystem service change globally, alongside land use changes, unsustainable use and exploitation of natural resources, global climate change and pollution (MEA 2005). Boyd et al. (2013) and others suggest a wide range of impacts, from tree pests and diseases on biodiversity, carbon sequestration, timber and wood fuel, flood alleviation, air quality, landscape change, recreation, health and cultural values (Boyd et al. 2013; Freer-Smith and Webber 2017; Potter and Urquhart 2017; Marzano et al. 2017), all of which will have implications for human well-being. Humans are also implicated in the increased spread of tree pests and pathogens—much of the recent literature highlights the increase in the global trade in commodities (including plant material, timber and wood products, wood fuel and wood packaging) and human movement as key pathways for new introductions (Brasier 2008; Freer-Smith and Webber 2017; Potter and Urquhart 2017).

Given the potential for substantial human well-being impacts from the effects of invasive species on trees and forest ecosystems and the role of humans in perpetuating their spread, it is surprising that until recently what we broadly define in this volume as 'the human dimensions of forest and tree health' has received very little scholarly attention. A search on Elsevier Scopus in December 2017 confirms that around 80% of the 25,663 journal articles on tree pests and diseases were published in agricultural and biological sciences publications, almost 30% in environmental science publications and 19% in biochemistry, genetics and molecular biology publications (see Fig. 2). Despite growing research interest in tree health over the past two decades, with 87% of the journal articles published during this period, much of the existing academic expertise on tree health stems from the natural sciences, notably in the fields of plant pathology, entomology and ecology (see Fig. 3). Less than 2% of published outputs can be classified as social science, with 40% of these published in the last 5 years and less than 0.5% from journals in the arts and humanities or economics. This is perhaps unsurprising, given policy imperatives for decision-making informed by scientific evidence and the need to justify the governance mechanisms currently being adopted. Yet, as Marzano et al. (2017) suggest, there is growing recognition from governments and practitioners worldwide that the social sciences and humanities have valuable contributions to make to addressing tree health issues. For instance, a deeper understanding of the social and human dimensions is influential in determining the success of control or eradication programmes (Crowley et al. 2017) or how tree pest risks are communicated to lay publics (Urquhart et al. 2017a, b).

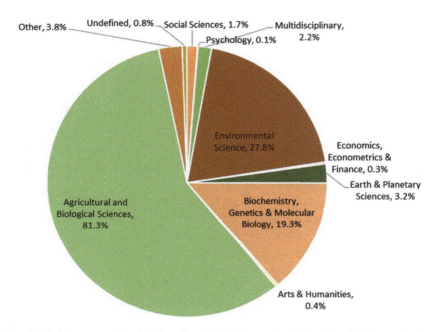

Fig. 2 Subject area distribution for 25,663 journal articles identified as relating to tree pests and diseases. Search terms used [tree OR forest AND pest OR disease OR pathogen], medicine, physics and related subject areas were excluded. *Note* Some articles may appear in more than one subject area; thus, the total is over 100% (*Source* Elsevier Scopus, 29 December 2017)

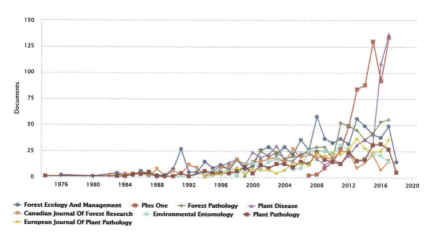

Fig. 3 The top 8 journals for publications on tree and forest health (*Source* Elsevier Scopus)

2   An Emerging Research Area

What is clear is that while there is emerging scholarly interest, finding a home for research on the human dimensions of tree and forest health can be difficult and to date such work has been published in an eclectic mix of journals, including *Landscape Ecology, Forest Policy and Economics* and *Environmental Science and Policy*. An aim of this book is to bring together a wide body of work from the social sciences and humanities in order to provide a point of reference for established and newly interested researchers in the area. This is the first book-length synthesis of the human dimensions of forest and tree health, and it is therefore appropriate to explain the background to the book's development and some underpinning discussions that are important for its framing.

In the first instance, *The Human Dimensions of Forest and Tree Health: Global Perspectives* is the product of the International Union of Forest Research Organizations' (IUFRO) newly formed working party '7.03.15 - Social dimensions of forest health', but has also been stimulated by a series of workshops hosted by the book's editors as part of the UK's Tree Health and Plant Biosecurity Initiative (THAPBI). The IUFRO working party, chaired by social anthropologist Dr. Mariella Marzano, represents a subgroup of Division 7 Forest Health of the IUFRO network (https://www.iufro.org/science/divisions/division-7/70000/70300/70315/). Division 7 includes research on the physiological and genetic interactions between trees and harmful biotic impacts, including invasive pests and diseases, and the impacts of air pollution on forest ecosystems. The 7.03.15 working party was formed in 2015 to explicitly recognise that tree health issues have a social dimension and that studying the impacts and implications of tree pests and diseases cannot be fully understood without consideration of their associated human dimensions. At the same time, Dr. Julie Urquhart and Professor Clive Potter, social scientists based at Imperial College London, hosted a 'Social Science for Tree Health and Plant Biosecurity Workshop' in September 2015 and (together with Dr. Marzano and Dr. Hilary Geoghegan) a 'Human Dimensions of Tree Health and Plant Biosecurity International Workshop' in August 2016. The workshops sought to bring together those social scientists working on a number of interdisciplinary tree health projects funded as part of the THAPBI programme with international researchers engaged in complementary activities in other countries. The aim was to provide a forum for sharing theoretical, applied and methodological insights from projects undertaken in a range of geographical contexts.

It is in this collaborative spirit that this book seeks to bring together arguments, relevant theoretical frameworks and the latest empirical research findings to consider the specifically human dimensions of the problem. A central theme of the book is to consider the role that social science can play in better understanding the social, economic and environmental impacts of such tree disease and pest outbreaks. This introductory chapter begins by setting the theoretical context for the book's central argument that in order to fully understand and manage the rise in new tree pest and disease outbreaks, traditional technical risk assessment tools and methodologies need to be integrated with approaches from a range of academic disciplines, such as economics, sociology, environmental psychology, cultural geography, environmental ethics, anthropology and history that consider the human dimensions of outbreaks, alongside the ecological and biological. We need to develop a deeper understanding of the complexity of macro-level governance dimensions, such as policy, regulatory and market forces that provide the context within which biosecurity and trade arrangements operate and that speak to the practical concerns of policy makers in the real world. With that in mind, quantitative assessments and economic approaches are also vital tools for evidence-based decision-making that can help inform global, regional or national assessments, including international trade, phytosanitary and biosecurity agreements.

At the same time, on a more micro-level, outbreaks are often geographically and contextually specific, with impacts that are not reducible to quantitative measures, but that are more amenable to qualitative methods that seek to understand how individuals or communities experience and respond to outbreaks in specific locations and in particular sociocultural landscapes. Here, understanding human–environment relationships and interactions is important, including cultural and indigenous responses to new pest or disease invasions. For those tasked with dealing with such outbreaks, bridging macro-level concerns with localised experiences or societal responses to new or emerging risks is a challenge. With this in mind, work conducted in recent years by the editors and others has considered how risks associated with tree pest and disease outbreaks are communicated, along with how publics and stakeholders can be engaged and enrolled in mitigating the impacts of outbreaks or reducing the risk of new introductions.

In the spirit of trying to capture the diverse human nature of tree and forest health issues, the book purposely includes contributions from a range of researchers working across the globe. Currently, the community of researchers actively engaged in social science research in tree health is small and is largely represented by the authors in this volume. With this in mind, a key aim of this book is to encourage researchers from across the social sciences, and arts and humanities, to engage in tree health research. It is our contention that scholars working across a range of disciplines in these fields can make valuable theoretical and empirical contributions to this growing research agenda. Methodologically speaking, the chapter authors will also explore how mixed qualitative and quantitative approaches can be used to highlight the challenges of assessing the human dimensions of tree and forest health. Our aim has been to put together a state-of-the-art collection of developing work in this field in order to lay the foundation for future work on the human dimension of tree and forest health.

Before introducing each of the contributions in this volume, we first set out three key challenges facing those tasked with undertaking research on the human dimensions of tree health.

3 Challenge 1: Developing Research Capacity and the Importance of Research Funding

Clearly, in order for researchers to engage in research, there has to be suitable and sustainable sources of funding. As an example, we describe the current funding landscape in the UK to illustrate the role that funders can play in developing strategic research capacity. Much of the social science research on tree health issues in the UK in recent years has been as a result of funding made available via several government-sponsored programmes. The first, the Tree Health and Plant Biosecurity Initiative, part of the Living with Environmental Change (LWEC) partnership, consists of a consortium of funders from across government agencies and the UK research councils established to develop a joint strategic research initiative to develop interdisciplinary research capacity to address tree health in the UK. Over three phases between 2013 and 2018, the initiative has funded 9 projects, a number having social science components, and one which specifically focused on social science: Understanding public risk concerns in relation to tree health (UNPICK), led by Professor Clive Potter at Imperial College London. The second is the Future Proofing Plant Health project, a programme of government research and development funded by the Department for Environment, Food and Rural Affairs (Defra) and involving Forest Research, Fera, Royal Botanic Gardens Kew, Natural England and the Joint Nature Conservation Committee (JNCC), running from 2014 to 2020. The programme involves primarily not only natural but also social science themes.

As a result of these programmes, social science research capacity has been developed in the UK, both through expanding the research portfolio of established academics and early career researchers, and the training of new researchers through Ph.D. studentships. However, it is imperative that funding continues to be made available for research in this field beyond the lifetime of these programmes if the research capacity that they have fostered is not to be lost. Clearly, budgets for research funding are finite, and priorities need to be determined, and at first glance, it may seem prudent to give precedence to those projects that offer more immediate or short-term results. Given that tree health issues are often slow to emerge and develop, it is important that longer-term funding programmes be secured to ensure the maintenance of research capacity into the future, given the prognosis for continued growth in the threat from invasive tree pests and diseases.

4   Challenge 2: Applying Social Science Research at the Policy–Practice Interface

Across tree health research, one of the key challenges is providing robust scientific evidence to inform policy and decision-making. This is not particular to social science research, and the importance of gathering scientific intelligence on new and emerging pests and diseases is crucial as part of pest risk assessments and horizon scanning. Delays or uncertainties in the scientific evidence can have severe consequences for how outbreaks are managed. For example, it is widely recognised that delays in correctly identifying the pathogen responsible for ash dieback, *Hymenoscyphus fraxineus*, hindered early regulatory action to protect the UK from infection (Urquhart et al. 2017a). For social scientists tasked with predicting sociocultural or other impacts of pest and disease outbreaks, there are particular challenges associated with predicting societal attitudes towards tree health risks that may lack relevant historical analogues, especially given the evolving nature of public attention to and concern about environmental issues.

Social science contributions to tree health research are strongly driven by the evidence needs of policy makers. Often there can be uncertainty over the credibility of qualitative research from a policy and scientific perspective, with a tacit belief in the robustness of quantitative methods over qualitative methods. While this is changing, and decision-makers are recognising the important role that qualitative methods can play in informing public policy, qualitative researchers may need to develop skills in how to 'translate' findings into a language that policy makers understand. This is likely to involve dialogue, trust and co-design of research programmes and outputs. Policy makers, too, are likely to need to reflect on their own biases and be open to new learning from disciplines that they have not engaged with previously.

5 Challenge 3: Identifying Research Priorities

This chapter has begun to make the case for developing research capacity around the human dimensions of tree and forest health. It highlights the need to bring together social, natural and economic expertise through interdisciplinary research in areas such as the critical analysis of environmental and biosecurity governance, understanding stakeholder and public perceptions and communication and engagement around tree health issues. The IUFRO working group organises itself around four broad themes, which provide a useful starting point for identifying research priorities that help to address fundamental questions around how to prevent new introductions and more effectively manage outbreaks, along with improving societal involvement in biosecurity through awareness-raising and changes in current practices. We outline these themes below, together with the relevant disciplines that may be able to contribute in these areas, but suggest that it is imperative that research priorities are identified through close collaboration between researchers and decision-makers, policy makers and practitioners.

5.1 Governance

Governance can be both formal, through legislation, policy and regulation, but also informal, such as voluntary codes of conduct. It can occur at the macro-level involving international organisations and agreements (e.g. WTO, SPS, CBD), regional (such as the European plant health regime), national or, at the micro-scale, it may involve local organisations and communities. Key questions include, for instance, how international trade regulation and national legislation across multiple sectors may influence what can be regulated and what mechanisms can be used to reduce the risk of invasive pest introductions. Further, how does governance at various scales influence biosecurity practice and the diverse ways in which key pathways (e.g. live plants) are currently regulated and implemented (e.g. through inspections and management). Researchers in policy studies, political ecology, political science and legal studies may offer contributions on the governance of tree health and biosecurity.

5.2 Stakeholder and Public Values, Perceptions and Behaviours

The tree health stakeholder landscape is notoriously complex and dynamic due to the range of public- and private-sector interests and influences, the number of potential pathways, the varying levels of awareness of tree health issues and the inherent uncertainties around which biosecurity practices and behaviours are most effective (Marzano et al. 2015). Values and perceptions are likely to play a key role in determining the actions and behaviours of stakeholders and publics. For example, practices that reduce the risk of introduction are often resisted due to a denial of the existence of a threat or beliefs that it is someone else's responsibility or due to a lack of trust in those tasked with undertaking control programmes. Understanding the complex nature of values, attitudes, perceptions and behaviours should involve researchers from across the social sciences and humanities, such as human and cultural geographers, environmental sociologists, social and behavioural psychologists, social anthropologists, historians and creative arts scholars.

5.3 Economic Values and Impacts

Economic analyses can provide useful information about the costs of improved biosecurity and the benefits of avoiding damage to forest ecosystems (Aukema et al. 2011; Holmes et al. 2009). Decisions about forest protection include the development and implementation of policies and practices to prevent the introduction of non-native pests, detect newly established pests and manage successful invasions. Economic analysis of biosecurity measures is challenged by a lack of knowledge about factors contributing to successful invasions, the efficacy of practices to detect and control pest outbreaks and the type and monetary value of damages caused by forest pests (Holmes et al. 2014). Important themes include the development of economic approaches to analysing biosecurity measures under conditions of risk and uncertainty, analysis of the spatial and temporal resolution needed to capture key elements of the joint ecological-economic system and the incentives required to motivate tree and forest health stakeholders to take protective actions. Researchers engaged in this theme are likely to be forest economists, environmental economists and natural resource economists, but there is also a need for these individuals to work closely with other social scientists in order to achieve an integrated understanding of drivers and impacts.

5.4 Risk Communication and Engagement

Action across all stages in the supply chain requires different sorts of engagement with different stakeholders and at different times. While public authorities (national and international) and other institutions often communicate about tree health issues, little attention is paid to the efficacy of such messages and whether they achieved the desired outcomes, such as information sharing or influencing behaviour change. Thus, we need a deeper understanding of how the perceptions, values and attitudes of stakeholders and publics, as outlined in Section 5.2 above, can help to inform engagement activities and communication strategies, and there may be useful lessons to learn from other sectors such as climate change, agriculture, human or animal health. Behavioural scientists, environmental sociologists, human geographers, social anthropologists, social psychologists as well as environmental ethics are likely to provide contributions. In addition, public relations and communications experts, including arts-based scholars, may offer useful insights, particularly in the design of awareness-raising or engagement programmes.

6 Contribution to the Volume

The chapters that follow were written in response to a call to members of the IUFRO working party and contributors to the two workshops outlined above. Authors were encouraged to contribute empirical or theoretical chapters that would be of interest to the research community, policy makers and professionals working in the tree health sector, as well as interested communities and citizens. As is clear from the table of contents, about half of the chapters are written by researchers from the UK, which largely reflects the geographical location of the books' editors, but also the enhanced research funding opportunities that have become available in the UK for interdisciplinary tree health research. As outlined earlier, substantial funding was made available by the UK Government and research councils from 2013 to 2018, largely in response to the ash dieback outbreak in the country in late 2012 and recognition of the increased risk from invasive tree pests and diseases and the need for empirical evidence to inform future biosecurity decision-making. However, the book's editors were also particularly keen to encourage contributions from across the globe and are delighted to include research from New Zealand (Lambert et al. and Allen et al.), the USA (Davis et al., Prentice et al., and Mattor et al.), Sweden (Keskitalo et al.), Romania (Dragoi) and Turkey (Gürsoy).

An indication of the diverse ways in which the authors have taken up the call for contributions is reflected in the considerable variety of theoretical, methodological and applied insights. The chapters also vary in the degree to which they focus on macro-level issues, such as broad consideration of the suite of economic assessment tools or regulatory mechanisms, conceptual frameworks for understanding human-environment relationships and specific attention on particular pests or diseases or geographical contexts.

While all of the authors make significant contributions to developing an understanding of the human dimensions of tree and forest health, Chapters 2–7 are primarily concerned with understanding how tree health issues are socially constructed, drawing on the historical, cultural, social and situated contexts of tree pest outbreaks. Chapter 2, by Williamson et al., provides a useful historical perspective on how the contemporary landscape in England has been shaped by past management decisions based on economic or social trends. The authors argue that although global trade and climate change are leading causes of the growing numbers of tree pests and diseases, the legacy of past management decisions has resulted in a treescape that is far from 'natural', dominated by just a few species and a fairly homogenous age structure, rendering woods and forests more vulnerable to new pests and diseases. Through an enlightening journey through historical farming and forestry texts from the sixteenth century onwards, the authors demonstrate that while poor health in trees is not a new phenomenon, the large-scale epidemics of the twentieth century were not commonplace in earlier times. The chapter concludes by suggesting that lessons can be learnt from the past that can inform future management and policy, especially in terms of developing a realistic perception of 'naturalness' for modern-day treescapes.

Williamson et al.'s historical account from England is followed by a complementary perspective from Turkey in Chapter 3. Here, Gürsoy provides a highly detailed anthropological account of how cultural and folklore visions of trees have shaped how forest villagers perceive trees and threats to their health, especially in the context of gardens, orchards and forests. Drawing on extensive ethnographic research in 12 forest villages across Turkey carried out between 2014 and 2017, the author contrasts local perceptions of tree disease to scientific and policy perceptions, often finding a lack of consensus about risks and priorities. In its emphasis on developing participatory approaches that bring together the local knowledge of forest villagers with scientific evidence to develop management approaches that are grounded in deliberation and consensus-building between all stakeholders, her contribution introduces a theme that is a central concern for many of the book's chapters. Chapter 4 by Prentice et al. develops this further in the context of the outbreak of mountain pine beetle (*Dendroctonus ponderosae*) in north central Colorado forests in the USA since 1996. By tracing the varying environmental narratives of both the local communities experiencing the outbreak and the regional organisations responsible for managing it, the authors reveal the distinct and sometimes overlapping subjective conceptions of how residents and managers perceive outbreaks. In this political ecology perspective, the authors concur with Gürsoy's anthropological example from Turkey that a deeper understanding of the diverse ways in which different actors frame outbreaks is needed in order to develop consensual and sustainable management solutions.

In Chapter 5, Lambert et al. present an example from New Zealand, where real efforts have been made to achieve a collaborative approach by explicitly integrating traditional Māori indigenous knowledge into modern-day biosecurity practice and forest management. The authors describe the successes and challenges faced by a joint agency programme, founded to manage the kauri dieback (*Phytophthora agathidicida*) outbreak in 2009, which involved Māori representation across all levels of governance and community engagement within the programme. This is contrasted with the more successful Māori Biosecurity Network, established to coordinate the involvement of Māori researchers, governance representatives and political lobbyists to combat the recent incursion of myrtle rust (*Austropuccinia psidii*) which threatens a range of taonga species.

In Chapter 6, Fellenor et al. make a valuable theoretical contribution to understanding some of the ontological questions that are raised when tree health threats are being spoken about. Through a rapid evidence review, the authors present a compelling argument that the growing concern with environmental degradation, including tree pests and diseases, is mediated by the Internet and user-generated content such as social media, which transform human interaction and knowledge about the world. This, they conclude, means that how forest health issues are perceived can be both a product of direct experience but is also represented as a 'digital artefact'. Following on from this, Urquhart et al. use the Social Amplification of Risk Framework (SARF) to assess how publics and experts perceive the threats from tree diseases in Chapter 7. Using the case of ash dieback in the UK, the chapter usefully demonstrates how the dynamic interactions between experts, policy makers, the media and publics influence how tree health risks are perceived, with implications for how risk managers communicate about such issues.

Three chapters, in particular, provide a useful analytical assessment in terms of the governance of tree health and the role of economics for informing policy and decision-making. In Chapter 8, Keskitalo et al. flag up the difficulties associated with the practical implementation of plant health regulations (e.g. Liebhold et al. 2012). With invasive alien species recognised by the Convention on Biological Diversity (CBD) as one of the major threats to biodiversity, this is an ontological issue for plant health regulation. Under the World Trade Organization's (WTO) Sanitary and Phytosanitary (SPS) Agreement, there needs to be 'proof of harm' before an organism can be regulated for, whereas the CBD promotes the precautionary principle, adhering to 'proof of safety' before allowing free trade in that organism. Thus, Keskitalo et al.'s finding—in the context of Swedish plant health legislation and its impact on forest plant nurseries—is that much of the focus for monitoring and detecting risk is placed on the different actors along the plant supply chain. This can be explained by the way the present WTO framework operates, resulting in only trade in high-risk organisms being targeted for preventative action. In Chapter 9, as part of a broad critique of the efficacy of economic research for informing public policy making, Jones suggests that there is a gap between outputs provided by academic economists and the needs of those tasked with developing plant health policies. In a similar vein, but focusing particularly on stated preferences approaches for tree health protection, Price outlines in Chapter 10 the possibilities and challenges of using such approaches for assessing the impacts of tree pests and pathogens and allocating resources for their control.

In addition to understanding how tree health issues are perceived and understood from the macro-level international governance regimes of the WTO through to national biosecurity policies and local community perspectives, the book's authors are also concerned with developing practical approaches that can address some of the issues raised in the preceding chapters. Thus, Chapters 11–15 present a series of studies that draw insights from diverse disciplines such as education, social learning and operant learning theory to develop much-needed collaborative approaches that facilitate stakeholder engagement and effective communication tools. In Chapter 11, Allen et al. used a participatory action research approach to develop a rubric as an assessment framework for post-border biosecurity management in New Zealand. The rubric proved a useful device for different practitioners and agencies to articulate the various social, technical and management dimensions, leading to a more systematic and outcomes-based approach to responding to the issue.

Marzano et al., in Chapter 12, are also concerned with dialogue between various stakeholders, in this case academics responsible for the development of technological innovations for the detection of pests and diseases and end users and commercial stakeholders to ensure they are 'fit for purpose'. Again, the authors adopted an action research approach, drawing on principles from social learning to develop a 'learning platform', but concluded that there are still numerous challenges for multi-stakeholder engagement for effective socio-technological development. Similarly Dragoi, in Chapter 13, also adopted a social learning approach by drawing on operant learning theory, to develop a training drill that could be used to encourage more effective and biodiversity-aware decision-making when Romanian forest managers are marking trees for sanitation felling or trees to maintain for biodiversity.

In Chapter 14, Mattor et al. focus on the specific issue of the effects of the mountain pine beetle outbreak on drinking water quality in North America and argue for improved communication and knowledge exchange between scientists and water managers. On a more community level, in Chapter 15, Davis et al. assess the role of 'forest collaborative groups' in eastern Oregon, USA, to provide a forum for dialogue and to mitigate the impacts of forest pests and other threats to forest health, such as wildfire.

The final two contributions, by Dandy et al. and Dyke et al., are very much concerned with conceptual issues, using 'non-human' approaches and environmental ethics as a lens through which to view tree health. In the first of these, in Chapter 16, Dandy et al. present three alternative accounts of the 2012 Asian longhorn beetle outbreak in Paddock Wood in the UK, as viewed through different ethical lenses. They argue that shifting the ethical focus from humans to non-humans (both the trees and the pests) may demand revised approaches to how outbreaks are managed and may result in different outcomes. In Chapter 17, Dyke et al. also argue for consideration of the non-human, arguing for a more relational approach to tree health management that sees trees as dynamic organisms that change across both time and space, thus requiring flexible and evolving approaches to tree heath management.

The book's conclusion in Chapter 18 argues that quantitative measures of the impacts of tree pest and disease outbreaks, such as economic assessments of impacts on commercial crops or willingness to pay for tree disease mitigation (Price), have to be coupled with other, less-tangible, qualitative aspects such as the meanings people attach to trees and the social and cultural connections they provide to individuals and communities (Prentice et al. and Urquhart et al.). Further, high-profile outbreaks across the globe are likely to have profound implications for the way future tree pest and disease threats need to be handled and communicated by government, its agencies and stakeholders. Thus, it is important to better understand the way in which people respond to notifications about outbreaks and engage with debates about biosecurity more broadly in order to inform risk communication and public engagement activities (Allen et al.). The book's case studies all engage in different ways with the central focus of the book on understanding the human dimensions of forest and tree health. Taken together, the chapters make theoretical, methodological and applied contributions to our understanding of this important subject area, and we hope provide a call to arms for researchers from across the social sciences and humanities to bring their own disciplinary perspectives and expertise to address the complexity that is the human dimensions of forest and tree health.

References

Aukema, J. E., Leung, B., Kovacs, K., Chivers, C., Britton, K. O., Englin, J., et al. (2011). Economic impacts of non-native forest insects in the continental United States. *PLoS One, 6*(9), e24587. https://doi.org/10.1371/journal.pone.0024587.

Boyd, I. L., Freer-Smith, P. H., Gilligan, C. A., & Godfray, H. C. J. (2013). The consequence of tree pests and diseases for ecosystem services. *Science, 342*(6160), 823–831. https://doi.org/10.1126/science.1235773.

Brasier, C. M. (2008). The biosecurity threat to the UK and global environment from international trade in plants. *Plant Pathology, 57*, 792–808. http://doi.org/10.1111/j.1365-3059.2008.01886.x.

Chafota, J., & Owen-Smith, N. (2009). Episodic severe damage to canopy trees by elephants: Interactions with fire, frost and rain. *Journal of Tropical Ecology, 25*(3), 341–345. https://doi.org/10.1017/S0266467409006051.

Crowley, S. L., Hinchliffe, S., & McDonald, R. A. (2017). Invasive species management will benefit from social impact assessment. *Journal of Applied Ecology, 54*, 351–357. http://doi.org/10.1111/1365-2664.12817.

Freer-Smith, P., & Webber, J. (2017). Tree pests and diseases: The threat to biodiversity and the delivery of ecosystem services. *Biodiversity and Conservation, 26*(13), 3167–3181. https://doi.org/10.1007/s10531-015-1019-0.

Gill, R. M. A. (1992). A review of damage by mammals in north temperate forests. 1. Deer. *Forestry, 65,* 145–169.

Holmes, T. P., Aukema, J., von Holle, B., Liebhold, A., & Sills, E. (2009). Economic impacts of invasive species in forests—Past, present, and future. *Annals of the New York Academy of Sciences, 1162,* 18–38. http://doi.org/10.1111/j.1749-6632.2009.04446.x.

Holmes, T. P., Aukema, J., Englin, J., Haight, R. G., Kovacs, K., & Leung, B. (2014). Economic analysis of biological invasions in forests. In S. Kant & J. R. R. Alavalapati (Eds.), *Handbook of forest resource economics* (pp. 369–386). New York: Routledge.

Liebhold, A. M., Brockerhoff, E. G., Garret, L. J., Parke, J. L., & Britton, K. O. (2012). Live plant imports: The major pathway for forest insect and pathogen invasions of the US. *Frontiers in Ecoogy and the Environment, 10*(3), 135–143. http://doi.org/10.1890/110198.

MacLeod, A., Pautasso, M., Jeger, M. J., & Haines-Young, R. (2010). Evolution of the international regulation of plant pests and the challenges for future plant health. *Food Security, 2,* 49–70. http://doi.org/10.1007/s12571-010-0054-7.

Marzano, M., Dandy, N., Bayliss, H. R., Porth, E., & Potter, C. (2015). Part of the solution? Stakeholder awareness, information and engagement in tree health issues. *Biological Invasions, 17*(7), 1961–1977. https://doi.org/10.1007/s10530-015-0850-2.

Marzano, M., Allen, W., Haight, R. G., Holmes, T. P., Keskitalo, E. C. H., Langer, E. R., et al. (2017, November). The role of the social sciences and economics in understanding and informing tree biosecurity policy and planning: A global summary and synthesis. *Biological Invasions, 19*(11), 3317–3332. https://doi.org/10.1007/s10530-017-1503-4.

MEA. (2005). *Ecosystems and human well-being: General synthesis. Millennium Ecosystem Assessment.* Washington, DC: Island Press.

Potter, C., & Urquhart, J. (2017, June). Tree disease and pest epidemics in the Anthropocene: An analysis of drivers, impacts and policy responses in the UK. *Forest Policy and Economics, 79,* 61–68. https://doi.org/10.1016/j.forpol.2016.06.024.

Raum, S., Bayliss, H., & Urquhart, J. (Under review). The potential impacts of non-native tree pests and diseases and their management on ecosystem services. *Forest Ecology and Management.*

Urquhart, J., Potter, C., Barnett, J., Fellenor, J., Mumford, J., & Quine, C. P. (2017a, November). Expert risk perceptions and the social amplification of risk: A case study in invasive tree pests and diseases. *Environmental Science & Policy, 77*, 172–178. https://doi.org/10.1016/j.envsci.2017.08.020.

Urquhart, J., Potter, C., Barnett, J., Fellenor, J., Mumford, J., Quine, C. P., & Bayliss, H. (2017b). Awareness, concern and willingness to adopt biosecure behaviours: Public perceptions of invasive tree pests and pathogens in the UK. *Biological Invasions, 19*(9), 2567–2582. https://doi.org/10.1007/s10530-017-1467-4.

2

English Tree Populations: Economics, Agency and the Problem of the "Natural"

Tom Williamson, Gerry Barnes and Toby Pillatt

Abbreviations

BRO	Bury St. Edmunds Record Office
ERO	Essex Record Office
HALS	Hertfordshire Archives and Local Studies
IRO	Ipswich Record Office
Lds RO	Leeds Record Office
NHRO	Northamptonshire Record Office
Nrth RO	Northallerton Record Office
NRO	Norfolk Record Office
PRO/TNA	Public Record Office/The National Archive
Shf RO	Sheffield Record Office

T. Williamson (✉) · G. Barnes
Landscape Group, School of History,
University of East Anglia, Norwich, UK

T. Pillatt
Department of Archaeology, University of Sheffield, Sheffield, UK

© The Author(s) 2018
J. Urquhart et al. (eds.), *The Human Dimensions of Forest and Tree Health*,
https://doi.org/10.1007/978-3-319-76956-1_2

1   Introduction: Trees and Disease in Historical Perspective

Dutch elm disease—caused by the fungus *Ophiostoma ulmi*, disseminated by beetles of the genus *Scolytus*—first arrived in England in the 1920s, but a more virulent strain—caused by *O. novo-ulmi*—appeared in the late 1970s and within a decade had effectively wiped out elm as a tree (Brasier 1991; Gibbs et al. 1994). A series of outbreaks has followed, including *Phytophthora ramorum*, leaf minor and canker in horse chestnut canker (*Cameraria ohridella* and *Pseudomonas syringae* pv.*aesculi*), oak processionary moth (*Thaumetopoea processionea*) and more recently ash dieback (*Hymenoscyphus fraxineus*) (Cheffings and Lawrence 2014). All are caused by invasive organisms—fungi, bacteria or insects—and have thus been seen as a consequence of globalisation, perhaps compounded by climate change (Brasier 2008). There are further threats of this kind on the horizon, including emerald ash borer (*Agrilus planipennis*), pine processionary moth (*Thaumetopoea pityocampa*), citrus longhorn beetle (*Anoplophora chinensis*) and xyella (*Xyella fastidiosa*). In addition, there are worries that tree health in England is suffering a more general deterioration, with recognition of such complex and diffuse conditions as "oak decline" (Denman and Webber 2009), manifested in progressive thinning of the crown and general ill health, leading to gradual death. An acute variant of this disease, leading to rapid death and with debated causes, has also been identified.

Current concerns need, however, to be placed within a broader historical context: only then can appropriate action be taken—or, in some cases, a more relaxed approach be adopted to arboreal ill health. The discussion that follows is mainly based on archival research (funded by the Arts and Humanities Research Council (AHRC) and the Department for Environment, Food and Rural Affairs (DEFRA)) which examined maps, tree surveys and forestry accounts from four English counties, chosen for their contrasting land use and agricultural histories: Hertfordshire, Norfolk, Northamptonshire and Yorkshire. This was supplemented with further and less detailed examination of the evidence from other areas, principally the counties of Essex, Herefordshire,

Kent, Suffolk and Shropshire. A systematic examination of farming and forestry literature from the seventeenth, eighteenth and nineteenth centuries was also undertaken. The purpose of this research was to characterise tree populations and their management in the period since the sixteenth century; to assess the extent to which trees suffered from endemic or epidemic disease in the past, and to identify any changes in management over time which might have contributed to rising levels of morbidity. One feature which rapidly became apparent as this research proceeded was that, while there is little evidence for large-scale epidemic disease prior to the twentieth century, poor health in trees is not in itself new. Oak, for example, has always been prone to fungal pests, such as *Aetiporus sulphureus* (cuboidal brown rot of heartwood) and *Stereum gausapatum* (pipe rot), while caterpillars of the indigenous micromoth *Tortrix viridana* have caused successive years of severe defoliation. The effects of the common cockchafer (the beetle *Melolontha melolontha*, also known as the May Bug) on trees appear to have been more serious in the past than they are today. In 1787, for example, the oaks around Doncaster "were entirely stripped by them" (Nichols 1795, 31), while fatal attacks on the roots of ash are also reported (Trans of Soc. For the Encouragement of Arts etc., 1795, cxcii). The condition known as "shake"—that is, internal splitting of the timber—in oak, sweet chestnut and other trees is widely reported in documents from the seventeenth century onwards, and the symptoms of "oak decline" are often described. At Prior Royd Wood near Sheffield in Yorkshire it was reported in the late eighteenth century that 206 trees were "nearly all dead top'd" (Shf RO ACM/MAPS/Shef/169), while on the Bolton Estate in Wensleydale in the same county, and period, "many of the trees were affected by crown dieback" (Dormor 2002, 222). "That dead-topped oaks are very common, cannot be disputed" (Pontey 1805, 130). Prior to the appearance of Dutch elm disease in the twentieth century, its vector *Scolytus* caused extensive damage on its own account, leading to the "decay and subsequent death of the finest Elms in the vicinity of London, particularly those in St James' and Hyde Parks" (Selby 1842, 114). Earlier generations clearly considered ill health in trees as normal. The terms used in a survey of timber at Staverton in

Northamptonshire in 1835, for example, include "decayed", "damaged", "small and very bad", "very bad", "decayed very bad" and "dead" (NHRO ZB 887). Moses Cook in the seventeenth century advised regular examination "of your Timber-trees, to see which are decaying … Why should any reasonable Man let his Trees stand in his Woods, or elsewhere, with dead Tops, hollow Trunks, Limbs falling down upon others and spoiling them, dropping upon young Seedlings under it, and killing them?" (Cook 1676, 163). Poor health was simply something to be tolerated, with diseased trees being felled as quickly as possible and sold for the best price possible.

In addition, although the increasing scale of global trade—in timber and live plant materials—is probably, together with climate change, the principal cause of the current increases in tree disease, not only in this country but across the world, it is worth emphasising that neither is new. Imports of timber from the Baltic were common in the Middle Ages—as early as 1273 the chapter of Norwich Cathedral dispatched "John the carpenter" to Hamburg to buy timber (Latham 1957, 28)—and were on a substantial scale by the sixteenth century. In November 1696, the Norfolk landowner Richard Godfrey lamented that frost was preventing the delivery of live fruit trees he had ordered from Holland (NRO Y/C 36/15/18), and live forest trees were also being imported on a significant scale by this date. Some time around 1700, John Bridges of Barton Seagrave in Northamptonshire planted "500 limes from Holland" (Morton 1712, 486). By the eighteenth century, substantial quantities of timber were arriving from the Americas, as well as from northern Europe; and the volume of imports rose steadily thereafter, reaching 2.6 million cubic metres by 1851 and 5.9 million by 1871—around a third of modern levels (Fitzgerald and Grenier 1992, 18). But it was probably the increased speed of transportation, rather than the quantities moved per se, that led to the arrival of oak mildew in the early years of the twentieth century, and to the first epidemic of Dutch elm disease soon afterwards. By the 1880s, the development of the screw propeller, the compound engine and the triple-expansion engine, made trans-oceanic shipping of bulk cargoes by steam economically feasible (Carlton 2012).

2 Historic Tree Populations: Density and Management

An historic approach invites us to consider the contribution made to current problems by long-term changes in the character of tree populations—in their species composition, management and age structure. This is important because—to use an obvious example—if ash was a rare tree, concern about ash dieback would be limited to a small number of botanists. It is important to assess how, and why, rural tree populations have developed in the ways that they have. It is often assumed that the particular kinds of tree we find in the countryside—the ubiquity of ash and oak, for example, and the relative rarity of trees like wild service—is a consequence of natural factors, but this is only partly true. For centuries, the overall numbers of trees, the relative numbers of different species and the ways in which trees are managed have all been embedded in social and economic structures. Tree populations have more the character of a human artefact, than of something essentially "natural".

Some "veteran" trees surviving in the modern landscape date back to the Middle Ages. But most are of early-modern date, and the overwhelming majority of our trees were planted in the eighteenth, nineteenth or twentieth centuries. The relatively recent character of tree populations is mirrored in the nature of our sources, for it is only from the seventeenth century that these provide really useful information about trees and their management. One striking feature they reveal is that, before the mid-nineteenth century, many districts boasted vast numbers of farmland trees. On the boulder clays of East Anglia, Essex and east Hertfordshire, for example, there were usually between 20 and 30 farmland trees per hectare, but often more. On a property at Thorndon in Suffolk, c.72 per hectare were present in 1742 (BRO BT1/1/16), while at Kelshall in north east Hertfordshire a farm contained 59 trees per hectare when surveyed in 1774 (HALS DE/Ha/B2112). Some properties, it is true, had fewer, but this was often the result of particular circumstances. At Saxtead in Suffolk in the 1720s, a tenant had taken down "so many trees that there was not enough wood left for the dairy, so that firing had to be fetched from up to four miles away", while another, at Cotton Hall, had "managed to strip much of the farm of its trees and hedges" (Theobald 2000, 9–10). Equally striking was the high proportion of trees which were managed as pollards: that is, regularly cropped at a height of two metres or so to produce a harvest of straight "poles". On the eastern boulder clays 70 and 80 per cent were usually managed in this way, but sometimes more, with figures of 92 per cent on a farm at Pulham in Norfolk in 1751 (NRO DN/MSC 2/22-25); 93 per cent at Kelshall in Hertfordshire in 1727 (HALS DE/Ha/B2112); and 94 per cent on a farm at Campsey Ash in Suffolk in 1807 (IRO HD11:475). At Curd Farm, Little Coggeshall Essex in 1734 there were only 46 mature timber trees, but no less than 3591 pollards (ERO D/Dc E15/2). Similar densities are recorded on the poor soils, formed in Eocene deposits, in the vicinity of London—on seven farms on the Broxbournebury estate in south Hertfordshire in 1784, for example, there were 3012 pollards but only 299 timber trees (HALS DE/Bb/E27)—and contemporary commentators record something similar in many western counties. In Herefordshire in 1792, "The Trees are much strip't and lopp'd by the Farmers" (*House of Commons Journal* 1792, 318), while in Shropshire they were "generally found most decayed in consequence of lopping" (Plymley 1803, 213).

This picture—of a countryside densely populated with trees, most of which were pollarded—did not apply everywhere. In particular, there were fewer trees in the "champion", open-field districts of the Midlands, largely because there were fewer hedges in which they could grow. Most of the arable lay in unhedged strips grouped into larger unhedged furlongs, which were in turn aggregated into two or three vast "fields". Nevertheless, these districts contained more trees than is often assumed. The hedges of the village "tofts"—the small enclosures behind the farmhouses—were usually densely planted. A survey of Milcombe in Oxfordshire, made in 1656, describes hundreds of trees in the village closes. One contained "126 trees of Ash and Elm … and 52 withes [willows] small and great and about as many new planted" (NHRO C(A) Box104 4 1656). In the wider landscape, trees grew on roadsides, on patches of waste, in hedges on parish boundaries and in the meadows beside rivers. When Irthlingborough in Northamptonshire was enclosed in 1808, landowners made claims relating to a total of 3,055 trees growing amidst the fields and meadows (NHRO ZA 906). Sixty-two per cent were willows, growing on the floodplain of the river Nene. The enclosure of open fields proceeded steadily in Midland districts through the seventeenth and eighteenth centuries, and as the numbers of hedges increased, so too did the density of farmland trees. A survey of land on the Duke of Powis' estate in Northamptonshire, made in 1758, records a total of 3004 trees on 770 acres—or 9.6 per hectare (NHRO ZB 1837), while on John Darker's estate in the same county in 1791 there were no less than 10.6 trees per hectare (NHRO YZ 2183). These figures, however, were significantly lower than those found in the old-enclosed districts described above.

In general, by the eighteenth century the lowest densities of trees, and the lowest proportions of pollards, were to be found in northern counties. Even enclosed parishes in the Vale of York only contained between 0.4 and 2.5 trees per hectare, averaging 1.4 (Nrth RO ZNS; Nrth RO ZIQ; Nrth RO ZDS M 2/12; Nrth RO ZMI; Lds RO WYL68/63). In some northern districts, it is true, rather more trees—and a higher proportion of pollards—existed, often for special reasons. In Cumberland in 1776, Thomas Pennant drew attention to the numerous ash pollards, cut for fodder (Pennant 1776), and in places, the remains of these populations can still be seen, growing against or, more rarely, within stone walls, especially in Borrowdale and Langdale. Holly pollards, again used as winter fodder, were a feature of many Pennine townships (Spray 1981). But by the eighteenth century, farmland trees (and especially pollards) were relatively thin on the ground in most northern areas.

3   Explaining Tree Density and Management

Many early writers railed against farmland trees, especially in high densities. The late seventeenth-century agricultural writer Timothy Nourse typically argued that "Corn never ripens so kindly, being under the Shade and Droppings of Trees; the Roots likewise of the Trees spreading to some distance from the Hedges, do rob the Earth of what should nourish the Grain" (Nourse 1699, 27). At Badwell Ash in Suffolk in 1762, it was said that the land was "capable of great improvement by destroying the timber and pollards that encumber the fields in many places" (BRO B E3/10/10.2/28). Most trees, as noted, were pollards and while the poles they produced had many uses they were mainly employed as domestic fuel. Not surprisingly, where farmland was sparsely treed, especially across much of northern England, there were usually other fuel sources. Peat, both from upland moors and lowland "mosses", was both burned locally and exported to major cities. Many of the inhabitants of seventeenth-century York burned peat brought from Inclesmoor, nearly 30 kilometres away (Hatcher 1993, 124), while peat was taken from the mosses of south-west Lancashire to supply Ormskirk and Liverpool (Langton 1979, 56–57). In addition, from an early date, many northern and western districts had access to coal. As early as the 1530s John Leland described how, although wood was plentiful across much of Yorkshire, many people were burning coal (Toulmin-Smith 1907). By 1790, coal had long been almost the only fuel consumed in Durham, Yorkshire, Nottinghamshire, Staffordshire, Lancashire and Cheshire. In the West Riding "The Use of Coal … has been universal, as far back as can be remembered", while in Staffordshire "Coals are, and have been universally used in this county" (*House of Commons Journal* 1792, 328–329).

In a similar manner, the increasing output of the Warwickshire, Nottinghamshire and Staffordshire mines—coupled with improvements in road transport and the construction of canals, which also allowed the greater dispersal of coal from the great north-eastern coal fields—explains why, as open fields in the "champion" Midlands were gradually enclosed through the seventeenth and eighteenth centuries, the new hedges were never as densely stocked with trees as those in most old-enclosed districts, in the Home Counties and East Anglia especially. These latter areas lay at a greater distance from coalfields. Although "sea coal" from the north east was, by the seventeenth century, widely burned in towns located on the coast, or on navigable rivers, elsewhere organic fuels had to be used. And where peat was in short supply, and heaths—with their combustible gorse and heather—were few, then the importance of firewood, and thus of pollards, was great.

What made the wood cut from pollards particularly important was the fact that early-modern England was poorly endowed with woods. Even in the well-wooded south east of the country, it was rare to find that more than 10 per cent of the surface area was devoted to woodland. In most champion counties, the figure was less than 5 per cent, as it was across the north of England. Estimates made by John Tuke at the end of the eighteenth century, for example, suggest that woodland occupied no more than two per cent of the land area of the North Riding of Yorkshire (Tuke 1800). Before the later eighteenth century, woods were almost all of "coppice-with-standards" type. The timber trees were mainly used for construction; it was the coppiced understory—cut down to ground level on a rotation of between eight and fourteen years—which potentially provided most fuel. Only in a few districts, however—especially in the vicinity of fuel-hungry London—were faggots[1] for burning a major product. Coppices mainly produced high-quality poles, used for making hurdles, fencing, hoops, tools and parts of vehicles, and to provide thatching and building materials. In many districts, only the twiggy residue appears to have been destined for burning. Indeed, variations in coppice management were often closely related to particular demands of local economies. Across much of the north of England, for example, and in the industrialising areas of the west Midlands, the coppice was usually dominated by oak and was cut on a very long rotation, of 20 or even 30 years. In Shropshire, "Large quantities of oak poles are used for different purposes in the coal-pits; as they are required to have some strength, they are seldom fallen before 24 years growth, and the bark (used in tanning leather) is an object of great importance…" (Plymley 1803, 219).

Indeed, bark from long-grown coppice poles, and stripped from felled timber trees, was a major source of profit in all areas, although especially industrialising ones. On the Millford Estate near Leeds in the eighteenth century bark accounted for around 20 per cent of the sale value of oak trees (Lds RO WYL500/939). At Hutton Rudby in the same county in the 1630s, the figure was as high as 33 per cent (Lds RO WYL100/EA/13/38).

4 Explaining Tree Species

Almost everywhere, oak, ash and elm accounted for between 85 and 100 per cent of farmland trees. This is remarkable given that there are at least 25 indigenous, or long-naturalised, species capable of growing into reasonably sized trees, with a height of ten metres or more. The contrast with the situation in remote prehistory, before the advent of farming, is striking: across southern and Midland England small-leafed lime (*Tilia cordata*), a rare tree by the seventeenth century, had been the most common species (Rackham 2006, 83–85). However, oak, ash and elm became the dominant trees in the farmed landscape of the post-Neolithic, not because they were well adapted to this environment naturally, but because they were deliberately planted. Mortimer, for example, described how "The best way of raising Trees in Hedges, is to plant them with the Quick", but he also gave advice on how to establish them "where Hedges are planted already, and Trees are wanting" (Mortimer 1707, 309). Even where trees were self-seeded they needed to be protected. Initial establishment might be the consequence of natural process, but survival to maturity was a function of human agency. An early eighteenth-century lease for a farm in Barnet in Hertfordshire typically instructed the tenant to "do every Thing in his Power for the Encouragement, and growth of the young Timber Shoots, under the Penalty of Twenty Shillings for every Shoot or Sapling which shall be wilfully hinder'd from growing" (HALS DE/B 983 E1).

Two main factors ensured the overwhelming popularity of these three trees: an ability to thrive in a wide variety of contexts, and the wide range of uses for their timber and wood. Oak was "The best Timber in the World for building Houses, Shipping, and other Necessary Uses" (Meager 1697, 110). It also made good firewood, excellent charcoal and clefts easily, making it suitable for floorboards and fencing while its bark, as we have noted, was employed in tanning. It could grow in most situations: "in any indifferent Land, good or bad, as Clay, Gravel, Sand, mixed, or unmixed Soils, dry, cold, warm or moist" (Meager 1697, 110). Ash was less useful as structural timber, but it had many other uses. Timothy Nourse thought it "a most useful wood to the

Coach-maker, Wheeler, Cooper, and a Number of other Artificers", and that it could be used for making fencing and bins, "for Spittle and Spade Trees, for Drocks and Spindles for Ploughs, for Hoops, for Helves, and Staves, for all Tools of Husbandry, as being tough, smooth and light" (Nourse 1699, 119). Its excellence as firewood was universally praised: "the sweetest of our forest fuelling, and the fittest for ladies chambers" (Evelyn 1664, 40); "Of all the wood that I know, there is none burns so well green, as the Ash" (Cook 1676, 76). And on top of this, ash grew rapidly and, like oak, was not very choosy about *where* it grew. Contemporaries agreed that it would flourish on "any sort of land", provided "it be not too stiff, wet and boggy", although in reality it seems to have been less prominent on more acidic soils (Mortimer 1707, 366).

Elm in its various forms also had many uses. It was "proper for Water-works, Mills, Soles of Wheels, Pipes, Aqueducts, Ship Keels and Planks beneath the Water Line … Axel trees, Kerbs Coppers … Chopping-Blocks … Dressers, and for Carvers work", as well as for making spades, shovels and harrows. But above all it made excellent boards and planks, for floorboards, external weatherboarding—and coffins (Nourse 1699, 115). Again, it could tolerate a wide range of conditions, and early writers singled out another advantage. It caused "the least offence to Corn, Pasture and Hedges of any Tree", in part because (unlike ash) its roots did not spread far, but also because it could be rigorously trimmed up as timber, so that it cast limited shade. Ellis thought that elms "don't damage any thing about them, as some other Trees do, whose Heads must not be trimmed up as these may" (Ellis 1741, 49).

A number of other species are recorded in early surveys, growing in small numbers in fields and hedgerows. These include maple, lime, hornbeam, rowan, aspen, black poplar, alder, sycamore, beech, holly, sweet chestnut, walnut, wild service, willows, crab and fruit trees like apple and cherry: indeed, only whitebeam and birch were regularly shunned by planters, appearing at very low levels only in some western and northern districts. In general, such "minority" species made up less (usually much less) than fifteen per cent of trees recorded, but there were exceptions. In certain districts, fruit trees might rival or outnumber oak, ash and elm in hedges—especially in parts of Herefordshire, Worcestershire and Shropshire, but also in Kent, Hertfordshire and Middlesex. A few farms in Essex and east Hertfordshire boasted diverse mixtures of trees: maple, lime, hornbeam and wild service made up a surprising 61 per cent of the trees on a farm at Navestock in south Essex in 1772 (ERO T/A 783). Such cases are rare, however, and in most places surveys reveal an overwhelming dominance of oak, ash and elm.

In most cases, the relative rarity of "minority" species was due to the fact that they had fewer uses, grew more slowly, or were more demanding in their requirements than oak, ash or elm. Some, however, were infrequent as *trees* because contemporaries thought they were better managed as coppice, in woods or hedges. Maple, for example, is widely recorded as a pollard and, more rarely, as a timber tree, but only infrequently did it account for more than 5 per cent of trees on a property. It was presumably common, as it is today, as a shrub in hedges, where it seeds relatively easily. It was also—usually in combination with hazel and/or ash—a frequent component of coppices, especially in the Midlands and south. Farmers and landowners evidently preferred to manage it as underwood, rather than as a pollard: if it self-seeded in a hedge, it was usually plashed or laid with the rest of the shrubs. Moses Cook in 1676 explained that:

> If you let it grow into Trees, it destroys the wood under it; for it leaves a clammy Honey-dew on its Leaves, which when it is washed off by Rains, and falls upon the Buds of those Trees under it, its Clamminess keeps those Buds from opening, and so by degrees it kills all the Wood under it; therefore suffer not high Trees or Pollards to grow in your Hedges, but fell them close to the Ground, and so it will thicken your Hedge, and not Spoil its Neighbours so much. (Cook 1676, 99)

The distribution of minority trees—and the relative importance of oak, ash and elm—displayed a measure of spatial variation, the consequence of a complex interplay of environmental and economic factors. Farmers and landowners, knowledgeable about the local environment, planted or encouraged trees which they knew would both flourish, and produce wood or timber of value or utility. But it was not only the trees of farmland which were shaped by such factors. Woods and wood-pastures (grazed woods on commons and in parks) also had their distinctive species, the result of choices made by land managers, or a side effect of management systems.

The timber trees in coppiced woods were mainly oaks, valuable as timber and able to self-seed and flourish under the canopy shade. Again and again we find a sharp contrast between the trees growing in woods, and those on adjacent farmland, clearly illustrating the highly artificial character of both populations. Three examples from Essex make the point well. On the Topping Hall estate in Hatfield Peverel oak made up 48 per cent of the farmland trees in 1791, but accounted for all but one of the 2000 trees growing in the Great Wood; at Finchingfield in 1773, oak constituted 57 per cent of the farmland timber but 100 per cent of the 968 trees in the four woods on the property; while at Little Baddow in 1777 it made up 65 per cent of the trees growing on the lands of Bicknaire Farm but 99.5 per cent of those in Bicknaire Wood (ERO D/DRa C4; ERO D/D Pg T8; ERO D/DRa C4). Even woodland coppices, which displayed much more variety in their composition, were not simply the "wild" vegetation of the places in question, tamed by management. The main coppiced species—ash, hazel and oak—were all of particular value for construction, tools, fencing and the like and the comparatively pure stands which often existed were in part the consequence of deliberate weeding and replanting. Boys in 1805 suggested that many coppices in Kent were regularly augmented with new plants simply because "wood, like everything else, decays and produces fewer poles every fall, unless they are replenished" (Boys 1805, 144). A lease for a wood in Wood Dalling, Norfolk, from 1612 bound the lessee to plant sallows in cleared spaces following felling (NRO BUL 2/3, 604X7); the tithe files of 1836 describe how there were 35 acres of coppice in Buckenham in the same county, "part of which has been newly planted with hazel" (PRO/TNA IR 29/5816); while Lowe described how on one Nottingham estate the hazel and thorns were stubbed up after the coppice was cut "and young ashes planted in their stead" (Lowe 1794, 34, 114). Rudge in 1813 described how ash was regularly replanted in the coppices in Gloucestershire and Vancouver in

1810 noted how, in Hampshire, ash shoots were plashed "in the vacant spaces" to form new plants (Vancouver 1810, 297); a similar practice is recorded in Surrey woods in 1809 (Stevenson 1809, 127). Coppiced woods, in short, were very far from being "natural" environments. They were intensively managed factories for the production of wood and timber, and their trees and coppiced stools were selected or manipulated accordingly.

In wood-pastures, oak was likewise usually the main tree but on commons and in deer parks in the Cotswolds and Chilterns, and on poor soils around London, beech was prominent, while in the latter district hornbeam was also present in vast numbers. No less than 24,000 hornbeam pollards were recorded on Cheshunt Common in south Hertfordshire in 1695: Rowe has argued that they were often deliberately planted on the wastes of this district by manorial lords in the early-modern period, responding to the high fuel prices in the proximity of London (Rowe 2015). Hornbeam wood had a range of specialised uses but it was mainly valued as firewood and as a source of charcoal. Beech and hornbeam have good resistance to grazing, especially when compared to ash or elm, and also produced mast which was consumed by deer and other livestock. How far their prominence in wood-pastures was a side effect of intensive grazing, how far they were deliberately planted, remains unclear, but as with woods the contrast with trees growing on the adjacent farmland was often sharp. At Drakes Hill Farm, Navestock, Essex, in 1772 hornbeam made up 16 per cent of the 419 mature trees growing in the fields and hedges, but 85 per cent of the 959 growing on the adjacent area of common land (ERO T/A 783).

5 The Age of Trees in the Past

Everywhere we look, early-modern tree populations were shaped by intensive management. One striking example is the way in which most timber trees were felled at a young age. Trees containing around 50 cubic feet of wood are, in general, likely to be around 80 years old, but most trees measured in surveys, or when felled, were much smaller than this. Of the 762 oaks (mainly in hedges) growing on an estate in Waltham Abbey in Essex in 1791, only 2 per cent were thought to contain more than 25 cubic feet, and none more than 40; all the 255 ash were thought to contain less than 15 cubic feet; and while some of the 197 elm were larger, one containing an estimate 40 cubic feet, most contained less than 15 (IRO HA 116/5/11/2). Many surveys, it is true, reveal larger trees, but they usually form a small minority, and Pringle, writing about Cumberland in 1794, noted that it was "general opinion in this and, I believe, in other counties that it is more profitable to fell wood at fifty or sixty years growth, than to let it stand for navy timber to 80 or 100" (Pringle 1794, 12). In part, such a practise was encouraged by the fact that bark was of better quality, and more easily peeled, from younger oak trees. But more importantly, before the development of industrial sawmills in the middle of the nineteenth century it was easier to select the size of timber for the job at hand, rather than to let a tree grow to a substantial size and then saw it up—especially as, from around sixty or seventy years, the growth rate of oak, in particular, begins to slow. It made more sense to fell trees at an appropriate size for gate posts, building repairs or whatever, and get others growing in their place.

It might be thought that pollards, which often formed the majority of farmland trees, were in general older, because they could continue to produce a reasonable crop of poles for centuries. But as Thomas Hale explained in 1756, "Pollards usually, after some Lopping, grow hollow and decay... The Produce of their Head is less, and of slower Growth". They should be taken down before the trunk rotted badly, and lost value; and the farmer should ensure a constant succession, by regularly replacing old pollards with young trees destined to be managed in the same manner (Hale 1756, 141). While neglect clearly allowed a proportion to reach a venerable old age—the veteran trees of today—these were exceptions. One eighteenth-century observer, railing against the dominance of old pollards in the hedges of East Anglia, commented disparagingly that these were "of every age, under perhaps two hundred years" (Middleton 1798, 345). In addition, we might note in passing that actively managed pollards were anyway maintained, in effect, in a state of permanent juvenescence (Read 2008, 251). In Lennon's words, because "the crown is constantly having to reform, pollarding can delay the emergence of the tree from the formative growth period… This can extend the natural lifespan of the tree significantly…" (Lennon 2009, 173). Compared with today, the countryside was filled with very young trees.

6 Changes in the Nineteenth and Twentieth Centuries

During the nineteenth century, the numbers of farmland trees in England declined steadily. Pollards gradually became redundant as better roads, the construction of canals and ultimately the spread of the rail network allowed coal to become the normal domestic and industrial fuel throughout the country. They were removed wholesale from farmland hedges. So too, in many areas, were timber trees, as a fashionable interest in agricultural improvement and a rising tide of imports (especially from the Baltic) ensured that forestry operations were increasingly concentrated in woods and plantations, so that felled hedgerow trees were not replaced.

Further changes followed in the twentieth century. As large landed estates experienced financial difficulties—or were broken up altogether—in the first half of the century, large numbers of trees were cut down. Much timber was also felled during the two World Wars, while post-war agricultural intensification and hedgerow removal, and the impact of Dutch elm disease, all took a terrible toll.

In most Midland and southern districts, tree densities were roughly halved in the nineteenth century and had more than halved again by the late 1970s (Williamson et al. 2017, 139–143). Since then, amenity planting and the growth of ash and maple in neglected hedges have, on some measures, reversed the decline: but it rather depends on what is being counted, for free-standing trees in hedges have continued to fall in numbers (Forestry Commission 2002).

Equally important, however, is the fact that, since the mid-nineteenth century, tree populations have become much less intensively managed, and the number of old trees in the landscape has in consequence increased markedly. The development of industrial sawmills and improvements in transport made it possible for more mature timber trees to be processed into smaller timber, leading to a rise in felling age. This was followed by an effective cessation of economic management as the increasing scale of timber imports made it less economically attractive to extract individual trees in hedges, and as post-war agricultural intensification encouraged the view that the countryside was for growing food, not trees. Where trees were allowed to remain in hedges, they thus gradually grew old and were not replaced when they died.

Increases in tree age were also arguably a consequence of social factors. The late nineteenth and early twentieth centuries saw the establishment of a number of organisations dedicated to the conservation of rural landscapes, open spaces and wildlife, including the Commons Preservation Society (1865); the Society for the Protection of Birds, later the RSPB (1889); the National Trust (1895); the Society for the Promotion of Nature Reserves (1912); and the Council for the Protection of Rural England (1926) (Evans 1992; Sheail 2002). Changes in the distribution of wealth and improvements in transport meant that middle-class urbanites, with little real experience of rural living, visited the countryside on a larger and larger scale, and increasingly settled in it, or in suburbs on its margins, and began to take an active interest in its conservation. The idea that the countryside was essentially "natural", which had been developing (alongside urbanisation and industrialisation) since the eighteenth century, now triumphed. Felling prominent hedgerow trees gradually came to be regarded, even by many landowners, not as a normal part of land management, but rather as a desecration. Such ideas were manifested with particular clarity, somewhat paradoxically, where countryside was being lost to urban or suburban development. It was proudly claimed that Letchworth Garden City, established in Hertfordshire in 1902, had been built on virgin farmland without the loss of a single tree (Rowe and Williamson 2013, 274). By the time of the Second World War, the idea—long promulgated by land use planners like Patrick Abercrombie and campaigners like Clough Williams Ellis—that state intervention was required to preserve the rural landscape from large-scale development was widely accepted, culminating in the Town and Country Planning Act of 1947

(Rowley 2006, 112–115). As well as introducing, for the first time, workable systems of spatial planning, this also established Tree Preservation Orders (TPOs), which allowed specimens deemed to be of particular value to be preserved from felling. Although largely applied in urban areas, TPOs represented the triumph of the new attitude to trees, as objects of the natural world to be preserved, rather than as economic objects to be husbanded and exploited.

The increasing age of farmland trees was manifested, in particular, in the growing incidence of "stag-headedness" or dieback. Photographs of the countryside dating from the late nineteenth century show, by modern standards, remarkably few stag-headed trees. Those taken in the post-war period, in contrast, show far more. By the 1950s and 60s, the ageing character of trees in the countryside was becoming a matter of concern. The *Report of the Committee on Hedgerow and Farm Timber* of 1955 estimated that over a third of hedgerow trees had girths in excess of 1.5 metres, that is, were at least sixty years old: the age by which, a century earlier, most would have been felled. The great gale of 1987 thinned a large number of old trees, but much remains, and while recent conservation and amenity planting have lessened the numerical dominance of old trees, in visual terms they often remain prominent in local landscapes.

7 Lessons from History?

It is within this broad historical context that we need to consider current concerns about tree health. Perhaps the main point to emphasise is that there is little that is "natural" about our farmland trees, a comment that applies more generally to our semi-natural woodlands and to a range of other valued habitats (Rackham 1986; Barnes and Williamson 2015; Fuller et al. 2017). Tree populations have, for centuries, been artefacts of management, and the same may well be true of some of the vaguer pathologies currently affecting English tree species, such as oak decline, a condition which principally affects trees a century or more in age. In historical terms, as the data discussed above should have made clear, these are over-mature trees, and these conditions may, to an extent, simply represent normal ageing, transformed into a "disease" by modern and unrealistic expectations of perpetual arboreal health. Equally important is the fact that, when tree populations were rigorously managed, few specimens would have exhibited symptoms of "decline" for very long, for they were simply taken as signs that useful growth was over. As Moses Cook put it in 1676:

> When a Tree is at its full Growth, there are several signs of its decay, which give you warning to fell it before it can be quite decay'd; as in an Oak, when the top boughs begin to die, then it begins to decay; in an Elm or Ash, if their Head dies, or if you see wet at any great Knot, which you may know by the side of the Tree being discolour'd below that place before it grows hollow …these are certain Signs the Tree begins to decay; but before it decays much, down with it, and hinder not your self. (Cook 1676, 171)

Although "oak decline" was only formally named and characterised in the 1920s, trees exhibiting the appropriate symptoms are referred to in early texts, but on an increasing scale from the nineteenth century. Curtis in 1892 described how "dead upper branches or 'stag-horn top', as it is usually called, is often met with… The manifestation needs but little remark, for it is apparent to all. The top branches die, the yearly growth is meagre, and the whole tree presents an enfeebled condition" (Curtis 1892, 25). It is noteworthy, however, that he drew attention to the prevalence of the condition, not on farmland or in woods, but "on lawns and pleasure grounds … and park lands"—that is, in locations where many trees were already, by the late nineteenth century, being retained beyond economic maturity, because they were primarily valued as ornaments to the landscape. The spread of the condition more widely, in other words, may simply reflect a decline in intensive management, and an increase in the proportion of over-mature trees in the countryside.

Of course, other changes in the rural environment over the last two centuries will have contributed to poor arboreal health. The increased scale of land drainage and water abstraction, and changing patterns of cultivation (with a shift to late summer cultivations and continuous courses of crops), have been noted by several authorities (e.g. Forestry Commission 1955). Less attention has been paid to the impact of the large-scale application of manufactured fertilisers, potentially an important influence given the suggestion that inorganic fertilisers can suppress the development of mycorrhizal fungi, on which tree health depends (e.g. Ryden et al. 2003). The amount of dead wood in the environment has also risen steeply over the last century or so, due to the decline of wood burning: in the past, any dead wood was rapidly gathered up by the local poor. The native buprestid beetle *Agrilus biguttatus*, thought by many to be a factor in acute oak "decline", was until recently considered a "red book" species, to be encouraged by the retention of fallen wood. Certainly, an earlier generation of foresters was clear about the potential threat posed by accumulations of decaying wood: "At the risk of repetition I would impress upon all foresters the necessity of cleaning up after every fall of timber, and the total destruction by fire of all dead organic matter" (Curtis 1892, 46). A decline in the practice of "quarantine felling", so regularly practised in earlier periods, may also be important: Curtis recommended it as the best way of dealing with fungal attacks, and with infestations of *Scolytus*. But the large numbers of old trees in the countryside, the main consequence of less rigorous management, may be key. Overall, the message from history may be, not so much that disease is a natural condition of trees, but that the most unnatural and most rigorously managed tree populations are also the most healthy ones. Forms of management that benefit rare saproxilic insects may not be so good for the health of trees themselves, and thus for the species which depend upon them; and difficult choices may need to be made by conservationists in the future.

But there are other important things that we can learn from history. It is clear that our present situation is uniquely serious. Elm has been lost from the landscape; ash, and possibly oak, are under threat. If we wish to ensure the continued presence of trees in the countryside, then we are obliged to plant a different and wider range of trees. At the same time, there is little doubt that the traditional dominance of oak, ash and elm in the countryside was mainly a consequence of economic rather than environmental factors. Recognising this essential artificiality of tree populations gives us more freedom in our choices of what we should plant in order to diversify and thus ensure future resilience, and history can suggest the kinds of species we should use. Some authorities have proposed the large-scale establishment of southern European varieties (such as downy oak (*Quercus pubescens*)), in anticipation of climate change, but given that many of our indigenous species have distributions extending far south into Europe this seems unnecessary. Instead, attention should turn to the "minority" trees, whose distributions—as we noted earlier—are often strongly regional in character. In Hertfordshire, for example, there were significant differences between the west of the county—the Chiltern dipslope, with soils largely formed in clay-with-flints and outwash gravels—and the east, with soils mainly formed in boulder clay. Before the mid-nineteenth century, cherry was regularly found, together with smaller amounts of apple, in the hedges of the west, together with aspen and beech. All were rare in the east of the county, where instead maple and hornbeam were present, with small quantities of black poplar on damper sites. Replicating, restoring and accentuating such patterns would ensure that a measure of regional diversity could be perpetuated into the future, providing a "sense of place" and a measure of historical continuity, things which might be lost if new species from abroad, or some indiscriminate "conservation mix" of indigenous species, were instead to be widely established. In addition, such "minority" trees are "tried and true" and likely to succeed in the localities in question. But we could also be bolder. In Hertfordshire, for example, attempts might be made to recreate the great wood-pastures of hornbeam, lost from the south of the county only relatively recently. There are arguments, too, for the large-scale planting of small leafed lime (*Tilia cordata*), largely banished from the landscape before the historic period. We need to plant very large numbers of trees, and we need to plant them now. But we need to think carefully about what we should plant, and where, and here the history of the landscape, and an awareness of its essentially anthropogenic character, ought to be one influence on our planning. Indeed, our habit of thinking of tree populations as primarily "natural" may be one of the problems we face when formulating future policy.

Quite how such ideas might, in practice, be implemented is a more complex question. Britain's exit from the EU, and more specifically from the Common Agricultural Policy, provides an opportunity for targeting grant aid towards large-scale replanting of a more diverse range of farmland trees, and also perhaps towards support for the more commercial, and more rigorous, management of farmland timber. But in addition, those currently involved in land management and conservation—county councils, trees wardens, the National Trust, the Forestry Commission, landowners—urgently need to be made aware of just how far our "natural" tree populations are, in reality, historically contingent; and of how some wildly shared aspects of current conservation policy, such as careful replication of their existing character in replanting programmes, or the retention on a large scale of over-mature trees and dead wood in the landscape, may be bringing as many problems as benefits.

Note

1. A bundle of sticks bound together as fuel.

References

Barnes, G., & Williamson, T. (2015). *Re-thinking ancient woodland.* Hatfield: University of Hertfordshire Press.
Boys, J. (1805). *General view of the agriculture of the county of Kent.* London.
Brasier, C. M. (1991). Ophiostoma novo-ulmi sp. nov., causative agent of current Dutch elm disease pandemics. *Mycopathologia, 115,* 151–161.
Brasier, C. M. (2008). The biosecurity threat to the UK and global environment from international trade in plants. *Plant Pathology, 57,* 792–808.
Carlton, J. (2012). *Marine propellers and propulsion.* London: Butterworth-Heinemann.
Cheffings, R., & Lawrence, C. M. (2014). *Chalara. A summary of the impacts of ash dieback on UK biodiversity, including the potential for long term monitoring and future research on management scenarios* (JNCC Report No. 501). Peterborough.
Cook, M. (1676). *On the manner of raising, ordering and improving forest-trees.* London.

Curtis, C. E. (1892). *The manifestation of disease in forest trees: The causes and remedies*. London.
Denman, S., & Webber, J. F. (2009). Oak declines—New definitions and new episodes in Britain. *Quarterly Journal of Forestry, 103*(4), 285–290.
Dormor, I. (2002). *Woodland management in two Yorkshire Dales since the fifteenth century* (Unpublished Ph.D. thesis). University of Leeds.
Ellis, W. (1741). *The timber tree improv'd: Or, the best practical methods of improving different lands with proper timber*. London.
Evans, D. (1992). *A history of nature conservation in Britain*. London: Routledge.
Evelyn, J. (1664). *Sylva, or a discourse of forest-trees*. London.
Fitzgerald, R., & Grenier, J. (1992). *Timber: A history of the Timber Trade Federation*. London: Batsford.
Forestry Commission. (1955). *Report of the committee on hedgerow and farm timber*. London: HMSO.
Forestry Commission. (2002). *National inventory of woodland and trees*. Edinburgh: Forestry Commission.
Fuller, R. J., Williamson, T., Barnes, G., & Dolman, P. M. (2017). Human activities and biodiversity opportunities in pre-industrial cultural landscapes: Relevance to conservation. *Journal of Applied Ecology, 54*(2), 459–469.
Gibbs J. N., Brasier C. M., & Webber J. F. (1994). *The biology of Dutch elm disease*. Edinburgh: Forestry Commission (Research Information Note 252).
Hale, T. (1756). *A compleat body of husbandry*. London.
Hatcher, J. (1993). *The history of the British coal industry, Volume 1. Before 1700: Towards the age of coal*. Oxford: Oxford University Press.
House of Commons. (1792). *House of Commons Journal, 47*. https://books.google.co.uk/books?id=M2RIAQAAMAAJ.
Langton, J. (1979). *Geographical change and industrial revolution: Coal mining in south-west Lancashire 1590–1799*. Cambridge: Cambridge University Press.
Latham, B. (1957). *Timber. Its development and distribution: An historical survey*. London: George Harrap and Co.
Lennon, B. (2009). Estimating the age of groups of trees in historic landscapes. *Arboricultural Journal, 32*, 167–188.
Lowe, R. (1794). *General view of the agriculture of the county of Nottingham*. London.

Meager, L. (1697). *The mystery of husbandry: Or, arable, pasture and wood-land improved.* London.

Middleton, J. (1798). *View of the agriculture of Middlesex.* London.

Mortimer, J. (1707). *The whole art of husbandry: Or, the way of managing and improving of land.* London.

Morton, J. (1712). *The natural history of Northamptonshire.* London.

Nichols, J. (1795). *The history and antiquities of the county of Leicester.* London.

Nourse, T. (1699). *Campania felix: Or, a discourse of the benefits and improvements of husbandry.* London.

Pennant, T. (1776). *A tour in Scotland, and voyage to the Hebrides; MDCCLXXII* (Part I). London.

Plymley, J. (1803). *General view of the agriculture of Shropshire.* London.

Pontey, W. (1805). *The forest pruner, or timber-owner's assistant.* London.

Pringle, A. (1794). *General view of the agriculture of the county of Westmoreland.* London.

Rackham, O. (1986). *The history of the countryside.* London: Dent.

Rackham, O. (2006). *Woodlands.* London: Collins.

Read, H. J. (2008). Pollards and pollarding in Europe. *British Wildlife, 19,* 250–259.

Rowe, A. (2015). Pollards: Living archaeology. In K. Lockyear (Ed.), *Archaeology in Hertfordshire: Recent research* (pp. 302–324). Hatfield: University of Hertfordshire Press.

Rowe, A., & Williamson, T. (2013). *Hertfordshire: A landscape history.* Hatfield: University of Hertfordshire Press.

Rowley, T. (2006). *The English landscape in the twentieth century.* London: Hambledon Continuum.

Rudge, T. (1807). *General view of the agriculture of the county of Gloucester,* London.

Ryden, L., Pawel, M., & Anderson, M. (2003). *Environmental science: Understanding, protecting and managing the environment of the Baltic Sea region.* Uppsala: Baltic University Press.

Selby, J. (1842). *A history of British forest trees.* London.

Sheail, J. (2002). *An environmental history of twentieth-century Britain.* London: Palgrave.

Smith, J. (1670). *England's improvement reviv'd.* London.

Spray, M. (1981). Holly as fodder in England. *Agricultural History Review, 29*(2), 97–110.

Stevenson, W. (1809). *General view of the agriculture of the county of Surrey.* London.

Theobald, J. (2000). *Changing landscapes, changing economies: Holdings in Woodland High Suffolk, 1600–1850* (Unpublished Ph.D. thesis). University of East Anglia, Norwich.

Toulmin-Smith, L. (Ed.). (1907). *The itinerary of John Leland in or about the years 1535–1543*. London: George Bell.

Tuke, J. (1800). *General view of the agriculture of the North Riding of Yorkshire*. London.

Vancouver, C. (1810). *General view of the agriculture of Hampshire*. London.

Williamson, T., Barnes, G., & Pillatt, T. (2017). *Trees in England: Management and disease since 1600*. Hatfield: University of Hertfordshire Press.

3

Local Knowledge on Tree Health in Forest Villages in Turkey

Akile Gürsoy

1 Introduction

1.1 Forests, Trees and Tree Health

It is estimated that about 30% of the world's land area is covered with forests (FAO 2016) but it is not clear whether this is an undisturbed primary forest, severely degraded or something in between. The global data do not allow us to see whether the forest is healthy or has been subject to attacks by pests, disease or forest fire, or damaged by wind or air pollution (Archard 2009). However, examining forest health is important in assessing the future of the forest and determining whether any remedial measures are needed. The ever-changing natural and man-made processes affecting forests are quite complex and require detailed

---

A. Gürsoy (✉)
Sociology Department, Faculty of Sciences and Letters,
Beykent University, Avalon Campus, İstanbul, Turkey

information not only on natural processes but also on human behaviour affecting forests (PROFOR and WORLD BANK GROUP 2017). Local observations and knowledge of tree health and well-being can provide extremely valuable sources of information that are crucial in assessing the picture on the ground.

There is concern that forest mortality caused by fires, invasive insects and pathogens has increased beyond the levels of twentieth-century experience (Millar and Stephenson 2015). The causes of these changes in nature and the consequences of environmental intervention have always been a matter of controversy among different observers and interest groups (Dobson and Eckersley 2006; Kagan 2009). Recent national forest policies also show that many interventions that were once thought to be beneficial and necessary, such as planting a monoculture of trees, do not provide optimum conditions for either tree survival or a balanced ecosystem (Angel 2008).

This chapter, based on anthropological research (2014–2017) on forest villages in Turkey sponsored by the Scientific and Technological Research Council of Turkey (TÜBİTAK), explores villagers' perception of forest and tree health. Perception of tree health is important since it reflects the villagers' attention and evaluation of the trees around them and it determines subsequent action to mitigate pests and take other decisions against poor tree health.

In addition to compiling visual documentation, qualitative and quantitative data were collected from selected villages across the country with the aim of understanding village livelihoods and human interaction with forests. As this chapter argues, the villagers relate to different trees in different ways and their perception of tree health very much depends on the value that they attribute to the tree in question. Currently, there is a lack of research on the social and cultural values of trees and how these may impact human behaviour related to tree health (see, for instance, Marzano et al. 2016).

This chapter first introduces the context for forests and trees in Turkey. This is followed by a consideration of the specific place of trees in private gardens and the value of trees in Turkish folklore. As state ownership and control of forests is a crucial issue in the Turkish context, a discussion of rules and regulations regarding forests and their impact on forest villagers is presented. Rapid urbanisation, however, has extended concerns related to forests from being rural concerns to also being urban matters. This social fact has meant that the categories of trees include not only trees in forests, gardens and orchards but also trees in urban parks and landscapes. Next, the chapter discusses research findings related to forest villagers' perceptions on selected tree diseases. This discussion shows the various controversies in interpreting the health of trees. Different explanations related to tree health and diseases are analysed with reference to local knowledge versus scientific knowledge.

1.2 Trees and Forests in Turkey

Turkey is rich in its forest wealth, particularly from the point of view of biodiversity. According to the Global Environment Facility (GEF) benefits index for biodiversity, the median is 1.5. The scale of the benefits index is (0–100), and Turkey has a 6.23 index value. In this respect, Finland for example has a 0.2 index, despite the fact that almost three-quarters of the country is covered by forests. Turkey ranks 43/192 countries, whereas Finland has a score of 0.17 and ranks 164/192 (The World Bank 2009). According to 2016 World Bank data, more than 15% of Turkey is covered with forests (see Table 1).

Situated at the centre of the junction joining the continents of Asia, Europe and Africa, and surrounded by three seas, namely the Black Sea in the north, the Mediterranean in the south and the Aegean in the west, the Anatolian peninsula experiences diverse and unique ecosystems. The forests range from predominantly oak forests in the northwestern parts of the country to predominantly pine forests in the south (De Planhol 1965; Çağlar 1992; Bingöl 2005). The country is known to have 11,000 plant species, and these almost total the number of plant species of the whole continent of Europe. Furthermore, 3708 of these species are considered endemic to the country. Diverse types of forests such as the *longos* wetland forests exist in western Turkey. With significant biodiversity, the health of these trees and forests in Turkey is of crucial ecological concern.

Table 1 Forest area, demographic and economic variables of selected countries (Source Compiled from 2016 The Little Green Data Book, World Bank Group, The World Bank, Washington, 2016)

Country	Population (millions)	Land area (1000 sq. Km)	GNI per capita (US$)[a]	Urban population (% of total)	Agricultural land (% of land area)	Forest area (% of land area)	Threatened species, higher plants
Brazil	206.1	8358	9844	85.4	33	59.0	516
China	1364.3	9388	6419	54.4	55	22.2	568
Finland	5.5	304	40,233	84.1	7	73.1	8
France	66.2	548	35,829	79.3	53	31	41
Germany	81.0	349	40,365	75.1	48	32.8	24
India	1295.3	2973	1352	32.4	61	23.8	385
Japan	127.1	365	29,768	93.0	12	68.5	24
Mozambique	27.2	786	511	31.9	64	48.2	84
New Zealand	4.5	263	36,053	86.3	42	38.6	23
Russian Fed.	143.8	16,377	10,677	73.9	13	49.8	59
Turkey	75.9	770	9762	72.9	50	15.2	104
UK	64.6	242	39,041	82.3	71	13	22
USA	318.9	9147	46,858	81.4	44	33.9	379

[a]Adjusted net national income per capita: equals gross national income minus consumption of fixed capital, energy depletion, mineral depletion, and net forest depletion, divided by midyear population (World Bank; data are for 2014)

Forested areas in the country have increased by almost one million hectares annually over the last three decades. This is mainly due to rural migration which has resulted in a flow of the population from rural areas into urban cities. As a result of this migration, unharvested land soon turns into forest land. Other reasons for the increase in forested areas may be attributed to increased public awareness of environmental issues, and afforestation and reforestation activities of the public and the private sectors.

The main types of trees in the forests are oak (*Quercus* spp, 13 different species); Calabrian pine (*Pinus brutia*); and Anatolian black pine (*Pinus nigra*) (Republic of Turkey, Ministry of Forestry and Water Affairs, General Directorate of Forestry 2015). In addition to natural and planted forests, Turkey enjoys numerous endemic and planted fruit trees. Planted fruit trees are mostly in orchards or in home gardens (Mataracı 2004; Namıkoğlu 2012).

Trees continue to provide an indispensable source of nutrition and valuable income in Turkey. Even in the most densely forested areas of the country, forest village livelihoods are a combination of different degrees of agriculture, animal husbandry and forestry. The place of fruit trees has always had significance for household economies. Therefore, when considering the health of trees, the overall picture would be incomplete if we exclude trees in gardens and orchards (Alkan and Toksoy 2008; Tan 2016; Çenkoğlu 2016; Bozok 2016; Erdoğan 2017; İdrisoğlu 2015).

1.3 The Place of Gardens

Discussing the place of gardens and fruit trees in the environmental history of humankind, Radkau argues that the study of the anti-wilderness (i.e. the garden with its fruit trees) reveals an especially intimate and, at the same time, creative relationship with nature since time immemorial (Radkau 2008). Radkau categorises the garden and the field as entities separate from the wilderness. The garden is the most intimate space, often separated and fenced, and, in addition to providing fruits and vegetables, is at the same time a space symbolically characterised as a private domain. This is still true of village gardens in rural Turkey, as well as private gardens in cities.[1]

Historically, compared to the fields, the garden, with its limited size, offered an enclosed laboratory-like space where peasants and aristocrats could experiment with natural fertilizers, pruning and other plant-growing practices. The closeness of the garden to human dwellings allowed daily observations throughout the year. The garden is important because, in conjunction with fruit trees, it is one of the most elementary forms of human interaction with nature. It provides a sense of sustainability, as well as an ideal of beautiful and "domesticated" nature. Radkau argues that there are many examples in agroforestry of the combination of garden-like cultivation and fruit trees which may provide inspiration for the sustainable development of agriculture and forestry in many parts of the Third World.

1.4 Trees in Turkish Culture and Folklore

Trees are much revered in Turkish history and mythology, with many deemed to be sacred. Attributing symbolic meanings to trees continued in Islamic times (Altan et al. 1998). In contemporary folklore, there are many legends attributed to trees in general and to monumental trees in particular (Gülersoy 1972; Üsküdar Municipality 2003; Öner et al. 2010). Certain trees are viewed as having a protective persona, ensuring the well-being of people connected to these trees, while others are perceived as having sacred qualities that ensure their own protection against those who plan to cut down or destroy them (Altan et al. 1998).

In republican Turkey, the founding leader Atatürk is known to have attributed much importance to trees and forests.[2] In the 1920s and 1930s, the country was predominantly rural. The dictum of "peasants are the masters of the nation"[3] rests on the acknowledgement that it is the villagers whose labour on the land provides the wealth of the country. The concept of the forest was integrated into the general framework of agricultural wealth, and into the understanding that trees and forests form the basis of the wealth of the country.

1.5 Trees and Forests Under Turkish Law

Today, there are about 10 million inhabitants living in the 21,000 villages that are administratively classified as "forest villages" in Turkey. According to the Forestry Law, inhabitants of villages classified as forest villages have the right to make a profit from timber from designated areas of the forest. The whole process of logging and cutting tree branches is strictly sanctioned and controlled through the agency of the Forestry Directories.

Despite mythic and symbolic values attributed to trees, protecting trees from human harm seems to be a legitimate concern that is often emphatically articulated by the administrative authorities (see Anadolu Agency 2017).[4]

It is worth noting that in Turkey, about 99% of the forest lands and the forest products are owned by the state. Only a very small percentage of forest land falls into private ownership or within the ownership of foundations (Gülöksüz 2010; Aygen 2002). Villages that are administratively classified as forest villages have a controlled usage of designated forest land. Tree felling and logging is strictly regulated and controlled by the state forestry administration.

The history of forestry in Turkey shows that forestry practices and policies in the Ottoman Empire manifested attempts to develop "rational" forest management, or scientific forestry in a dominantly agrarian setting, where industrial and technological progress was only in the making. Dursun argues that centralisation had a direct impact on the development of forestry in both the Ottoman era and Republican Turkey (Dursun 2007).

1.6 Urban Expansion and the Fate of Forests

Even though forests were traditionally associated with villages and rural life, rapid urbanisation in Turkey has meant that forest and tree health concerns have extended to urban areas (Atmış 2007; Ata 2007; Erdönmez and Özden 2009; Göl et al. 2011; Atmış et al. 2007).

It should also be remembered that most city dwellers in Turkey are recent or one generation back migrants from villages. Many sociological analyses in Turkey have described this urbanisation process as "the ruralisation of towns" and have noted that city spaces are inhabited by people whose cognitive approach to life is more rural than urban (Suzuki 1964, 1966; Karpat 1967; Özbay 2015). Appadurai draws attention to the changing social, territorial and cultural reproduction of group identity under times of rapid dislocation of people in internal or out of the country migration. He makes the point that groups are no longer tightly territorialised, spatially bounded, historically unselfconscious, or culturally homogenous (Appadurai 1991). The question still remains as to how these dislocations affect traditions of perception related to nature in general and to trees in particular. Villagers who once lived in proximity with their trees in their garden now have a different relationship to trees in the city parks or in nearby public woods and forests. Indeed, social science analyses of rural to urban migration have often overlooked the changes related to migrants' relationship to nature in general and to trees in particular.

Urban livelihoods mean a different daily relationship with nature. Once they move to the towns, the vast majority of migrant villagers no longer have a garden. They can no longer walk to their orchards or fields. They live in urban apartments that are much more constricted compared to village life. Yet, some migrants routinely go back and visit their villages and can compare the difference between rural and urban environments. Our research has shown that many villagers who come to temporarily live in towns to take care of their grandchildren or who leave the village for health reasons have voiced their preference for the spacious, fresh-aired village life.

1.7 Categories of Trees

It is in the village that villagers build gardens. Trees in their gardens and orchards are seen to be within their own direct responsibility. They will buy insecticides, compare notes with other villagers and consult the Forestry Department if they see any ailment in the trees in their garden. They take pride in the fruits and vegetables that grow there.

Similarly, they are observant and concerned about the cash-bringing and fruit-bearing trees in their orchards. The health of trees in the forest, however, is of importance only to the extent that the tree has some direct economic value for them.

In her groundbreaking categorisation of food, with her theories of symbolic boundary maintenance, anthropologist Mary Douglas has tried to understand why and how some animals and foods are considered edible whereas others are not (Douglas 1966). Based on recent anthropological research from forest villages in Turkey, I would like to suggest that for the villagers, the garden, the orchard and the forest are circles of three distinct spaces that embody different meanings. Trees in the garden are like household pets, trees in orchards are like domesticated animals, and trees in the forest signify wild animals in nature (see Table 2).

Why and how we care for some and are more distant and inattentive towards others depends on how they are placed in these symbolic spaces. From the point of view of the villagers, this categorisation affects their perception of trees and tree health. Trees in the garden and in orchards are considered more their responsibility, whereas their relationship to the forest is ambivalent. On the one hand, there is a discourse that owns up to the forest and says that the forest is their biggest wealth and asset. They would feel enraged if their rights related to the forest were taken away from them, if they no longer had access to using the forest. On the other hand, they also say that they are not responsible for the trees in the forest. Since the state legally owns the forests, it is up to the state to take the necessary measures for ensuring the healthy survival of trees there.

2   Research on Forest Villages: Methodology

Ethnographic research was carried out (2014–2017) in 12 selected forest villages covering the 12 *NUTS-1* districts of Turkey, each region representing a geographic and socio-economic entity[5] (see Fig. 1). For the purposes of this research, one village was selected from each of these regions. Thus, the research covered a total of 12 forest villages across the country.

Table 2 Locations and categories of trees

Territory	Category	Relationship to humans	Tree protection
The garden (private gardens)	Domesticated	Close proximity, has cash and/or sentimental value	Private
The field, orchards	Domesticated	Cash value	Mostly private
Planted forests	Domesticated	Cash value	State controlled
Urban public parks	Domesticated	Touristic and recreational value	State controlled
National and natural parks, reservations	Wild nature	Touristic and recreational value	State controlled
Planted forests	Controlled wilderness	Cash value, touristic and recreational value	State controlled
Natural forests	Wilderness	May have cash value	State controlled

3 Local Knowledge on Tree Health in Forest Villages in Turkey

Fig. 1 Map of Turkey with 12 NUTS districts. I would like to thank Turan Asan from the Graphics Design Department of Beykent University, for the design of the map

Selected graduate students and social scientists carried out participant observation in these villages.[6] Each student stayed in one of these villages for a duration of one to three months. In addition to participant observation related to village life and the villagers' interaction with the forest, in each village semi-structured interviews were conducted with the village headman and with a random sample of households permanently dwelling in the village. In all, the data cover semi-structured interviews with 330 households and 12 village headmen. Discussions with forestry officials, publications and brochures of environmentalist NGOs, and discussions that have taken place in the social media have also been analysed for the purposes of our research.

The research team[7] visited each selected village and interacted with the village headman and village men and women. One of the primary concerns of the research project was to understand the relationship of the villagers with the forest that surrounds the village and the perceptions of forest health from the point of local folk aetiology. The project was of an explorative social anthropological nature, with the overall objective of understanding and assessing the role of forests in the economic livelihood of villages (see Figs. 2 and 3).

Fig. 2 Forest village in TR7 Central Anatolia (Erdoğan 2017)

Fig. 3 Forest village in TR9 Eastern Black Sea region (Türkeli 2014)

3 Findings

The research confirmed national findings related to demographic change in these villages: all the villages have lost their young population. This has meant that village livelihoods and economies have become more stagnant and the potential of making use of the agricultural land and the forest has been traded off for wages in towns and cities (Çakın 2015; Doğru 2016; Emlik 2017; Tüncer 2016). The exodus of the younger generations is also reflected in village settlement landscapes. Dilapidated houses, the remains of buildings, once inhabited by three or four generations all living together are now left empty. Such "ghost" houses left in ruins give the impression that the villages are deserted. However, the villages also contain newly built villa-type modern houses that are constructed with modern technology. These are mostly built by migrant villagers working in Europe and wanting to have a future retirement house. The villages are invariably characterised by the fact that the population increases in the summer and the holidays when urban migrants are able to come and visit their home village.

The fact that the majority of the younger generations are no longer in the village and no longer involved in agriculture and forestry as they once were means that local knowledge on forests and wildlife (plants and animals) is steadily disappearing. Fifty years from now, extensive indigenous knowledge on herbs and other natural life will have significantly diminished, if not perished. This is also true of knowledge on perceptions of forest and tree health.

One frequently articulated cause of tree ill health is climate change. All villagers agree that they are experiencing some kind of climate change, which they perceive to be very much part of their daily lives. Perceptions of climate change are frequently associated in the local discourse with poor tree health. Villagers have been vocal about various environmental problems such as global warming (hotter summers, less snow on the ground), changes in seasonal temperatures and erratic rainfall, drying up of streams and fountains. This change manifests itself in repeated accounts:

In the past, there would be longer periods of snow on the ground. We would have snow in mid-November until the end of March, and it would be 2 metres in height. Now it snows and the snow disappears in a week. It snows again, but the snow is about 50 cm on the ground. It is not as cold as it used to be. We have more rain and more storms. There is more humidity. Temperatures are erratic. (Villager from TR9 Eastern Black Sea Region 2014)

3.1 Examples of Tree Diseases and Local Perceptions

The villagers seem to be mostly concerned about the health of the fruit trees in their own gardens. For most villagers, the health of the trees in the forest is a matter of observation but not so much a matter of direct concern. They seem to believe that the forest will be there forever and that it will continue to thrive despite the presence of tree diseases and other environmental challenges. For the villagers, the forest was there at their birth and will continue to be there regardless of the changes occurring within the forest and around it. For example, the pine processionary moth (*Thaumetopeo pityocampa* Schiffemüller) causes damage to pine forests in the Mediterranean, Aegean and Marmara regions. The pest is visible with its manifestations on infected trees (see Fig. 4). However, none of the villagers mentioned this pest without being specifically asked about it. They confirmed seeing the infested trees and they knew about the pest. Some of them had joined the teams working to collect and burn the bags before the eggs hatch and pass the larval stage. Still, this tree pest was not a priority for them. This finding confirms our analysis that the villagers are more concerned about trees that have a direct economic return for them.

On the other hand, the inhabitants of the Princes Islands in İstanbul have taken civic action to combat the pest that is affecting the trees on the island. The Forestry Administration of the Princes Islands has released 10,000 "gladiator insects" to combat the pest in a pilot area on the islands. The directorate decreed a statement saying the harmful insect spread unprecedentedly due to climate change and

Fig. 4 *Thaumetopea pityocampa (Çam kese böceği)* found on all types of pine trees across Turkey. The moth larvae affects primarily *Kızılçam (Pinus brutia)*, but also other pines such as *Karaçam (Pinus nigra)*, *Sarıçam (Pinus silvestris)*, *"Pinus maritima"*, *"Pinus halepensis"*, *Fıstık Çamı (Pinus pinea)*, *Lübnan Sediri (Cedrus libani)* and sometimes *"Juniperus excelsa"* (Gürsoy, TR8 Western Black Sea region, September 2015)

that in combatting the pest, priority is given to methods that protect the ecological balance. The gladiator insect (*Calosoma sycophanta* L.) was first grown in the laboratories on the island in 2003 under the initiative of the members of a civil organisation, the *Cultural Foundation of the Princes Islands* (*Milli Gazette*, 1 May 2012). It is important to remember that the pine trees of the Princes Islands are renowned for their beauty, in addition to their recreational and touristic attraction and value.

Citizens in other parts of urban Turkey have also their voiced concern and have sought for remedy against the pine processionary moth.

> I want to ask friends who may know: caterpillars have made nests on pine trees. However, I could not find the means to rid the tree from them, and now the branches of the tree have become terrible. Will the tree renew its branches or will it wither away? What can I do to save the tree? P.S. The tree is quite tall and I pruned the lower branches so that it does not harm the fruits and the vegetables. (Inquiry from the city of Hatay, Social Media, 17.7.2006)

Today, there is general consensus in scientific literature that there are three major causes of tree and forest ill health in Turkey. These are (1) forest fires; (2) invasive insects; (3) tree diseases due to pathogens (Çepel 2008). From the point of view of the villagers, there is yet another major biotic cause of poor tree health: harmful mammals and birds. In fact, the issue in question is not the tree per se, but the crops that the tree yields. It seems that for many villagers a tree is valuable and worthy to the extent that it adds to their livelihood and economy. One threat they mention is mammals' attacks, such as bears and wild boar. The villagers will go to great lengths to protect their trees from these predators (see Figs. 5 and 6). It was observed that one villager in the Marmara region turned his tractor into a tent in order to keep watch overnight against such predators (Kaplan 2015).

In the Eastern Black Sea region of Turkey, villagers are known to guard their garden trees from the attacks of bears (Fig. 7). They attach CDs or similar gadgets on the branches of the trees. Apparently, the

Fig. 5 Watch tents and sound-making bottles placed in fields to protect the crops and the fruit trees from bears and wild boar, Western Turkey, August (Kaplan 2015)

bears and birds interpret these shiny objects as reflections of rifles and they do not come near the tree which they would have otherwise plundered for its fruit or for the bee hives that are placed on or naturally found on it. When interviewed about causes of tree ill health and damage, many villagers have mentioned the lethal harm done to trees by bears and wild boar (Türkeli 2014; Çenkoğlu 2016; Kaplan 2015).

3.2 Controversies in Interpreting the Health of Trees

Our research has shown that there are different kinds of knowledge: (1) that of the villagers, based on their daily observations and intergenerational accumulated knowledge of their environment (i.e. traditional indigenous knowledge); and (2) that of the forest engineers, based on

Fig. 6 Watch tents and sound-making bottles placed in fields to protect the crops and the fruit trees from bears and wild boar, Western Turkey, August (Kaplan 2015)

scientific learning, reasoning and periodic observation. A study of the latter would merit a study of the science of forestry. Anthropological studies acknowledge that there are different categories of "knowledge" in societies and studies of indigenous knowledge (IK), traditional ecological knowledge (TEK) and studies of technology and science (STS). These approaches in essence focus on different sociocultural settings and different "kinds" of knowledge. However, there is a call on students of IK, TEK and STS that they "talk to each other" (Knudsen 2009, p. 9; Atran 1996). Knudsen challenges the view that addresses knowledge as separate "systems" or "traditions", arguing for more sensitivity to how knowledge is constructed or reconstructed in opposition to other knowledge. Our research has shown that there are differences in the views of villagers and forestry officers as regards causes and consequences of poor tree and forest health.

Fig. 7 Pear tree damaged by bear, TR4 Eastern Marmara region (Kaplan 2015)

Presently, many villagers living in forest villages attribute tree ill health and forest deterioration to certain causes. These, however, do not always correspond to official, "scientific" views of forest engineers and forest officers of the Forestry Administration. The importance of understanding local knowledge and experience; and the importance of a dialogue between local understandings and scientific knowledge and perspectives is fundamental for making advances in tree health.

Maybe the foremost controversy is related to the forest management techniques that are employed. In forest management one method that is extensively employed is the Finnish style of tree cutting called "thinning" (*seyreltme*) which involves cutting the old and unhealthy branches of trees and leaving the younger and healthier ones in a given forest area. This method also includes the removal of some trees to make space for the growth of others. Space is opened in the forest. Villagers say: "Humans should be able to walk and air needs to circulate between trees". The previously widespread method was called clear felling (*traşlama*), which literally means shaving. In this method all trees were cut down in a certain area and the plot would be left 15–20 years for regrowth. Forest regrowth under this latter method produces same age trees (see Görcelioğlu 2003).

In the Finnish style of forest management with the selection method, forests are visually ever present. The forest may not be dense but the landscape shows uninterrupted forest scenery. Many villagers claim that with the thinning method, trees lose their vigour and health and that the young trees are weak. When it snows, trees break and collapse on each other. Many younger off shoots begin to rot before reaching adulthood. Roots rot due to lack of sunshine. This method also poorly affects the soil. The oil dripping from the machines used to cut the tree trunks and branches fall on the soil and pollute and negatively affect new growth of plants. Furthermore, the thinned forest does not provide sufficient cover for wild animals. Wild animals do not have enough covered places to form nests or to hide from predators. The villagers claim that the clear felling method allowed for healthier regrowth. Cutting all trees at once allows young trees to grow faster. With plenty of sunshine, the healthiest trees flourish. The forest becomes a thick, robust forest and wildlife thrives better in a denser forest.

This seems to be an unresolved controversy and on the whole, the forest engineers who are present in the discussions with the villagers do not accept the above arguments. They agree that young trees growing under this system do not see enough sunshine. Engineers are ambivalent about roots rotting due to frequent trimming and cutting. They do not accept the claim that newly growing trees are weak. They argue that the tree trunks are thicker and they believe that the best resistance to snow is with the selection method. Finally, engineers add that it is in clear cutting that wildlife becomes homeless. As for the advantages of the clear cutting method, forest engineers agree that cutting all the trees at once allows young tees to grow faster. There is also consensus that this method ensures plenty of sunshine for the young trees and that the healthiest ones flourish. The forest becomes a thick, robust forest but this is only after a certain time and wildlife certainly suffers until the new forest grows to a height.

Different points of view emerge in nearly all aspects of forestry and tree health. Some villagers believe that the forest should be left to expand by itself instead of planting trees to increase forest land. The controversies are not restricted to the different forest management techniques. There are different perceptions related to the effects of artificial lakes (*gölets*) and man-made dams: villagers claim that dampness, dew and foggy-misty weather caused by nearby dams negatively affect the health of trees, especially fruit trees and their leaves. Again, this view is not always supported by agricultural and forest engineers. Villagers also claim that mobile phone transmitters and electromagnetic emissions have increased, negatively affecting the natural surroundings. The villagers believe that bees and birds have suffered. Both villagers and the forestry administration agree that different types of insects have increased. The regional forestry administration and the village council are debating the causes and the remedies.

Similarly, in the Eastern Black Sea Region where there is a rich cover of natural forest including spruce, fir, different types of pine, all kinds of conifer trees as well as indigenous oak, poplar, wild fruit trees, walnut trees (the most profitable ones being walnut trees, pines and spruces), villagers say that spruce trees are drying up from their top. This condition is also affecting young trees and this ailment has accelerated since 2010. Villagers think that this is due to the dams in the area that cause humidity and negatively affect pine trees (see Fig. 8).

Fig. 8 Branches of spruce trees believed to be infected due to humidity, TR9 Eastern Black Sea region (Türkeli 2014)

Differences of opinion exist between local inhabitants and the forestry administration regarding the remedy for tree diseases. One such case is that of the fungal disease threatening the box tree. The box (boxwood) tree (*Buxaceae* from the *Buxus* family) is known to be a valuable endemic tree that has both economic and folkloric value in the Black Sea Region of Turkey. However, presently this tree is suffering due to climatic changes and human activities such as indiscriminate cutting that also increase the spread of tree infection. In view of this situation, the Regional Forestry Administration called in 2012 for a bid to have all infected trees cut down to prevent the infection of other trees in the vicinity. However, nobody volunteered to undertake this task. The Forestry administration justified its decision by saying that the wood of the dried up trees would be turned into economic profit. The administration is of the view that all boxwood trees within their regional district (which includes gene protection trees and national forest reservations) should be cut down. Local inhabitants and members of nature preservation societies, however, have observed that new leaves have grown on trees that were thought to have died out.

The difference of opinion also extends to views related to the spread of the disease. Members of the environment protection societies observe that whereas boxwood blight *Cylindrocladium pseudonaviculatum* does not need any such opening to penetrate and infect the tree, *V. buxi* enters the tree from cuts and wounds on the tree. Therefore, if the trees are pruned or cut down, this will lead to the further spread of the disease. The famous *Fırtına Valley* forests on the Black Sea Coast of Turkey are very dense natural forests. Members of nature preservation societies are also concerned that an attempt to cut the boxwood trees will mean that people will enter and trek through densely forested areas, most probably increasing the spread of the disease. An additional concern is related to the economic use of the tree. Branches and leaves of boxwood are consumed by commercial florists as decorative green fillings when making bouquets of flowers. The use of the tree branches in this manner possibly greatly contributes to the spread of the disease (Rural Environment and Forestry Problems Research Foundation 2013).

The stone pine tree (*fıstık çamı*) is again a most valued tree because of its fruitful cones. As part of their maintenance, the trees are pruned and sprayed with insecticide. In the past years this tree suffered from a particular disease. Villagers deem this to be most distressing: "Initially the tree flowers well with lots of cones, but then these do not remain to mature on the tree. At a certain height of the tree, the cones fall off. The income of the village has indeed fallen", one villager interviewed said. Another villager added, "I am not scientifically knowledgeable. The forestry administration is fighting this but it is not enough" (Villagers from the Mediterranean Region). The expectation of villagers is that scientists should engage to help them understand the problem in order to get rid of this tree disease (Arslan and Şahin 2016).

Another very common tree ailment is the European mistletoe, (*Viscum album* ssp. *austriacum (Wiesb.) Vollman*) which is locally named *gökçe otu* or *ökse otu*. This parasitic weed grows on the trunks and branches of all kinds of trees. It latches on and sucks the tree, draining its essential minerals and eventually causing it to dry out. Villagers value these parasitic weeds which they use to feed their sheep and goats, claiming that, apart from being nutritious, the plant regulates the bowel movements of the animals. However, the weed is known to hinder the healthy development of trees. The European mistletoe causes the drying up of the branches of the trees. Villagers see it as the responsibility of the Forestry Department to fight this tree parasite. On the other hand, the Forestry Department says that villagers, having shifted the responsibility to the Forestry Department and do not take the necessary precautions that they could and should do themselves.

Villagers believe that it is not only villagers who harm the trees by improper logging activities. The forestry administration too can make mistakes, and they too should be punished for their wrong deeds against trees and forests. They believe that the villagers more than the forestry administration protect the forest and that cooperation and mutual understanding is needed for the sake of the forest. Villagers believe that it is the older generation compared to the younger ones; it is men compared to women; local inhabitants compared to foreigners and villagers compared to forestry administrators who understand more about trees and the forest. The reasons for these convictions are based on the criteria of being more experienced, having more accumulated knowledge, spending more time in the forest and seeing at first hand the results of certain practices and applications (see Barlı 2006; Sağlam and Öztürk 2008; Tolunay and Alkan 2008).

4   Concluding Remarks

This chapter has tried to demonstrate that in the context of Turkey where forests and forest products are predominantly owned and regulated by the state; villagers take more responsibility for the trees in their own gardens and orchards. Ethnographic research has suggested that the villagers' perception and concerns over tree health very much depends on their proximity to and their relationship with the tree in question. The economic value of the tree increases concerns over its health and villagers are more aware about the condition of these trees.

There is a vast array of known or unknown causes underlying poor tree health and tree mortality. The villagers' perception of tree health provides valuable information for the challenges we face in understanding the causes of tree ailments and determining the success of interventions. The villagers feel that since they have lived for long periods of time near forests and since they spend more time with and observe the trees, they are the ones more knowledgeable and understanding about tree health. At the same time, they feel the need to consult scientific points of view and want to see forestry officials actively engaged in combatting tree diseases that affect the trees that are valuable for the villagers.

In Turkey, it is the Ministry of Forestry and Water Affairs that has the duty to protect, improve and operate the forests. The Ministry also has the responsibility of conserving nature as well as improving and operating the protected areas. The vision of the Ministry is stated to be "a reputable and strong organization which rationally manages forests and water resources; conserves and improves nature" (Eroğlu 2016). At the same time, it is within the strategic understanding of the Ministry that there should be a participatory approach in their work, giving priority to the "satisfaction of the people and organizations it serves and making efforts to ensure that a sensible, participatory, transparent, integrated and sustainable understanding of management predominates in all its applications" (Eroğlu 2016).

Thus, both on the part of the villagers, and also on the part of the central Forestry administration, there is an understanding of a need for collaborative work and mutual cooperation in comprehending tree diseases and working towards tree health. Working strategies are needed to ensure that such participatory approaches are developed and sustained to integrate the voice and the needs of the villagers and the concerns and endeavours of the Forestry Ministry. Although it is not new to state that participatory action that includes the voice of the villagers is needed, such a dialogue is particularly important in cases where there are different interpretations of natural phenomena.

Clearly, folk aetiology and scientific reasoning need to "talk to each other" and explain how these different modes of understanding have influenced each other. The cross-cutting domain where villagers, forestry officials and social scientists meet and construct their views is a complex ground that very much depends on the positioning of one group vis-a-vis the others. Views and convictions on tree health are negotiated in such a social and natural environment where different groups interact with each other, and ultimately affect tree health and survival.

Acknowledgements The research reported in this chapter was produced as a social sciences research project, entitled "Social anthropological research on forest villages in Turkey", sponsored by the Scientific and Technological Research Council of Turkey (TÜBİTAK) (Grant Number SOBAG-1001–114K125).

Notes

1. In Turkish history, Ottoman Sultans constructed many private gardens, however not all of them were designed as a setting for the impressive state ceremonies but rather as places where the sultans and their family could spend enjoyable hours or even days in privacy (Atasoy 2002).
2. http://www.isteataturk.com/haber/1111/yalovada-millet-ciftliginde-yuruyen-koskun-insaatinda-24071930.
3. In Turkish: "*Köylü milletin efendisidir*".
4. http://aa.com.tr/en/turkey/worlds-5th-oldest-yew-tree-discovered-in-nturkey/596600.
5. *nomenclature d'unités territoriales statistiques.*
6. The ethnographic village studies were conducted by Dr. Nihan Bozok, Büşra Şahin, Cennet Erdoğan, Emrah Tüncer, Handan Türkeli, Gizem İdrisoğlu, İbrahim Arslan, Pelinsu Çakın, Sena Çenkoğlu, Tangör Tan, Ünsal Uzan and Volkan Kaplan. Reference is made to their unpublished research reports in the chapter.
7. The multidisciplinary research team is composed of anthropologist Dr. Akile Gürsoy (Principal investigator), forest engineer, silviculture/horticulture specialist Dr. Cemil Ata, anthropologists Drs. Yüksel Kırımlı, Ebrar Akıncı, Abdurrahman Yılmaz, demographer Dr. Aykut Toros, historian Dr. Selçuk Dursun, sociologists Drs. Nihan Bozok and Ozan Zeynek.

References

Alkan, S., & Toksoy, D. (2008). Economic livelihood in forest villages: The example of Trabzon (in Turkish). *Kastamonu Üniversitesi Orman Fakültesi Dergisi, 8*(1), 37–46.

Altan, Y., Akgül, H., Uğur, Y., & Solak, H. (1998). *The monumental trees of Manisa and their place in folk culture* (in Turkish). Manisa: Celal Bayar University Publications. ISBN 975-6829-00-1.

Anadolu Agency. (2017). http://aa.com.tr/en/turkey/worlds-5th-oldest-yew-tree-discovered-in-nturkey/596600.

Angel, H. (2008). *Green China*. UK: Stacey International.

Appadurai, A. (1991). Global ethnospaces—Notes and queries for a transnational anthropology. In R. G. Fox (Ed.), *Recapturing anthropology: Working in the present*. Sante Fe: School of American Research.

Archard, F. (2009). *Vital forest graphics*. UNEP, FAO, UNFF.

Arslan, İ., & Şahin B. (2016). *Ethnographic village report for TR6 Mediterranean Region*, Turkey, (TÜBİTAK Project No. 114K125).

Ata, C. (2007, September 13–15). Arboratums and urban forests (in Turkish). *Proceedings of Urban İstanbul Symposium*, İstanbul.

Atasoy, N. (2002). *A garden for the Sultan—Gardens and flowers in the Ottoman culture*. Istanbul: Aygaz.

Atmış, E., Özden, S., & Wietze, L. (2007). Urbanization pressures on the natural forests in Turkey: An overview. *Urban Forestry & Urban Greening, 6*(2), 83–92.

Atran, S. (1996). *Cognitive foundations of natural history—Toward an anthropology of science*. Cambridge, UK: Cambridge University Press.

Aygen, D. (2002). *Forest law and related articles* (in Turkish). Ankara: Adil Yayınevi.

Barlı, Ö. (2006). Analytical approach for analyzing and providing solutions for the conflicts among forest stakeholders across Turkey. *Forest Policy and Economics, 9*, 219–236.

Bingöl, İ. H. (2005). *Forests and forest management in our country* (in Turkish). Ankara: Baran Ofset.

Bozok, N. (2016). *Ethnographic village report for TR3 Aegean Region, Turkey* (TÜBİTAK Project No. 114K125).

Çağlar, Y. (1992). *Forests and forest management in Turkey* (in Turkish). Cep Üniversitesi, İletişim Yayınları.

Çakın, P. (2015). *Ethnographic village, Turkey* (TÜBİTAK Project No. 114K125).
Çenkoğlu, S. (2016). *Ethnographic village report for TR5 Western Anatolia Region, Turkey* (TÜBİTAK Project No. 114K125).
Çepel, N. (2008). *Forests and functional values we have destroyed and forest mortality in our time* (İn Turkish). İstanbul: TEMA Vakfı Yayınları.
De Planhol, X. (1965). Les nomades, la stepe et la forêt en Anatolie. *Geographische Zeitshchrift, 53*(2–3), 291–308.
Dobson, A., & Eckersley, R. (2006). *Political theory and the ecological challenge.* Cambridge, UK: Cambridge University Press.
Doğru, S. (2016). *Ethnographic village report for TRA Northwestern Anatolia region, Turkey* (TÜBİTAK Project No. 114K125).
Douglas, M. (1966). *Purity and danger, an analysis of concepts of pollution and taboo.* UK: Routledge and Keegan Paul.
Dursun, S. (2007). *Forest and the state: History of forestry and forest administration in the Otoman Empire* (Unpublished Ph.D. thesis). Sabancı University.
Emlik, H. (2017). *Ethnographic village report for TRC South East Anatolia region, Turkey* (TÜBİTAK Project No. 114K125).
Erdoğan, C. (2017). *Ethnographic village report for TR7 Central Anatolia, Turkey* (TÜBİTAK Project No. 114K125).
Erdönmez, C., & Özden, S. (2009). Relations between rural development projects and urban migration: The Köykent Project in Turkey. *Ciencia Rural, 39*(6), 1873. Santa Maria.
Eroğlu, V. (2016). *Republic of Turkey, Ministry of Forestry and Water Affairs.* www.ogm.gov.tr.
FAO. (2016). *Global forest resources assessment 2015. How are the worlds' forests changing?* (2nd ed.). Rome: UN Food and Agriculture Organization.
Göl, C., Özden, S., & Yılmaz, H. (2011). Interactions between rural migration and land use change in the forest villages in the Gökçay Watershed. *Turkish Journal of Agriculture and Forestry, 35,* 247–257.
Görcelioğlu, E. (2003). *New environmental program for Finnish forestry* (in Turkish), TEMA Foundation, No. 41.
Gülersoy, Ç. (1972). *Monumental trees of İstanbul* (İn Turkish). İstanbul: Türkiye Turing ve Otomobil Kurumu.
Gülöksüz, E. (2010). Ownership rights in the forest, historic and contemporary developments (in Turkish). *Proceedings of Ownership of Land, Symposium.* Ankara: Memleket Yayınları.
İdrisoğlu, G. (2015). *Ethnographic village report for TR1 İstanbul region, Turkey* (TÜBİTAK Project No. 114K125).

Kagan, J. (2009). *The three cultures—Natural sciences, social sciences and the humanities in the 21st century*. Cambridge, UK: Cambridge University Press.

Kaplan, V. (2015). *Ethnographic village report for TR4 Eastern Marmara Region, Turkey* (TÜBİTAK Project No. 114K125).

Karpat, K. (1967). *The Gecekondu: Rural migration and urbanization*. Cambridge: Cambridge University Press.

Knudsen, S. (2009). *Fishers and scientists in modern Turkey, the management of natural resources, knowledge and identity on the Eastern Black Sea Coast* (Vol. 8), Studies in Environmental Anthropology and Ethnobiology. New York: Berghahn Books.

Marzano, M., Dandy, N., Papazova-Anakieva, I., Avtzis, D., Connolly, T., Eschen, R., et al. (2016, September). Assessing awareness of tree pests and pathogens amongst tree professionals: A pan-European perspective. *Forest Policy and Economics, 70*, 164–171.

Mataracı, T. (2004). *Trees, the exotic trees and shrubs of Marmara Region (Turkey)* (in Turkish) (TEMA, Publication No. 39).

Millar, C. I., & Stephenson, N. L. (2015). Temperate forest health in an era of emerging megadisturbance. *Science, 349*(6250), 823–826.

Namıkoğlu, N. G. (2012). *The trees and shrubs of Turkey* (in Turkish). İstanbul: NTV Publications.

Öner, N., Özden, S., & Üstüner, B. (2010). Relationship between a natural monumental stand in Turkey and local beliefs. *Journal of Environmental Biology, 31*, 149–155.

Özbay, F. (2015). *Family, city and population—Past and present* (in Turkish). İstanbul: İletişim Publications.

PROFOR Innovation and Action for Forests and World Bank Group, Poverty, Forest Dependence and Migration in the Forest Communities of Turkey, www.profor.info. Washington, June 2017.

Radkau, J. (2008). *Nature and power* (2002 German). Cambridge, UK: Cambridge University Press.

Republic of Turkey, Ministry of Forestry and Water Affairs, General Directorate of Forestry. (2012). *Forests of Turkey*. Ankara. ISBN 978-605-393-044-0.

Republic of Turkey, Ministry of Forestry and Water Affairs, General Directorate of Forestry. (2015). *Forest wealth of Turkey 2015* (in Turkish). www.ogm.gov.tr.

Rural Environment and Forestry Problems Research Foundation. (2013). *A rural environmental story: What happened to boxwood trees* (in Turkish).

Ankara: Kırsal Çevre ve Ormancılık Sorunları Araştırma Derneği Yayını No. 16.

Sağlam, B., & Öztürk, A. (2008). The relationship between forest villagers and forestry administration in increasing activities for the protection of the forest: The example of Artvin Regional Directorate (in Turkish). *Orman Fakültesi Dergisi, 8*(2), 131–143. Kastamonu Üni.

Suzuki, P. (1964). Encounters with İstanbul: Urban peasants and village peasants. *International Journal of Comparative Sociology, 5,* 208–216.

Suzuki, P. (1966). Peasants without plows: Some Anatolians in İstanbul. *Rural Sociology, 31*(4), 428–438.

Tan, T. (2016). *Ethnographic village report for TR2 western Marmara Region, Turkey* (TÜBİTAK Project No. 114K1250).

The World Bank. (2009). *The little green data book.* Washington: The World Bank.

The World Bank. (2016). *The little green data book.* Washington: The World Bank.

Tolunay, A., & Alkan, H. (2008). Intervention to the misuse of land by the forest villages: A case study from Turkey. *Ekoloji, 17*(68), 1–10.

Tüncer, E. (2016). *Ethnographic village report for TRB Central Anatolia Region, Turkey* (TÜBİTAK Project No. 114K125).

Türkeli, H. (2014). *Ethnographic report for TR9 eastern Black Sea region, Turkey* (TÜBİTAK Project No. 114K125).

Üsküdar Municipality (Üsküdar Belediyesi). (2003). *Impressive witnesses over centuries: From verses to our hearts—The monumental trees of Üsküdar* (in Turkish).

4

Mountain Pine Beetles and Ecological Imaginaries: The Social Construction of Forest Insect Disturbance

Elizabeth W. Prentice, Hua Qin and Courtney G. Flint

1   Introduction

Forest insect disturbances bear attributions at multiple scales, from the practices of local extractive industries to the politics of state and federal forest management to global climate change (Bentz et al. 2010; Dale et al. 2000, 2001; Müller 2011; Petersen and Stuart 2014). In the late 1990s and early 2000s, an outbreak of mountain pine beetle (MPB) (*Dendroctonus ponderosae*) swept through north central Colorado forests with an unprecedented scope and intensity, leaving massive swaths of red, beetle-killed trees throughout the landscape and precipitating community responses reflective of their unique economic bases, histories,

---

E. W. Prentice (✉) · H. Qin
Division of Applied Social Sciences, University of Missouri,
Columbia, MO, USA

C. G. Flint
Department of Sociology, Social Work & Anthropology,
Utah State University, Logan, UT, USA

and biophysical attributes (Colorado State Forest Service [CSFS] 2016). Such a setting provides an opportunity to examine how institutional forces pattern experiences of the natural world and responses to an ecological event or disturbance (Flint et al. 2012).

While other visual blights such as litter or pollution from industry are clearly attributable to human activity, the rust blanketed mountainsides appear perplexing to say the least. In the eyes of many local residents, the mountains are just not supposed to look this way, and yet there is no single perpetrator to blame. Reconciling such a shocking image with the knowledge that it is largely a natural event, demands attention to the way that such a disturbance is influenced by various political, market, and community factors.

Community responses to outbreaks are structured by local economies and the priorities of community members with access to local decision-making power, the policies of management officials and politicians, and the biophysical characteristics of the region itself (Flint et al. 2012). With the knowledge of how the responses of local communities have been patterned by broader socioeconomic and historical forces, it is important to also examine the competing conceptions of nature inherent in such responses and the discursive practices surrounding ecological events. While other studies of the human dimensions of insect disturbances explore how landscape and socioeconomic heterogeneity inform community responses to outbreaks (Flint et al. 2012; Qin and Flint 2010; Qin 2016), there has been insufficient attention to the ecological imaginaries that underpin these responses, and the way that culturally embedded conceptions of the natural world create context for the construction of and response to environmental events.

This work engages a political ecology perspective in order to consider the role and power of environmental narratives, and the various environmental identities of actor groups that emerge in relation to prevailing institutional power structures and to a constructed environmental problem. Since fixed assumptions about what the relationship between the human and non-human world "should be" become dogmatic and may even lend themselves to becoming the basis for exclusion and marginalization, this process of discursive analysis can be useful as a means of detaching knowledge claims from institutional structures of privilege, and evaluating environmental narratives on the basis of the subjectivities that simultaneously construct and are constructed by them.

2 Mountain Pine Beetles and Ecological Relationships

To adequately evaluate how communities and other human actor groups participate in unique constructions of MPB disturbances, it is first necessary to understand the ecology of the MPB and how it functions in the context of Colorado forest communities as a distinct, non-human actor group.

The MPB issue is endemic to Colorado forests. Moreover, insect disturbances play a critical role in maintaining forest ecosystems and are crucial to the process of forest succession (Dale et al. 2000). What has proved variable in outbreaks throughout North America in recent decades has been both the scope and the intensity of outbreaks (Dale et al. 2000; Petersen and Stuart 2014). Insect outbreaks that are deemed to exceed their natural range of variation can alter natural processes, such as nutrient cycles, and precipitate or further the extent of future disturbances such as wildfires, which represent a major challenge and destructive force for Colorado communities (Dale et al. 2000). Ironically, it is the suppression of wildfire and resultant processes of regeneration that is associated with the unique susceptibility to insect outbreaks seen in dense, homogenous forests (CSFS 2005).

Adult MPB preferentially seek out mature lodgepole (*Pinus contorta*) and ponderosa pine (*Pinus ponderosae*) trees to bore into, mate, and deposit eggs. Once in a tree, beetles emit aggregation pheromones that attract other beetles, and after many beetles have attacked the tree, they emit an anti-aggregation pheromone to ensure their eggs have sufficient resources (Petersen and Stuart 2014; Raffa et al. 2008). The beetles carry a fungus that stains the wood of the affected trees blue and obstructs the trees' water-transporting vessels while beetle larvae eat the trees' inner bark (CSFS 2005). Forests with an abundance of mature trees provide an ideal opportunity for beetles to reproduce and reach epidemic levels (Petersen and Stuart 2014).

Aside from the direct effects to beetle populations associated with the age of the available trees, epidemic-level outbreaks of MPB are attributed to more complex and broader climate-related variables, including drought and warming average temperatures (Bentz et al. 2010). Even with an abundance of prime host material, extended periods of cold temperatures have historically served as an effective regulator of MPB populations (Carroll et al. 2003). Warming temperatures speed up the life cycles of beetles, leading to increased numbers of individuals in populations and reducing generation times. In the context of disturbance events of unprecedented scope and intensity, the resilience boundaries of the rest of the forest ecosystem are under threat (Bentz et al. 2010).

3 Human Dimensions of Forest Insect Disturbances

Exploration of the human dimensions of forest insect disturbance represents a burgeoning area in the study of human–environmental interactions. Existing work examines the linkages between disturbances and broader socioecological systems at multiple scales; considering how local disturbances are informed by global forces including the market and changing climate (Bentz et al. 2010; Dale et al. 2000, 2001; Petersen and Stuart 2014), and how disturbances affect the attitudes and risk perceptions of local communities (Chang et al. 2009; Flint and Luloff 2007; McFarlane and Wilson 2008; Müller and Job 2009; Parkins and MacKendrick 2007; Qin and Flint 2012; Qin et al. 2015).

There is a discernible link between global climate change and forest disturbances including insect outbreaks (Bentz et al. 2010; Dale et al. 2000, 2001). While disturbances have always served an important role in shaping the composition and functional processes of forests, the frequency, intensity, and magnitude of recent disturbances are traced to changes in climate. Given the tendency of disturbances to interact and cascade within ecosystems, their increased prevalence represents a unique and unprecedented challenge for forest-dependent communities (Dale et al. 2001).

While the link with global climate change is important to consider, it is insufficient as a sole explanation for insect disturbances as it neglects the specific socioeconomic, political, and cultural forces that contribute to disturbance events and human community responses. Recognizing this deficiency, Petersen and Stuart (2014) engage a political ecology perspective to consider an MPB outbreak in British Columbia, which was regarded as the most severe beetle infestation in recorded North American history. They argue that too simplistic a link has been drawn between global warming and bark beetle outbreaks, effectively removing attention from global market pressures and local extractive industry practices that exacerbate outbreaks. In the case of British Columbia's outbreak, Petersen and Stuart argue that the scope and intensity was due to the timber industry's privileging of short-term economic gains by seeking to harvest unaffected old growth stands over the less profitable, beetle-affected timber, thus inhibiting the potential for timely regeneration and exacerbating the effects of the outbreak to forest communities.

Within the broader global context, much of the existing work in the human dimensions of forest insect disturbances is focused on community perspectives and responses (Chang et al. 2009; Flint and Luloff 2007; Kooistra and Hall 2014; McFarlane and Wilson 2008, 2012; Müller and Job 2009; Parkins and MacKendrick 2007; Porth et al. 2015; Qin and Flint 2012). Community contexts illustrate the collective and variable responses to natural events and the specific effects of tree health management on a local level. Disturbance-affected communities offer sites at which it is possible to see the interaction between local residents and management entities and the relations of trust and power implicit in these interactions (Porth et al. 2015). The community interactional perspective is engaged to explore the spruce bark beetle (*Dendroctonus rufipennis*) outbreak on the Kenai Peninsula in Alaska and the MPB outbreak in Northern Colorado (Flint 2006; Flint and Luloff 2007; Flint et al. 2012; Qin and Flint 2010; Qin et al. 2015). In these cases, risk perceptions and attributions vary by affected community, illustrating each community's unique and collective experience of the disturbance (Flint et al. 2012). As insect disturbances usher in further disturbance regimes, the shifting priorities of community members indicate the extent to which environmental problems and vulnerabilities are socially constructed as "social, institutional, environmental and cultural processes shape the way society experiences risk," with these forces and processes coming to coalesce at the community level (Flint 2006, p. 1598).

The fluidity in value-laden community experiences of, and responses to, insect disturbances is globally visible in a synthesis of forest insect disturbances in Canada, the USA, and the Bavarian forests in Germany (Flint et al. 2009). The synthesis notes the extent to which disturbances, as all ecological events, are viewed through complex cultural and economic lenses. Combined with local political and legal frameworks that work to create and constrain management opportunities, these broader forces ultimately determine how communities are moved to respond in the face of threats (Flint and Luloff 2007; Tomlinson et al. 2015). This process is inherently dynamic and shown to change over time as a community's economic history, amenity status, and biophysical features inform perceptions of risk, responses to disturbances, and ultimately a specific ecological imaginary, or idea about what the landscape should look like (Flint et al. 2012; Qin et al. 2015). Illustrating a similar fluidity, in national studies in the UK, attitudes about management are shown to vary by demographic factors, with men and older people favoring more aggressive management and women and young people more targeted management (Fuller et al. 2016).

Though the political ecology perspective is relatively underutilized in the study of the human dimensions of insect disturbances, Müller (2011) focuses on landscape as a cultural object and the symbolic meaning of disturbances and the sociocultural reverberations of bark beetle outbreaks. His case study explores the symbiosis between landscape ecology and sociopolitical forces in discussing how the image of the disturbed landscape becomes symbolic of the political process of forest management, and the deep cultural, identity-based significance of landscapes to local groups.

Inherent in both the social forces that precipitate forest insect disturbances and the perceptions of and responses to those disturbances are embodied conceptions and experiences of nature, or environmental subjectivities that coalesce at the community level. Considering the existing literature on the human dimensions of forest insect disturbances and the important role of communities in filtering both individual experiences and systemic forces, there is a need to further understand and empirically investigate the political and psychosocial, structurally entrenched forces that inform community perceptions of nature and emergent environmental narratives.

4   Environmental Narratives

Environmental discourses and the individual narratives they are composed of are fundamentally discourses of power, producing particular environmental subjectivities and individual-level experiences of environmental phenomena in the course of their exercise. This perspective, which draws on a Foucauldian interpretation of disciplinary power, has been referred to as "green governmentality" (Rutherford 2011, p. xvi). It seeks to interrogate the sort of stories that are being told about an environmental event, in this case an outbreak of MPB, and more importantly the consequences of those tellings for understanding how individuals and communities interact with environmental disturbances, and the various structural factors that inform the attribution of causality. Such discourses and narratives then, in line with a Foucauldian understanding of power, are fundamentally productive (rather than repressive) forces, producing understandings and subjectivities. Furthermore, local narratives and the broader global discourses they uphold are not simply stories, but are created relationally by the interaction of individuals and communities with their material and institutional contexts.

Within this conceptual framework, this work seeks to investigate how the MPB outbreak in northern Colorado during the early 2000s functioned as a site for the emergence and deployment of various environmental narratives, and how these narratives are nested within broader institutional and power arrangements within the area and globally. When evidently natural ecological events prove disruptive to the habitual flow of local society, an official narrative of explanation emerges alongside competing local narratives that are framed according to the vital interests of different actor groups, which are themselves positioned in complex and contested political, socioeconomic, and ecological contexts (Bixler 2013). Jasanoff (2010) writes that the competition between scientific and local narratives and conceptions of the non-human world are contentious because they separate the epistemic from the normative, or global fact from local value and issue a totalising image of reality without due consideration of the nuanced, complex, and culturally embedded investments communities have made in constructing reality as they know it. This work seeks to begin to untangle some of these nuanced investments in reality and how they vie with "official" explanations for recognition as truth.

Communities in the MPB-affected areas offer sites to capture the interaction and competing influence of various socioeconomic and political influences in determining emergent environmental narratives. The local community is recognized as situated at the nexus of broader society, and the lived, local environment, and community contexts, or the socioeconomic and biophysical features of local places, provide filters for individual experiences (Qin and Flint 2010). The local community is the conduit whereby individual social identities are established through social interaction and the place where people encounter the physical environment as part of their lived reality. It is where life is "least abstract," and where broader issues of market-driven inequality and power discrepancies that occur on a national or international scale converge to define individual experience (Wilkinson 1991, p. 24). Furthermore, the community's locality is the plane at which the effects of disturbances are most immediately felt and where successful work to mitigate their damages is most visible (Bridger and Luloff 1999). The social and biophysical community context effectively mediates both the construction of and response to an environmental event.

In the interactional framework, the physical place of a community constitutes the "spatial manifestation of a fundamental organization of interdependencies among people" (Wilkinson 1991, p. 53). While abstraction of the particular to the generalizable may be regarded as the method by which science achieves its universality and weight, communities are undeniably specific: constituted by specific people living grounded and particular lives in identifiable places (Jasanoff 2010).

With such an understanding, communities provide a compelling context in which to consider environmental identities and the exchange between official and local narratives, and the local-level implications of systemic power structures.

5 Study Area and Methods

Between 1996 and 2007, a massive outbreak of MPB (*D. ponderosae*) swept through north central Colorado, killing upwards of 3.4 million acres of primarily lodgepole and ponderosa pine trees and priming the landscape for further ecological disturbances in the forms of invasive species and fire (CSFS 2016; Qin and Flint 2010; Qin 2016). Although MPB are endemic to Colorado forests, this particular outbreak was unprecedented in both spatial extent and tree mortality; leaving massive swaths of dead, rust-colored trees throughout the northern part of the state and affecting communities throughout the region (Flint et al. 2012).

The research combines an interpretive approach based on understanding contextualized values, meanings, and representations of experiences conveyed in organization documents with results of interviews and a household survey in nine study communities in north-central Colorado: Breckenridge, Dillon, Frisco, Granby, Kremmling, Silverthorne, Steamboat Springs, Vail, and Walden. The analysis draws on a tri-part knowledge framework to integrate scientific, local and professional perspectives to address research questions, interpret the implications of the findings, and frame their local and extra-local applications. The various methodological components are integrated for their complementarity rather than for strict triangulation (Greene 2007). In other words, different forms of data drawing on different knowledge networks are integrated to tell a holistic story of the social construction of the MPB forest disturbance.

The empirical basis for this work begins with an analysis of the environmental narratives conveyed in organizational documents of purposively selected community and regional actor groups including CSFS, the Colorado Timber Industry Association (CTIA), and the Sierra Club and Wilderness Society, the two preeminent environmental advocacy organizations in the area. Organisations were selected to represent diverse institutional perspectives on the local environment and to assess the linkages between institutional narratives and the environmental subjectivities of community residents.

Colorado Forest Service annual reports for the outbreak period, CTIA outbreak period newsletters where MPB is discussed, and beetle-related literature disseminated by the Sierra Club and Wilderness Society during the 5-year period around the survey time, 2004–2008 (near the MPB outbreak peak period), were analyzed to assess: (1) whether, and the extent to which, the outbreak is problematized, (2) the attribution of the outbreak, and (3) the proposed course of action. Documents were collected for the indicated time period, read for discussion of MPB (one among many perceived threats to the health of Colorado forests), and those areas where the MPB outbreak was discussed were coded according to the above criteria. As nearly 60% of land in the Colorado mountains is owned and managed by the state or federal government (Riebsame et al. 1996), the Forest Service reports offer the dominant narrative of the outbreak and hold a position of authority in determining action plans for the majority of the affected area. Thus, much attention in this work is devoted to analyzing these reports. The Forest Service's narrative is countered by the voices of industry and environmental groups.

The area affected by the outbreak is home to a spectrum of socioeconomic and amenity characteristics, containing two distinct clusters of communities: high amenity resort communities with a large proportion of high socioeconomic status absentee property owners, and lower-amenity, lower socioeconomic status communities characterized by their recent roots in extractive industries and agriculture (Flint et al. 2012). To contextualize the outbreak in the specific places of communities and balance the critique of disembodied interests operating at the state and regional scales, interviews with community members and the results of a survey from the study communities were analyzed to further assess the features of narrative framing and environmental subjectivity listed above.[1]

A total of 165 key informant interviews were conducted in the Summer of 2006 to explore the range of experiences across the study area. To draw on multiple perspectives, key informants included individuals from schools, businesses, libraries, government, clergy, fire or police, community organizations, logging industries, environmental organizations, forest management agencies, and newspapers. These interviewees included both longtime residents and newcomers. The interviews focused on interviewees' attribution of the MPB outbreak, perceptions of land management entities, how the community experienced and responded to the outbreak, and the extent to which they felt their community was able to coalesce and act collectively. Interviews were recorded, transcribed, and thematically coded. Quotations that typify emergent themes are included in the results below.

A mail survey was developed based on the preceding key informant interviews and was administered to 4027 randomly selected households in nine study communities in the Spring of 2007, with a total of 1346 surveys completed and returned. The survey included questions that focused on respondents' environmental subjectivities and the features of narrative framing listed above, including perceptions of forest risks, faith in the forest industry, and trust in forest management (e.g., agreement/disagreement with statements dealing with inherent versus use value of forests and citizen representation in management decision making), support for forest industry options (e.g., biomass/biofuels power generation and small-scale timber processing), as well as satisfaction with land managers (e.g., private individuals and landowners, local fire departments, city and county governments, the US Forest Service). Risk perception was measured by asking respondents how concerned they were about a series of forest risks for their community, including fire, decline in wildlife habitat, increased erosion and runoff, loss of forests as an economic resource, loss of scenic/aesthetic quality, and loss of community identity tied to the forest (possible responses ranged from 1 (not concerned) to 5 (extremely concerned)). The survey also assessed attitudes about the values of forests and forest management. The level of agreement or disagreement was measured with a series of thirteen statements on a scale from 1 (strongly disagree) to 5 (strongly agree). Examples of statements include "forests should be managed to meet as many human needs as possible," "forests should be left to grow, develop and succumb to natural forces without being managed by humans," "the present rate of logging is too great to sustain our forest in the future," and "forestry practices generally produce few long-term negative effects on the environment." Respondents were also asked to indicate their attitudes from 1 (strongly oppose) to 5 (strongly support) about different forest industry options and levels of satisfaction from 1 (very dissatisfied) to 5 (very satisfied) with main natural resource management entities. The survey data also included information on the main sociodemographic characteristics of respondents such as age, gender, ethnicity, education, annual household income, political views, and employment in the forestry or agricultural sector. Key variables were explored with one-factor analysis of variance to assess variations across the study communities. In the comparison of newer and longer-term residents on major survey variables, two-tailed independent t-test, two-sided Mann–Whitney U test, and chi-square tests were used for numerical, ordinal, and categorical variables, respectively.

6 Results

6.1 Organizational Narratives

6.1.1 Colorado State Forest Service

In 2004 the Colorado Forest Service's annual report specifically focused on the ecology, condition, and management of ponderosa pine forests. At this time aerial surveys had recorded approximately 1.2 million trees killed by the MPB—nearly one hundred times the mortality at the beginning of the outbreak, in the mid-1990s (CSFS 2005). The MPB outbreak in affected areas was called "the most damaging insect and disease situation affecting Colorado's state and private lands," and the report emphasises the increasing insect populations and activity periods and the drought conditions that made trees particularly susceptible (CSFS 2005).

In this and subsequent years, the Colorado Forest Service annual reports make calls for thinning and diversifying forest stands to pre-settlement densities and diameter distributions as a substitute for the natural processes of forest succession that have been suppressed since settlement in the late 1800s (CSFS 2005). The MPB outbreak is held up as an example of the consequences of reactive rather than proactive management. In the 2005 issue specifically devoted to the health and management of aspen forests, the authors write that, "unlike the mountain pine beetle situation, we still have the opportunity to be proactive in the management of Colorado's trademark aspen forests" and later, "less than a quarter of Colorado's lodgepole pine trees are small enough to be resistant to MPB. Without forest management, future landscapes will be vulnerable to another widespread outbreak" (CSFS 2005). In this way, the outbreak is framed to encourage wider public acceptance of a more aggressive management paradigm in Colorado forests. This treatment of the MPB outbreak as an example of inadequate management appears consistently throughout the report, as other disturbances are linked to the MPB and the limitations of curbing an outbreak once it is underway reinforce the Forest Service's belief in proactive, aggressive management. An admonishing tone emerges at the end of a letter from the chairperson of the Colorado Forest Advisory Board appearing in the issue:

> As members of Colorado's Forestry Advisory Board, we encourage all Coloradoans to better understand the natural processes and human decisions that influence the condition of our forests – and support proactive treatments that improve that condition before negative impacts occur. (CSFS 2005, Introduction)

At the height of the MPB outbreak in Colorado in 2006 and 2007, the Forest Service's annual report on the health of Colorado forests details the extent and anticipated effects of the outbreak (CSFS 2008). The outbreak is contextualized as part of a complex set of issues threatening the future of Colorado forests, including forest fragmentation due to rapidly increasing development, fire suppression, and climate change. The authors of the report write that:

Two features of the current outbreaks appear to be unprecedented: (i) mountain pine beetle is now killing lodgepole pine at higher elevations than previously seen; and (ii) several different species of bark beetles are undergoing outbreaks at the same time, simultaneously affecting several different forest types and regions of the state. (CSFS 2008, p. 6)

The most emphasized risks associated with the outbreak by the Forest Service are the loss of clean air and clean water, particularly for the increasingly populated Front Range metropolitan area which relies on watersheds in affected areas for drinking water, and the loss of revenue for local, forest-dependent economies. The unprecedented scope and intensity of the outbreak is attributed to warmer temperatures associated with climate change, and a lack of effective forest management which has resulted in overgrown forests of older, less resilient trees (CSFS 2008).

While the outbreak is problematized in its own right (trees are dying and trees are vital to forest ecosystems), the Forest Service's report repeatedly emphasizes the precipitous effects that the outbreak can have in creating ripe conditions for wildfires that are predicted to exceed historic levels and intensities. The report contains images of huge swaths of beetle-killed forest alongside images of thriving young pine and aspen trees in actively managed areas.

A clear link between beetle kill, fire, and the potential for drinking water contamination in an area with a booming amenity migrant population is also emphasized. As described in the 2006 report, most of the MPB activity is located at the headwaters of Colorado's and many other Western state's drinking water supplies. To this end, the Forest Service promotes the need for more aggressive management throughout Colorado forests, including "harvesting timber, removing poor quality or low-value trees, forest thinning, prescribed fire and regulating development within fire prone forest types" (CSFS 2008, p. 6).

With a growing sense of urgency, the changing image of the landscape is noted as cause for concern. The 2006 report reads, "the resulting landscapes may not meet society's desires and needs and could be even less appealing than those created by the current mountain pine beetle epidemic" (CSFS 2006, p. 3). While the outbreak is a natural event, the consequent image of the landscape conflicts with the prevailing ecological imaginary, or idea of what the land should look like.

In the 2007 report, a management paradigm favoring a higher degree of intervention is framed as the most near-term solution for beetle-related issues, with a special role to be played by industry. The state has never had a large forest timber industry, and in 2007 only around 5% of available timber was being actively harvested, with only 5 mills in the state employing more than 50 people (CSFS 2008). Although sustainable harvesting is framed by the Forest Service as an integral part of working to regenerate forests and add diversity to the landscape, in 2007 at least 90% of all wood products used in Colorado were imported from other states or foreign countries (CSFS 2008). According to the Forest Service, obstacles to the implementation of more sustainable harvesting include funding shortfalls, a lack of processing facilities, and a lack of social acceptance for the necessary harvesting.

6.1.2 Colorado Timber Industry Association

CTIA describes themselves as a trade association that advocates for Colorado's forest products, companies, and for scientific, sustainable forest management (CTIA 2016). It is composed of nearly 50 forest product and logging companies from throughout the state.

As with the Forest Service annual reports, CTIA newsletters were collected for the outbreak period and examined for discussion of the MPB outbreak. In the Spring 2006 edition of the association newsletter, the president describes how he is often confronted with the question, "where is the timber industry and why aren't they cleaning up this big bug mess?" (CTIA 2006, p. 2). In response, he writes:

> The same people who spent 25 years trying to put me out of business have been spending the last 5 years trying to work me to death! When we the people chose not to properly manage the forest, Mother Nature takes over and many of those who pressured the Forest Service not to allow any tree cutting seem to be changing their tune. (CTIA 2006, p. 2)

The CTIA newsletter cites an as yet undiscussed contributor to the dense forest stands that were instrumental in the scope of the outbreak: the Forest Service's decision to limit pre-commercial thinning of lodgepole pine trees. According to the organization, this decision can be traced to the US Fish and Wildlife Service's classifying the lynx as an endangered species. As the Endangered Species Act requires the maintenance of critical habitat for designated species, this decision meant the prioritization of lynx management over other aspects of forest management. The industry argues that the Forest Service has "abdicated their forest management responsibilities to wildlife biologists and the US Fish and Wildlife Service" (CTIA 2006, p. 4). In the eyes of the timber industry, the best way to avoid future insect outbreaks is to reverse this prioritization of lynx management over timber management and thinning of regenerated lodgepole pine stands (CTIA 2006).

The Winter 2006 edition of the association newsletter emphasizes the role to be played by industry in maintaining the forest as "Mother Nature's healthy alpine garden" (p. 2). Accordingly, the bark beetle epidemic is framed as symptomatic of an unhealthy, unmanaged forest. The CTIA president writes in his newsletter message, "we must realize that we are a tool to be used to prevent the over aged, overstocked, and generally unhealthy conditions which have promoted such outbreaks as the present bark beetle epidemic" (CTIA 2007, p. 2). This narrative appears consistently throughout the text: that the beetles are the consequence of a diminished industry presence and that the timber industry is the true caretaker of Colorado forests.

The newsletter also contains a comic depiction of a forester, equipped with a chainsaw, pressing a stethoscope to the trunk of a tree inscribed with the words *National Forest*. The caption reads: "You're in terrible health!! You have heart rot, root rot, bugs and more! Who's taking care of you?!" (CTIA 2006, p. 6).

Later in the newsletter, the executive director of CTIA laments the plummet of North American lumber markets that began in the Summer of 2006. While Colorado markets were not as affected by the downturn as others, he contextualizes the market downturn within the Forest Service's call for the increased role of industry saying:

The fall in lumber prices has coincided with increased public support for increasing timber harvest levels to respond to overall forest health concerns, especially the mountain pine beetle and spruce bark beetle epidemics in Colorado's national forests. But the current lumber markets make it harder for sawmills to respond as aggressively as they, or the public, would like. (CTIA 2006, p. 6)

As a fundamentally economic interest, the constraints of the market necessarily inform the critical position the timber industry adopts in responding to the outbreak. This management constraint is added to those presented by the Forest Service.

6.1.3 The Sierra Club and Wilderness Society

Throughout the timber industry's newsletters, there is a clear frustration with the environmentalist, preservationist ethos that has informed the management of Colorado forests. A dominant voice of this environmentalist perspective, and one directly criticized by the industry, is that of the Sierra Club. In a special newsletter that examines the MPB outbreak, the Sierra Club emphasizes that the bark beetle is native and that insect disturbances play an important role in forest succession. In contrast to the Forest Service and the timber industry, the Sierra Club argues that fire suppression has *not* altered the frequency of fires or the density of the forests. Instead, to account for the scope and intensity of the MPB outbreak, they point to more global environmental phenomena, specifically drought and warmer temperatures (Bidwell 2008).

While the Forest Service raised alarms in their reports about beetle-killed trees being a catalyst for catastrophic wildfires, the Sierra Club argues that the risk posed is minimal, and at the most merely one of many fire threats faced in Colorado forests (Bidwell 2008). Looking forward, the thinning of forests is deemed an impractical response and a risk factor for crucial wildlife habitat.

The Wilderness Society describes itself as the leading American conservation organization working to protect wilderness areas. Though it is a leading environmentalist voice, there is relatively little literature devoted to the MPB outbreak, indicating that despite the alarm raised by the Forest Service and timber industry, it is not of great concern to environmental entities. In an article about the MPB outbreak, the author distills the organization's position on the outbreak, the extent to which it constitutes a problem and the proposed course of action into several talking points, which emphasise that despite the scope and scale of the outbreak the forests are resilient and sufficiently diverse to endure, and that beetle-killed trees do not pose any significant or particular danger in terms of erosion or fire (Aplet 2009).

The prevailing tone from both organizations is one that lacks the alarm and outrage apparent in the literature of the Forest Service and the timber industry, ultimately arguing that the beetle outbreak is a natural event, and the forests will "recover relatively quickly" (Aplet 2009).

6.2 Community Perspectives

While the above-organizational narratives illustrate the interaction of various local and regional interests with the MPB outbreak, the histories, biophysical, and socioeconomic contexts of the communities themselves produce distinct ecological imaginaries and environmental narratives. Interviews and findings from the 2007 survey of MPB-affected communities illustrate how experiences of the outbreak are informed by community contexts (Flint et al. 2012). Survey findings indicated that respondents in the lower-amenity communities of Granby, Kremmling, and Walden were relatively older, of lower income and education and had resided in the communities for a longer period of time than residents of the other communities. Relative to other communities in the beetle-affected area in north central Colorado, Vail, Steamboat Springs, Frisco, Breckenridge, Silverthorne, and Dillon are distinguished by high average household income, high educational attainment, low levels of employment in forest management, forest industry or agriculture, and relatively liberal political views.

This clustering according to sociodemographic and economic indicators corresponds to a clustering of attitudes and ascriptions to particular paradigmatic views about the health and appropriate management of the forests: a more preservationist, minimal intervention approach among the more affluent and liberal communities, and an approach that supports a greater degree of intervention and utilization of industry options among the less affluent, more timber-dependent communities. The relationship between Walden, Kremmling, and Granby and the timber industry proves a strong one with regard to levels of trust in land managers and perceptions of outbreak response options. Survey respondents from these communities were highly supportive of pursuing all industry options, including biomass and biofuels power generation, large- and small-scale timber processing and niche marketing/production of wood products, and were characterized by high levels of trust in private logging companies, relative satisfaction with the work of local land managers and markedly low levels of trust in environmental organizations and the Forest Service.

The role of the timber industry in defining community perspectives and approaches to MPB is clear in interviews with residents of these communities. As a Walden resident described:

> It's what we have been raised in, we know more about managing the forests than half of the people living in the city. And we respect the land.

These sentiments are echoed by residents in similar communities, with a Kremmling resident saying:

> …our roots are in logging and our roots are in timbering. So we feel that the government has ignored this issue to the point where it's gotten to the point of an epidemic and now uncontrollable […] They're all tree-hugging bastards. I'm a tree-hugger. I love trees, there's a need for them, but they don't look at the all-around picture.

There is a tangible and at some points visceral frustration with the outbreak as an unnecessary consequence of the diminished role of loggers and industry in forest management. Many residents who were interviewed saw the outbreak as a direct effect of the decline of the logging industry and the ascension of a management paradigm of minimal intervention and preservationist attitudes that reflect the priorities and interests of more amenity-oriented and affluent communities.

In reflecting on the economy and general quality of life in these communities with less amenity orientation, clear correlations are drawn between the health of the forest and the socioeconomic stability of the town, which one Walden resident describes as, "a real crisis area." The closing of local sawmills followed by the closing of the railroad in the early 1990s was referred to as major catalysts for the economic downturn, and multiple residents refer to the challenges associated with keeping public schools open. One resident summarized the challenge of remaining in Walden saying:

> We've got 3 kids and found ourselves many, many, many times at the end of the month with not enough money to pay bills and thought, you know, this is a great place to live, but you can't eat the scenery.

A nearer-term solution for residents in these communities was removing affected trees as swiftly as possible. Looking more long term, residents saw the potential expansion of the forest products industry as something important for the vitality of the forest, the town economy, and to keep young people from leaving when they graduate high school.

By contrast, more amenity-oriented communities had considerably lower levels of faith in the forest industry and relatively higher levels of trust in prevailing management regimes. Looking at community variations in support for forest industry options in responding to the outbreak, these respondents were generally less supportive, and particularly opposed to large-scale timber processing. For resort towns, the aesthetic loss associated with beetle kill was frequently cited as a problem for vacationers, and for residents who depend on tourism revenue. As a resident in Vail described, "it's really the visual as opposed to the potential danger." When it comes to devising a plan for dealing with the outbreak, residents in these communities generally favored a more restrained approach to management, with one Steamboat Springs resident saying:

> I don't think anybody likes to see logging trucks go into the wilderness, because we're all really avid outdoor enthusiasts here and we like to enjoy our forests.

While individual responses to the outbreak, such as taking specific action on private land or attending community informational meetings, were only moderately variable between communities, the differences in attributions of the outbreak, feelings of trust in local and state management, and support for industry options suggest fundamentally different experiences and vulnerabilities. Residents with histories in extractive industries felt constrained and marginalized in decision making, ultimately seeing the outbreak as a consequence of their diminished role in forest management. Nevertheless, in resort towns where the landscape has been commodified to fuel a tourism industry and draw amenity migrants, the aesthetic loss associated with beetle kill was a dominant concern among respondents.

For more amenity-oriented community participants, perspectives focused on the economic ramifications and uncertainty caused by the MPB outbreak, and these interviewees noted the way that responses throughout the region were economically constrained. A Vail resident pointed out that those with the means to do so can engage in more mitigation work:

> It's really driven by both economics, size of the organization and its ability to address issues. If you've got a poor homeowners association with a lower economic scale, they are less likely to do something. If you've got a homeowners association that is in a trendy mountainside tree surrounded environment, they are probably a little more attuned to what needs to be done. More buck to bang with.

A Breckenridge resident said, "Our economic base is basically tourism and we're 70% national forest land in the county. Anything that affects 70% of the county is obviously going to be a very important thing in the county." Noting that not all people appreciated risks, a Vail resident said, "There's so many billions of dollars of infrastructure at risk that people don't seem to be aware of although I think they're getting there."

Additionally, better relationships between local residents and resource management agencies were described in higher amenity communities, including more understanding of the limitations faced by local forest managers:

> We have a good collaboration with the Forest Service. They have the technical…they virtually have no dollars to help with actual cutting, but they have helped us a lot with the technical aspects of it. (Vail)

> No local community will be able to get anything done. I don't even think any single state will be able to get anything done. The only way we will see something done is if the affected western states pull together. (Breckenridge)

6.3 Newcomers and Old-timers

A further area of difference among residents' experiences of the outbreak and perceptions of appropriate responses was the time they had resided in the affected communities. As a natural amenity destination, northern Colorado has seen a marked influx of migrants in recent decades. US Census data for the five non-metropolitan counties in the study area (Eagle, Summit, Grand, Routt, and Jackson) show that local population increased more than four times from 22,673 to 119,937 between 1970 and 2010 (Qin 2016; US Census Bureau 1970, 2010). This influx of new residents implies an influx of unique, culturally situated attitudes about the local environment.

A longtime resident of Steamboat Springs spoke of the changing demographics of the community:

> 30 years ago when I first moved to Steamboat […] We got together and had potlucks and made songs about the ski area and the coal mines. We were just poor and we didn't really care. For $50 a month, you could have a place to stay. Now, you're lucky to find something for $400 a month. So, as we sold our town as a commodity not a community, there's a huge monster comp up here, we have simply discounted the future. We discounted our kids, so they can't even live here, because we're a single economy environment […] we sell our community, with family values to the tourists as a commodity.

Resort town status also means unique obstacles to eliciting a cohesive community response to the outbreak. In the eyes of longer-term residents, the increasingly fragmented socioeconomic base of the town is problematic for trying to catalyze community action. Residents commented on how second homeowners were less aware of the causal complexities of the beetle outbreak, had less investment in local life, and indicated that those who vacationed in the town were less likely to be supportive of management entities taking aggressive steps to mitigate fire risks like cutting trees or having controlled burns:

> It takes a lot of time for a second homeowner to understand the social and economic and environmental issues here because they're only here two or three weeks out of the year, and while they're here they want to ski… In the older days, even the rich people met a lot with the working people and the poor people. Nowadays, it's divided.

In reference to the MPB outbreak, 1980 was used as a cutoff date to compare the attitudes of longtime and newer full-time residents in the analysis of the community survey data. This cutoff meant that "oldtimers" already lived in the area prior to or in the early stage of the recent amenity in-migration and would have lived in the communities for more than 15 years at the start of the outbreak. As shown in Table 1, differences between the two groups were highly significant, illustrating both a demographic division and differences in environmental attitudes. On average, longtime residents were older, less educated, had lower household incomes, were more politically conservative, and were twice as likely to be employed in forestry-related occupations or agricultural production as compared to newcomers.

In terms of perceptional differences, newer residents had higher levels of perceived forest risks, less faith in the forest industry, and relatively more trust in forest management than longer-term residents. Newcomers were also less satisfied with local land management entities (private individuals and landowners, local fire departments, private logging companies, developers, and private homeowners associations), but comparatively more satisfied (or less dissatisfied) with government land managers (city and county governmental, the CSFS, the Bureau of Land Management, and the US Forest Service). Related to industry options for dealing with beetle-killed trees, newcomers were generally

Table 1 Differences between newer- and longer-term residents in sociodemographic and perceptional variables. Given as means of variables except for gender, ethnicity, and the two employment measures. No significant difference was found between the two resident groups in terms of gender or ethnical composition and support for biomass/biofuels power generation. Both categories included relatively more male than female respondents, were mostly white, and generally supported this forest industry option

Variable	Newer-term residents (Max $N = 894$)	Longer-term residents (Max $N = 323$)
Sociodemographic characteristics[a]		
Age	49.48***	57.56***
Gender	40.4% female	44.6% female
Ethnicity	95.6% white	96.5% white
Education	4.51***	3.91***
Household income	5.39(*)	5.14(*)
Political view	2.93***	3.29***
Forestry employment	13.2% yes***	27.3% yes***
Agricultural employment	20.0% yes***	41.3% yes***
Composite perceptional indicators[b]		
Risk perception index	3.67**	3.80**
Faith in the forest industry	2.65***	3.12***
Trust in forest management	2.65***	2.29***
Satisfaction with local land managers	2.88**	3.01**
Satisfaction with governmental land managers	2.71***	2.49***
Support for biomass/biofuels power generation	3.67	3.74
Support for large-scale timber processing	2.59***	3.24***
Support for small-scale timber processing	3.52***	3.98***
Support for niche marketing/ production of wood products	3.74***	3.97***

(*)$p < 0.10$, *$p < 0.05$, **$p < 0.01$, ***$p < 0.001$
[a]Variable measurement: gender (male or female), ethnicity (white or non-white), education (from "1" less than a high school degree to "6" advanced degree, i.e., Master's, JD, Ph.D.), household income (from "1" less than $15,000 to "8" $150,000 or more), political view (from "1" liberal to "5" conservative), and employment in the forestry/agricultural sector (yes or no)
[b]Computed as the averages of responses (on 1–5 Likert scales) to relevant survey questions following exploratory factor analysis. See Sect. 5 for further detail

less supportive of small-scale timber processing and niche marketing/ production of wood products, and much more opposed to large-scale timber processing. These differences, occurring across the study area, illustrate the extent to which the local environment is constituted by and interacts with varying culturally and historically situated identities and interests.

7   Discussion and Conclusions

In analysing the above narratives, special attention was paid to the way organizations and respondents took part in the active construction of the pine beetle outbreak and the extent to which it constituted an environmental problem. Analyzing narratives from diverse stakeholders allows for the emergence of distinct story lines and attributions that can be linked to larger global environmental discourses. Such stories elucidate the interconnections and interactions between biophysical, social, economic, and political realms and structures (Bixler 2013). People from each organization and each community demonstrated particular understandings of the local environment and an emergent, socioeconomically and politically nested narrative of explanation.

With the exception of environmental organizations, consistently within their narratives, the Colorado Forest Service, the CTIA, and the less amenity-oriented communities faulted the restricted role of forest managers and industry in maintaining forest equilibrium; linking this diminished management role to the ideals of politically powerful, and largely newcomer residents. This is evidenced by the industry's complaints about decades of public pressure to diminish harvesting and later by the assertion that the Forest Service abdicated its role in managing forests to the Fish & Wildlife Service's efforts to leave forests undisturbed to protect the endangered lynx. Such an ascendant preservationist ethos is common in the American West, where amenity migrants are increasingly seeking a pristine, commoditized landscape (Walker and Fortmann 2003).

The Forest Service narrative is one that promotes the need for more active management, but is constrained by bureaucracy and public wariness about what such management entails. The industry narrative is one of systemic marginalization in the wake of market constraints and shifting public opinions about what sorts of activities should be permitted in Colorado forests. Within affected communities, the narratives surrounding the beetle outbreak are structured by socioeconomic characteristics and by varying ecological imaginaries, or conceptions of what constitutes a legitimate image of the landscape. This conflict is common in Western lands, which are increasingly sought out by amenity migrants seeking to consume an "imagined idyllic landscape" (Walker and Fortmann 2003). The conception of a humanless and pristine nature is starkly at odds with the ecological imaginaries and environmental subjectivities of longtime residents currently or historically engaged in extractive industries, as indicated in attitudes about the cause of the outbreak, appropriate levels of management, and the role of industry in maintaining Colorado forests. For them, the relationships with and expectations of the land are based around "work, management and ongoing transformation" (Robbins 2011, p. 206). Such a conflict is represented in the interaction of new migrants and long-time residents with the MPB outbreak. These groups vary significantly in terms of socioeconomic indicators but also in terms of attitudes about the roles of forest industry and management, levels of satisfaction with land managers and support for industry options moving forward. These differences indicate broader, culturally situated differences in beliefs about what should constitute people's relationship with the environment, and who can be trusted in critical decision making.

While other political critiques of MPB outbreaks attribute them to the prioritization of economic gains through overharvesting (see Petersen and Stuart 2014), the role of the logging industry in Colorado seems to have been constrained by the relatively privileged attitudes about what forested landscapes should look like, and what kinds of use are deemed socially desirable due to an increasingly tourism and amenity-based economy. Given their economic histories, residents of less amenity-oriented, resource extractive communities have experienced the loss of a livelihood opportunity in the timber industry given the changing economy and the emergence of a specific, powerful ecological imaginary. Those in amenity-based resort communities are threatened by the loss of a particular, commodified image of the forest inconsistent with beetle-affected landscape.

Moving beyond the level of environmental subjectivities and ecological imaginaries, it is possible to discern linkages between the more systemic causal factors of narratives. Narratives consistently attributed the scope and intensity of the outbreak to insufficient management and global warming, yet in the context of the local logging industry's decline, and a period of massive population influx for the Colorado Front Range, at least 90% of all wood products used in Colorado were imported from other states or foreign countries, constituting an enormous expenditure of fossil fuels (CSFS 2008). Such an example shows the extent to which causal factors overlap and are fueled by the commodification of a particular ecological imaginary.

In conclusion, this work has sought to engage in a discourse and narrative analysis of the MPB outbreak in northcentral Colorado to consider the relationships between power, environmental narratives, and a constructed environmental problem. Intrinsic to these narratives are distinct and sometimes overlapping conceptions about what natural spaces should look like and what sorts of activities should constitute people's relationship with the environment. These narratives reveal the contested nature of nature in the discursive practices of actor groups. Tracing the narratives and the framing of environmental issues is an important part of developing empathy for different needs and vulnerabilities with respect to the environment, and can help shed light on how broader structures are implicated in environmental subjectivities. This sensitivity to unique environmental subjectivities and vulnerabilities is essential to the development of management regimes that are considerate and inclusive and ultimately, sustainable.

Acknowledgements This research was supported by the Decision, Risk and Management Sciences Program of the National Science Foundation (Award #1733990). The analysis drew partially on data from a previous research project funded by the Pacific Northwest Research Station and Region 2, US Forest Service. Helpful comments on an earlier draft from Dr. James Sanford (Sandy) Rikoon are also sincerely appreciated.

Note

1. For detailed description of interview and survey methodology, see Qin and Flint (2010) and Flint et al. (2012).

References

Aplet, G. (2009). *Understanding the mountain pine beetle: Seven facts you need to know*. The Wilderness Society. Retrieved from http://wilderness.org/blog/understanding-mountain-pine-beetle-seven-facts-you-need-know.

Bentz, B. J., Régnière, J., Fettig, C. J., Hansen, E. M., Hayes, J. L., Hicke, J. A., & Seybold, S. J. (2010). Climate change and bark beetles of the western United States and Canada: Direct and indirect effects. *BioScience, 60*(8), 602–613.

Bidwell, R. (2008). What's going on with the mountain pine beetle? *Rocky Mountain Chapter of the Sierra Club: Peak & Prairie, 42*(3), 6.

Bixler, R. P. (2013). The political ecology of local environmental narratives: Power, knowledge, and mountain caribou conservation. *Journal of Political Ecology, 20*, 273–285.

Bridger, J. C., & Luloff, A. E. (1999). Toward an interactional approach to sustainable community development. *Journal of Rural Studies, 15*(4), 377–387.

Carroll, A. L., Taylor, S. W., Régnière, J., & Safranyik, L. (2003). Effect of climate change on range expansion by the mountain pine beetle in British Columbia. In T. L. Shore, J. E. Brooks, & J. E. Stone (Eds.), *Mountain Pine Beetle Symposium: Challenges and Solutions*, October 30–31. Kelowna, BC. Natural Resources Canada, Information Report BC-X-399, Victoria, pp. 223–232.

Chang, W. Y., Lantz, V. A., & MacLean, D. A. (2009). Public attitudes about forest pest outbreaks and control: Case studies in two Canadian provinces. *Forest Ecology and Management, 257*(4), 1333–1343.

Colorado State Forest Service (CSFS). (2005). *2004 report on the health of Colorado's forests*. Fort Collins: Colorado State Forest Service.
Colorado State Forest Service (CSFS). (2006). *2005 report on the health of Colorado's forests*. Fort Collins: Colorado State Forest Service.
Colorado State Forest Service (CSFS). (2008). *2007 report on the health of Colorado's forests*. Fort Collins: Colorado State Forest Service.
Colorado State Forest Service (CSFS). (2016). *2015 report on the health of Colorado's forests*. Fort Collins: Colorado State Forest Service.
Colorado Timber Industry Association. (2006). Colorado Timber Industry Association. *Timber Times Newsletter*.
Colorado Timber Industry Association. (2007). Colorado Timber Industry Association. *Timber Times Newsletter*.
Colorado Timber Industry Association. (2016). Colorado Timber Industry Association. *Timber Times Newsletter*.
Dale, V. H., Joyce, L. A., McNulty, S., & Neilson, R. P. (2000). The interplay between climate change, forests, and disturbances. *Science of the Total Environment, 262*(3), 201–204.
Dale, V. H., Joyce, L. A., McNulty, S., Neilson, R. P., Ayres, M. P., Flannigan, M. D., et al. (2001). Climate change and forest disturbance. *BioScience, 51*(9), 723–734.
Flint, C. G. (2006). Community perspectives on spruce beetle impacts on the Kenai Peninsula Alaska. *Forest Ecology and Management, 227*(3), 207–218.
Flint, C. G., & Luloff, A. E. (2007). Community activeness in response to forest disturbance in Alaska. *Society and Natural Resources, 20*(5), 431–450.
Flint, C. G., McFarlane, B., & Müller, M. (2009). Human dimensions of forest disturbance by insects: An international synthesis. *Environmental Management, 43*(6), 1174–1186.
Flint, C., Qin, H., & Ganning, J. P. (2012). Linking local perceptions to the biophysical and amenity contexts of forest disturbance in Colorado. *Environmental Management, 49*(3), 553–569.
Fuller, L., Marzano, M., Peace, A., Quine, C. P., & Dandy, N. (2016). Public acceptance of tree health management: Results of a national survey in the UK. *Environmental Science & Policy, 59*, 18–25.
Greene, J. C. (2007). *Mixed methods in social inquiry*. San Francisco: Wiley.
Jasanoff, S. (2010). A new climate for society. *Theory, Culture & Society, 27*(2–3), 233–253.
Kooistra, C. M., & Hall, T. E. (2014). Understanding public support for forest management and economic development options after a mountain pine beetle outbreak. *Journal of Forestry, 112*(2), 221–229.

McFarlane, B. L., & Wilson, D. O. T. (2008). Perceptions of ecological risk associated with mountain pine beetle (*Dendroctonus ponderosae*) infestations in Banff and Kootenay National Parks of Canada. *Risk Analysis, 28*(1), 203–212.

McFarlane, B. L., Parking, J. R., & Watson, D. O. T. (2012). Risk, knowledge and trust in managing forest insect disturbance. *Canadian Journal of Forest Research, 42*(4), 710–719.

Müller, M. (2011). How natural disturbance triggers political conflict: Bark beetles and the meaning of landscape in the Bavarian Forest. *Global Environmental Change, 21*(3), 935–946.

Müller, M., & Job, H. (2009). Managing natural disturbances in protected areas: Tourists' attitude toward the bark beetle in a German national park. *Biological Conservation, 142*(2), 375–383.

Parkins, J. R., & MacKendrick, N. A. (2007). Assessing community vulnerability: A study of the mountain pine beetle outbreak in British Columbia, Canada. *Global Environmental Change, 17*(3–4), 460–471.

Petersen, B., & Stuart, D. (2014). Explanations of a changing landscape: A critical examination of the British Columbia bark beetle epidemic. *Environment and Planning A, 46*(3), 598–613.

Porth, E., Dandy, N. D., & Marzano, M. (2015). "My garden is the one with no trees": Residential lived experiences of the 2012 Asian longhorn beetle eradication program in Kent, England. *Human Ecology, 43*(5), 669–679.

Qin, H. (2016). Comparing newer and longer-term residents' perceptions and actions in response to forest insect disturbance on Alaska's Kenai Peninsula: A longitudinal perspective. *Journal of Rural Studies, 39*, 51–62.

Qin, H., & Flint, C. G. (2010). Capturing community context of human response to forest disturbance by insects: A multi-method assessment. *Human Ecology, 38*(4), 567–579.

Qin, H., & Flint, C. G. (2012). Integrating rural livelihoods and community interaction into migration and environment research. *Society & Natural Resources, 25*(10), 1056–1065.

Qin, H., Flint, C. G., & Luloff, A. E. (2015). Tracing temporal changes in the human dimensions of forest insect disturbance on the Kenai Peninsula Alaska. *Human Ecology, 43*(1), 43–59.

Raffa, K. F., Aukema, B. H., Bentz, B. J., Carroll, A. L., Hicke, J. A., Turner, M. G., & Romme, W. H. (2008). Cross-scale drivers of natural disturbances prone to anthropogenic amplification: The dynamics of bark beetle eruptions. *Bioscience, 58*(6), 501–517.

Riebsame, W. E., Gosnell, H., & Theobald, D. M. (1996). Land use and landscape change in the Colorado mountains I: Theory, scale, and pattern. *Mountain Research and Development, 16*(4), 395–405.

Robbins, P. (2011). *Political ecology: A critical introduction* (2nd ed.). West Sussex: Wiley.

Rutherford, S. (2011). *Governing the wild: Ecotours of power*. Minneapolis: University of Minnesota Press.

Tomlinson, I., Potter, C., & Bayliss, H. (2015). Managing tree pests and diseases in urban settings: The case of Oak Processionary Moth in London, 2006–2012. *Urban Forestry & Urban Greening, 14*(2), 286–292.

US Census Bureau. (1970). *Census of population and housing, 1970*. Washington, DC: US Census Bureau.

US Census Bureau. (2010). *Census of population and housing, 2010*. Washington, DC: US Census Bureau.

Walker, P., & Fortmann, L. (2003). Whose landscape? A political ecology of the 'exurban' Sierra. *Cultural Geographies, 10*(4), 469–491.

Wilkinson, K. P. (1991). *The community in rural America*. Westport, CT: Greenwood Publishing Group.

5

Indigenous Biosecurity: Māori Responses to Kauri Dieback and Myrtle Rust in Aotearoa New Zealand

Simon Lambert, Nick Waipara, Amanda Black, Melanie Mark-Shadbolt and Waitangi Wood

1 Introduction

The New Zealand economy relies predominantly on the primary sector, which contributes over 50% of the country's total export earnings and accounts for over 7% of GDP (New Zealand Treasury 2012). Being an Island nation in the South West Pacific, New Zealand's native flora and fauna are highly endemic, many having evolved in isolation over 65 million years. Both GDP and the conservation of native flora and fauna are dependent on having manageable levels of pests and diseases, something that is becoming increasingly difficult with the unprecedented levels of global movements of materials and people (McGeoch et al. 2010). Despite biosecurity issues being critical to New Zealand's

---

S. Lambert (✉)
Department of Indigenous Studies,
University of Saskatchewan, Saskatoon, Canada

N. Waipara
Auckland Council, Kauri Dieback Programme,
Auckland, New Zealand

© The Author(s) 2018
J. Urquhart et al. (eds.), *The Human Dimensions of Forest and Tree Health*,
https://doi.org/10.1007/978-3-319-76956-1_5

biological heritage, policy and management systems have yet to realise and embed the priorities of Māori who are theoretically the government's formal partner since the signing of the Treaty of Waitangi in 1840.

There are a growing number of cases in New Zealand where Indigenous Knowledge (IK) contests mainstream science for recognition, support and implementation, although the implementation of this is still problematic (see Prussing and Newbury 2016). In New Zealand, Māori-sourced IK, referred to as *mātauranga* Māori, has an increasingly important role in environmental management, including protection of biological heritage from biosecurity risks and threats. This chapter discusses two case studies of collaboration between Māori and non-Māori in the biosecurity space, resulting in (some) empowerment of Māori and more efficient biosecurity strategies and programmes.

This chapter proceeds with Mead's (2003) all-encompassing definition of *mātauranga Māori* as Māori knowledge and philosophy, thus allowing a contrast with 'Western' science and philosophy. It is acknowledged that both these philosophical bases (*mātauranga Māori* being one of many examples of IK) are dynamic and expanding. *Mātauranga Māori* also has an intimate connection to *Kaupapa Māori* (Māori methodology) as both a means to progress research with Māori (Smith 1999; Cunningham 1998) and as the fundamental expression of Māori culture within mainstream research (Pihama et al. 2002). We position *Kaupapa Māori* as an array of research principles for engaging with Māori in, for example, protecting kauri and other species valued by Māori. These principles are, of course, not limited to Māori-focused research and could be said to be fundamental to any research that relies on human participants (see, e.g. Piddington 1960; Whyte 1981). The justification for professional (and therefore ethical) acknowledgement

---

A. Black · M. Mark-Shadbolt
Bio-Protection Research Centre,
Lincoln University, Lincoln, New Zealand

W. Wood
Tangata Whenua Rōpū, Kauri Dieback Programme,
Aotearoa, New Zealand

of Kaupapa Māori (Māori methodology) is that these principles have grown from explicitly localised responses to the perceptions and realities of what Russell Bishop terms 'epistemological racism' (Bishop 1999). The grounding in Māori lives, from the use of Māori words and terms to the social and cultural engagement that occurs specific to Māori people and the spaces that they control, presupposes both the legitimacy of Māori knowledge and methodologies.

This chapter presents two case studies of Indigenous biosecurity action from Aotearoa New Zealand. The first concerns the giant conifer, *Agathis australis* (New Zealand Kauri), a *taonga* (treasured, sacred) plant to all New Zealanders and especially for Māori on whose lands these gigantic trees grow. The resilience and health of remnant kauri forests and dependent ecosystems are under increasing threat from the disease phenomenon Kauri Dieback (*Phytophthora agathidicida*). A seminal joint agency programme that included Māori from governance to community engagement was initiated in 2009. Eight years on this programme is still in existence, although it is yet to realise the potential of Māori knowledge and customs to manage successfully Kauri Dieback.

More successful collaboration has been achieved in the second case study where Māori are involved in extensive efforts to combat the recent incursion of Myrtle Rust (*Austropuccinia psidii*) which threatens a range of *taonga* species. Central to this case has been the establishment of a Māori Biosecurity Network that supports the involvement of Māori researchers, governance representatives and political lobbyists.

2   The Use of Indigenous Knowledge in Forest Conservation and Biosecurity Management

IK has an as yet unknown value to contemporary forest biosecurity, but such knowledge is increasingly recognised for the opportunities it offers states and jurisdictions that are prepared to accept and resource indigenous participation in this increasingly important and dynamic research area. Given that environmental concerns are increasingly couched in terms of political-economic concerns and environmental sustainability, IK discourse represents a convergence of state, corporate and community interests competing for resources in such vital areas as biosecurity. The World Bank has estimated that around 60 million Indigenous peoples are heavily dependent upon forests for their livelihood while an additional 350 million are dependent on them for their income and subsistence (World Bank Group Forest Action Plan FY16-20 2016). Many Indigenous communities will therefore have vested interests in the protection and health of forests, the management of which is mainly subject to a legacy of colonial management.

International literature on Indigenous communities and conservation is dominated by Western paradigms of conservation but includes examples of researchers working alongside and documenting IK for the purpose of gaining insight into aspects of ecology and natural history (Walter and Hamilton 2014; Camara-Leret et al. 2014; de Freitas et al. 2015). Studies examining alternative (including Indigenous) approaches to forest conservation document positive impacts of co-managing forests, including minimising the loss of biodiversity (Souto et al. 2014; Singh et al. 2015). The need for more inclusive approaches to biosecurity research and forest conservation, in partnership with relevant Indigenous communities, is perhaps critical to ensuring the long-term health of many tree species and forest ecosystems such as those found in New Zealand's kauri forests.

2.1 The Adoption of Māori Knowledge for Forest Conservation

Māori, like other Indigenous peoples, have developed customary practices to sustainably manage their lands and resources. However, the adoption of *mātauranga Māori* (Māori knowledge) in mainstream conservation ventures is often limited and mainly focused on the customary harvest of species for food, such as the *kereru* (New Zealand Wood Pigeon, *Hemiphaga novaeseelandiae*) and *titi* (Sooty Shearwater, *Puffinus griseus*) (Moller et al. 2009; Lyver et al. 2008, 2009), and the customary harvest of flax (*harakeke* in Māori, *Phormium tenax*) and seaweed, *karengo* (*Bangiaceae* spp.) (McCallum and Carr 2012; O'Connell-Milne and Hepburn 2015).

The use of Māori knowledge in New Zealand forest conservation is not particularly visible in research and policy (Walker et al. 2013), and discourse around the use and interpretation of Māori knowledge is often limited to scattered Māori representation in governance roles. While this is an ongoing issue, the longer-term strategic goals of government must expand to include the operationalisation of Māori methods and research priorities in forest health.

3 The Discovery of Kauri Dieback

Kauri are an ancient tree species now reduced to a fragment of their pre-colonial habitat and threatened with extinction from an introduced plant pathogen (*P. agathidicida*) (Waipara et al. 2013). Only recently has the soil-borne pathogen responsible for 'Kauri Dieback' been taxonomically described and named as *P. agathidicida* (Weir et al. 2015). The pathogen initially infects kauri through its roots before progressing to an aggressive collar rot resulting in large basal trunk lesions, then canopy defoliation and eventually death (Bellgard et al. 2016) (Fig. 1). All size and age classes of kauri are susceptible to infection and death. Details on the origin and introduction of Kauri Dieback to New Zealand are still unknown. One hypothesis is that the disease may have initially established through imported seedlings, plant and soil materials from the Oceania region destined for a kauri nursery in Waipoua Forest (Beauchamp and Waipara 2014). It is then thought to have spread to Great Barrier Island and other sites through New Zealand Forest Service plantings from the 1950s, along with secondary spread by domesticated cattle or feral pigs. The initial misidentification as *Phytophthora heveae* (Gadgil 1974) was corrected by Beever et al. (2009), and the working name '*Phytophthora taxon agathis*' or 'PTA' was used up until 2015. The current distribution of Kauri Dieback is shown in Fig. 2.

Fig. 1 Dead mature kauri tree >500 years old. Commonly referred to as 'stag head' (*Source* Kauri Dieback Programme, www.kauridieback.co.nz)

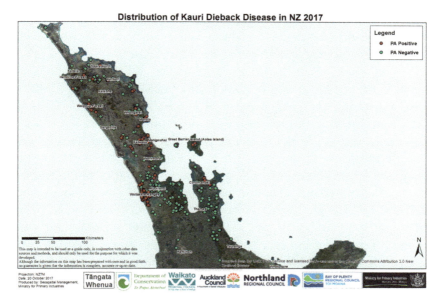

Fig. 2 The distribution of Kauri Dieback disease in New Zealand 2017 (*Source* Kauri Dieback Programme, www.kauridieback.co.nz/more/where-has-it-been-detected/)

3.1 Management Strategies for Kauri Dieback

Human activities, including the transfer of contaminated soils between nurseries, recreational use of kauri forests, and track building and maintenance practices, have been all correlated to the spread and incidence of the disease (Bellgard et al. 2016). Beever et al. (2009) and Horner et al. (2014) demonstrated how susceptible kauri are to infection and how easily infectious propagules, such as oospores, are transmitted from infected to non-infected plants. The pathway of oospore infection is through soil pore water and into the roots of healthy plants; hence, root health and protection of the root zones have a significant effect on the susceptibility of trees to infection (Beauchamp 2013; Waipara et al. 2013).

In October 2008, Kauri Dieback was declared a pest of national priority (an 'Unwanted Organism') under New Zealand's Biosecurity Act, and a biosecurity response was initiated by MAF Biosecurity New Zealand (now Ministry for Primary Industries), Tangata Whenua (local Māori), Department of Conservation (DoC) and Local Authorities (Regional Councils) within the natural range of kauri (Waipara et al. 2013). In 2009, a long-term management (LTM) programme was implemented to mitigate the disease. As per standard crisis response management models, the early focus was on the pathogen itself, with surveillance programmes set up and containment methods put in place to restrict the movement of soil. These methods are only commonplace on government-owned land and include phytosanitary measures to reduce soil-borne spread, such as footwear wash stations containing Sterigene (a disinfectant), vector control (feral pig and goat eradication), upgrading recreational visitor walking tracks (boardwalks) and closing public access to some high-value kauri areas, including imposing a *rāhui* (restriction) by local Māori to certain areas (Fig. 3). As of 2017, there is still a lack of measures that can effectively stop the spread of Kauri Dieback which has led to recent recommendations for restricting access and/or closures to threatened kauri forest areas such as the heavily infected kauri stands of the Waitakere Ranges in West Auckland (Hill et al. 2017).

3.2 Impacts on Māori of Kauri Dieback

Very few studies exist on the impacts of plant diseases on cultural identity, which highlights the importance of these two case studies. Harris (2006) acknowledges the devastating impact Potato Blight (*Phytophthora infestans*) had on Māori in 1905–1906, and Beever et al. (2007) identified many potential pre-border pests and diseases that could damage species highly valued by Māori and therefore pose risks of cultural impacts for Māori and their *kaitiakitanga* (guardianship) roles over particular species (Coffin et al. 2009).

In the case of New Zealand kauri, Nuttall et al. (2010) outline the cultural significance on Māori of the remaining ancient stands of kauri forests. More than 75% of remaining kauri forests lie within the

Fig. 3 Signage designed to raise public awareness of Kauri Dieback and the key hygiene measures in place to help reduce the spread of the pathogen (*Source* Kauri Dieback Programme, www.kauridieback.co.nz)

Northland region, mostly as fragmented remnants of ancient forests or regenerating successional stands, and for Māori on these territories kauri grow, the tree is the centrepiece of cultural and spiritual beliefs. Of these forests, Waipoua, within Te Roroa tribal lands, is home to the famous 1500-year-old Tāne Mahuta (Fig. 4), standing at 51.5 metres tall and having a girth of 13.8 metres. (Waipoua Forest is also home to the second and third largest kauri). In this territory, local Māori often referred to kauri in speeches, cultural performances and proverbs; the fundamental importance is expressed in the proverb *'Ko te kauri ko au, Ko te au ko kauri* - I am the kauri, the kauri is me'.

The health of Waipoua Forests is inextricably linked to by Te Roroa Māori to the *mauri* (spirit, essence) and *mana* (respect, authority, status, spiritual power) of their communities, elders and succeeding generations. For Te Roroa, the presence of Kauri Dieback represents yet another negative colonial impact, comparable to the land and population losses of the 1800s where the iwi was essentially landless with little or no resources and struggling to practice traditional concepts.

3.3 Use of Cultural Health Indicators for Kauri Forest Management

The application of Māori knowledge for kauri conservation is outlined in three reports: *'Te Roroa Kauri Dieback effects assessment'* (Nuttall et al. 2010); *'Kauri dieback cultural indicators'* (Shortland 2011), and a report commissioned by the Kauri Dieback Programme (KDP) on kauri cultural health indicators (CHI) (Chetham and Shortland 2013). Both Shortland (2011) and Chetham and Shortland (2013) outline a rationale and framework for Kauri Dieback based entirely on *mātauranga Māori*, using a holistic approach based on the domains of *Atua* (spiritual guardians) and recommending the inclusion of the monitoring of other species within the kauri forests; surrounding environmental conditions (soil characteristics, leaf litter, decaying wood detritus); the proximity of significant water bodies, levels of sunlight, human activities; and tree condition. This approach reflects the desire of Indigenous communities to combine selected ecological variables with community

Fig. 4 Tāne Mahuta, Waipoua Forest, Northland, New Zealand (*Photograph source* Alastair Jamieson, Wild Earth Media, and Auckland Council)

spiritual experiences of their forests. Attributes such as culturally framed spirituality are difficult for historical academic disciplines to assess within standard scientific ecological impact assessments of trees and, for example, the spread of Kauri Dieback. However, Māori insist that such an approach is essential to capture the wider well-being of their forest systems.

3.4 The Role of Māori in Managing Kauri Forest Health

In 2009, prompted by Māori advocacy, central government resourced engagement with Māori communities throughout the kauri districts to determine their role in the newly proposed joint government agency response, the KDP.[1] This was seen as a significant event in granting partnership status and resulted in a governance and management structure that includes Māori (Fig. 5). This was the first example of a long-term pest management programme attempting to incorporate a partnership model with Māori in New Zealand in accordance with the Treaty of Waitangi.

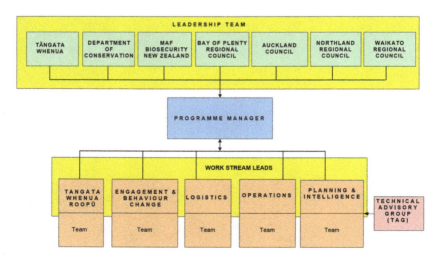

Fig. 5 Governance and management structure for the Kauri Dieback Programme (*Source* Kauri Dieback Programme, www.kauridieback.co.nz)

After a series of meetings, participating Māori established the Tangata Whenua Roopu (TWR) as a Māori reference group, comprised of representatives of those Māori whose lands included kauri (Wilson 2009). The TWR determined they would modify current biosecurity management through culturally framed methods and the use of Māori knowledge to manage or resolve the threat of PTA. In addition, the TWR committed to ensuring effective engagement in the PTA long-term management (LTM) plan, aiming to have local Māori continue to manage PTA beyond the LTM conclusion. At the outset, TWR expressed their expectation that Māori knowledge was fundamental to resolving Kauri Dieback management. In support of the purpose articulated by the TWR, programme partners[2] recognised the TWR as a key partner. They also formally acknowledged Māori as *kaitiaki*, guardians, of kauri and as landowners in their own right.

In April 2010, the TWR commissioned as part of its focus a cultural impact assessment on kauri (Chetham and Shortland 2013). It was also determined that the KDP programme would include increasing the capability of Māori in additional management-related activities such as surveillance, long-term monitoring and research. The TWR has representation in operations (operational management of Kauri Dieback); planning and intelligence (informing the programme with technical expertise and underpinning science); and engagement and behaviour change (including communication, media, public awareness and compliance with programmes key messaging). This model is the first case in which Māori have been represented at all levels of a management programme, and this has been captured in the KDP programmes Strategic Overview Goal Two (Ministry for Primary Industries, n.d.-a), 'Building Knowledge and Tools', in which Māori knowledge was embedded. The goal was then to harmonise mainstream science with *mātauranga Māori* through a plan that identified: (1) how *mātauranga* Māori (Māori knowledge) research, tools and monitoring will be implemented; (2) priority knowledge gaps that needed to be addressed; (3) how advice from experts will be obtained and utilised; and (4) arrangements to provide assurance and demonstrate that scientific evidence and analysis are sought, obtained, interpreted, used and communicated appropriately within the programme (ibid., p. 17). The expected benefits of implementing these goals include greater confidence that Māori would be harnessing the right advice and that decision-making is founded on robust scientific and cultural knowledge; enhanced knowledge of how to manage Kauri Dieback is obtained; and knowledge is gained from and used by those who are guardians of kauri.

The commissioning of a cultural impact assessment has helped initiate the operationalisation of Māori knowledge in the KDP, including the development of a Kauri Cultural Health Index and potential sites to pilot these indices for the detection of Kauri Dieback. The KDP has included funding specifically for the development of forest health indicators using Māori knowledge, with three broad areas of scientific and community interest chosen: Ngahere (forest), Oneone (earth, soil) and Water (wai). The model has the potential to produce successful synergistic social and conservation benefits for kauri forests.

However, these efforts to introduce an IK base into contemporary forest biosecurity were met with strong resistance and a general lack of recognition by some forestry managers and agencies. A change in membership and leadership has seen the programme delayed. Frustration for participating Māori and missed opportunities for better biosecurity outcomes describe the Indigenous experience in this case, although participants are continuing to advocate for future opportunities to ensure the status of Māori knowledge in New Zealand forest management.

4   The Māori Biosecurity Network: Te Tira Whakamātaki (TTW)

Informed by the experiences of Māori trying to address Kauri Dieback and aware of the continuing absence of a Māori voice in wider biosecurity issues, a group of Māori researchers travelled around New Zealand in 2015 and 2016 and met with interested individual Māori and collectives whose interests were across a range of commercial and environmental sectors. With funding from the Ministry of Business, Employment and Innovation (MBIE) and New Zealand's Biological Heritage National Science Challenge (NZBHNSC) discussions took place on the necessity for a national network that could focus on the need for Māori to have a voice in the biosecurity system. At these meetings, Kauri Dieback was presented as an existing biosecurity threat, and Myrtle Rust was used as an exemplar of a disease that would likely prove to be a biosecurity threat at some point in the near future (Te Tira Whakamātaki 2017; NZ Biological Heritage National Science Challenge, n.d.-a, b).

An important aim of these *hui* (meetings) was to engage with Māori pre-incursion and to develop processes to frame effective responses to current and future incursions based in large part on Māori knowledge. Additionally, the network wanted to make better use of data, including data sourced by or from Māori, and the insights and experiences of participating Māori, many of whom had established networks vital for understanding and combating the threats of pests and diseases affecting Māori bio-cultural interests.

The Māori Biosecurity Network has been vocal in their concern about the exclusion of Māori from the biosecurity system in New Zealand, as well as the existence of multiple strategies across several Ministries that overlap and are reactive, creating additional costs in administration and management and duplication. The network argues that 'Māori are in the best position to remind Ministries and agencies that a holistic view to fixing our biodiversity issues needs to be taken' (Mark-Shadbolt 2017a). The important role of the Māori Biosecurity Network in organising and overseeing a Māori response to a significant biosecurity incursion is discussed next.

5 The Discovery of Myrtle Rust in New Zealand

Myrtle Rust (*A. psidii*) is a devastating fungal plant disease. It is indigenous to South and Central America and the Caribbean (Teulon et al. 2015) but has spread to many other regions, including New Guinea and Australia, where it is threatening the extinction of several plant species of significance to Aboriginal Australians (Robinson et al. 2016). It was discovered in Hawai'i in 2005 and has since devastated the 'Ōhi'a tree which is an important species for native Hawaiians (Uchida et al. 2006). Since its arrival in Australia in 2010, its host range has doubled to over 346 known Myrtle Rust hosts globally, and at least nine native New Zealand Myrtaceae species that are cultivated in Australia have been confirmed as being infected (Teulon et al. 2015). Myrtle Rust spores can easily spread large distances by wind and can also be transported on clothing, equipment, insects, rain splashes and probably also cyclones. Impacts of the pathogen have ranged from superficial temporary infections to devastating outbreaks.

The first identification of Myrtle Rust in a New Zealand territory was on Raoul Island (Rangitahua), part of the Kermadec Island group situated a thousand kilometres north of the mainland's North Island. At the time, the newly established Māori Biosecurity Network released a press statement in which they argued, 'as [Myrtle Rust] has now reached our outer islands, we need to be vigilant and we need a plan' (Te Tira Whakamātaki 2017). The Network also offered their support and their knowledge (*mātauranga*) to the Ministry for Primary Industries (MPI) and DoC to help with the response (Te Tira Whakamātaki 2017). The offer was made because while the severity of the disease's effects in New Zealand was unknown, what was known was the disease's likelihood to infect native New Zealand Myrtaceae species.

Since that initial discovery on Raoul Island (Rangitahua), Myrtle Rust has been discovered in mainland New Zealand, initially in Northland in May of 2017, and then further south in Waikato, Taranaki, Auckland and Te Puke (Ministry for Primary Industries 2017).

5.1 The Impacts of Myrtle Rust on Māori

Indigenous Myrtaceae species utilised by Māori for medicine, construction and food, and that are susceptible to Myrtle Rust, include kānuka (*Kunzea ericoides*), mānuka (*Leptospermum scoparium*), ramarama (*Lophomyrtus bullata*), rohutu (*Lophomyrtus obcordata*) and various rātā species (*Metrosideros* spp., particularly *M. excelsa*, the pōhutukawa or New Zealand Christmas Tree). Other introduced species which Māori utilise, such as feijoa (*Acca sellowiana*) and several eucalyptus varieties (*Eucalyptus* spp.), may also be vulnerable (Teulon et al. 2015).

Māori have increasing commercial interests in mānuka honey, pharmaceuticals and cosmetics, and the loss of flowers and new growth could potentially have significant implications for these industries (Teulon et al. 2015).

> Honey production for both pōhutukawa and mānuka may be significantly affected in terms of both productivity and quality. Similarly, the quality of medicinal (traditional/rongoa and modern) products from key species may also be compromised. Impacts in this area may very much depend on which elite honey and medicinal plant biotypes are affected by Myrtle Rust. (Teulon et al. 2015, p. 70)

While the future distribution and impacts of Myrtle Rust are relatively unknown, it can be assumed that all Myrtaceae species in New Zealand are at risk and the impacts could be devastating. However, the potential sociocultural consequences for Māori are yet to be fully understood or addressed.

5.2 Management Strategies for Myrtle Rust

The current New Zealand government strategy for managing the spread of Myrtle Rust is focused on identifying outbreaks, spraying infected plants to halt the spread of the disease, removing infected plants and then burying them (Ministry for Primary Industries 2017). A Technical Advisory Group (TAG) made up of science experts and industry representatives was established to support the Ministry of Primary Industries to make decisions around the response; the Australian members of the group delivered a strong message to New Zealand: aim for eradication. However, with the increasing number of finds, it is anticipated that central government will move the Myrtle Rust response into one of LTM. The focus will then shift from eradication to long-term management, and central government's efforts and resourcing will be diverted to research and management.

The Māori Biosecurity Network, guided by *iwi* (tribes), *hapū* (subtribes) and *whānau* (individual Māori families), believed strongly in aiming for eradication of the disease and argued that Māori *kaitiaki* (as local environmental guardians) were the best 'eyes on the ground', and with their community, networks were 'best placed to identify the first signs of the disease on mainland Aotearoa' (Te Tira Whakamātaki 2017). The Network believed that eradication and containment, even if only regionally, were achievable if Māori knew how to recognise Myrtle Rust, report suspected discoveries in a timely manner and were allowed to be involved in the strategies designed to halt the spread (Te Tira Whakamātaki 2017). Additionally, the Network has argued for greater engagement with Māori at all levels, noting that a failure to properly engage will create tensions and hinder an effective long-term response. Evidence of this tension was reported on Radio New Zealand by McSweeny (2017) who noted 'iwi members were heavily critical of the way the ministry engaged with them over the incursion and voiced their condemnation at the Thursday meeting to MPI officials'.

At the time of publishing, the Māori Biosecurity Network was continuing to offer support to researchers and government agencies in the development of management strategies (Te Tira Whakamātaki 2017). However, despite support from numerous research organisations, there has been little uptake from either MPI or DoC, the two key government agencies. Accordingly, the Māori Biosecurity Network has been forced to develop its own Myrtle Rust management strategy. The Network's strategy is based on the articulated aspirations of over 300 Māori they consulted with between May and October 2017. The Network's response to those aspirations is discussed next.

5.3 The Use of Cultural Health Indicators and the Role of Māori in Managing Vulnerable Species and Ecosystems

One of the founding motivations for establishing the Māori Biosecurity Network was that the inclusion of Māori in biosecurity management was important because if Māori were informed by the latest research about incoming pests and diseases they would be better prepared, more easily mobilised and able to take an active role in the protection of the species and sites of significance to them. This view, which is a very traditional role, was also evident in the Māori response to Kauri Dieback. In both cases, Māori expressed a desire for tools based on their knowledge and for surveillance training and accreditation to be developed. Additionally, they have requested that the proprietary rights of Māori over particular plants and plant materials be considered and protected. In particular are concerns at the lack of, or ad hoc, engagement by agencies collecting seeds and germplasm 'under urgency' (for ex situ preservation and conservation of susceptible plants) without robust prior cultural safety agreements with local tribes.

While, in the Kauri Dieback space, Māori have and continue to struggle to get Māori management strategies recognised, resourced and/or implemented, the Māori Biosecurity Network achieved quick successes in the implementation of responses to Myrtle Rust. Within five months of the first mainland incursion, the Māori Biosecurity Network had trained over 100 *kaitiaki* to identify Myrtaceae plants and Myrtle Rust, and report suspected Myrtle Rust finds. They, along with other partners, had also released a smartphone application that also assisted in the identification of both Myrtaceae species and Myrtle Rust, while providing a platform for live reporting of suspected finds (New Zealand's Biological Heritage National Science Challenge, n.d.-a, b).

The incorporation of *mātauranga Māori* in the response to Myrtle Rust, while better than Kauri Dieback, has been limited to date. The approach by government has mainly been one of engagement, and the development of CHI is still at an early stage. However, *kaitiaki* (guardians) are already developing indicators or ideas on how to mitigate the effects of the disease; for example, they have expressed a desire to plant *ramarama* (*Lophomyrtus bullata*) either near sites of significance to take the brunt of the infection, or close by as sentinels. More time and resourcing are needed to find and refine further indicators.

6 Discussion

The special relationship that mana whenua (local Māori) have with kauri was recognised with the inception of the KDP in 2010, a joint agency response that included central government, regional agencies and Māori community groups. This was the first 'true' partnership between Māori and government, as per the expectations of the Treaty of Waitangi, which was established to manage any biosecurity incursion but particularly the devastating impacts of *P. agathidicida* on kauri forests.

In trying to address Kauri Dieback, Māori have struggled in their attempts to collaborate with researchers and government officials, both local and national. By arguing for a new role in biosecurity management which would integrate Māori knowledge in any effort to understand and combat the disease, local Māori and their supporters found themselves challenged by mainstream scientists and regulators. So far, only Western-style management methods have been implemented: phytosanitary measures to reduce soil-borne spread; vector control; upgrading walking tracks; and closing public access to some areas. Research outside of the programme is underway on how scientists can better collaborate with IK holders to produce solutions to mitigate the effects of Kauri Dieback.

The 2017 arrival of Myrtle Rust poses a significant threat to several native plants including culturally and commercially significant species such as *mānuka* (*Leptospermum scoparium*), source of the highly lucrative mānuka honey (Department of Conservation 2017). The near-cotemporaneous establishment of Te Tira Whakamātaki, as a Māori-centric 'network of the willing' (Mark-Shadbolt 2017a), was fortuitous. With members including Māori researchers and wider support from mainstream allies, the network has both scientific and political credibility. It is important to note that Te Tira Whakamātaki receives no direct funding but instead leverages off the existing research and programmes of its Executive; indeed, the leadership made a conscious decision not to accept money or contracts unless it was very clear about the purpose of that funding (Mark-Shadbolt 2017b). Their argument has been that by accepting money from the government results in government assuming the right to dictate or control the conversations, results, data generation and measures of success; at least two government agencies were accused of claiming Te Tira Whakamātaki events (community meetings) as their own achievements.

Te Tira Whakamātaki, the Māori Biosecurity Network, has argued that Māori roles of environmental guardianship are the best option to access the forests efficiently and with minimal disturbance to other species. As government officials and agents are not undertaking extensive surveillance in the wild, it is often only these *kaitiaki* who also know the sites of significance that need to be inspected and observed. This saves costs, ensures sufficient geographical coverage, secures Māori their rightful role as Treaty partners and supports Māori aspirations for their economic, environmental and cultural well-being.

Worldwide, there are undoubtedly many other biosecurity events that would benefit from local IK and the empowered participation of Indigenous representatives and their communities. Researchers, officials and the private sector must take seriously the rights of Indigenous peoples to determine their self-development and elevate the ethical engagement with Indigenous communities as a priority in the biosecurity of the world's forests.

7 Concluding Summary

The implementation of alternative models of partnership with Indigenous communities as demonstrated by the KDP and Te Tira Whakamātaki has resulted in the involvement of Indigenous representatives across research governance, strategy and field operations. In this chapter, we have argued that the adoption of IK and indigenous practices and the empowered participation of Indigenous environmental managers and their communities are vital for the sustainable management and long-term protection of many of the world's forests. In Aotearoa New Zealand, the inclusion of *kaitiaki* (Māori guardians) and the adoption of Māori practices such as *kaitiakitanga* (guardianship) can enhance and inform the long-term protection of kauri ecosystems and Myrtaceae species across the country. Such a collaborative approach provides efficiencies in national and local biosecurity strategies and tactics and, importantly, enables the fulfilment of Indigenous aspirations of economic, environmental and cultural well-being.

Acronyms

DoC	Department of Conservation
CHI	Cultural Health Indicators
KDP	Kauri Dieback Programme
LTM	Long-Term Management
MAF	Ministry of Agriculture & Fisheries, now MPI
MBIE	Ministry for Business, Innovation & Employment
MPI	Ministry for Primary Industries: Manatū Ahu Matua
NZBHNSC	New Zealand's Biological Heritage National Science Challenge
PTA	*Phytophthora* Taxon Agathis
RMA	Resource & Management Act 1993
TAG	Technical Advisory Group
TTW	Te Tira Whakamātaki, the Māori Biosecurity Network
TWR	Tangata Whenua Roopu

Glossary

Atua	Spiritual guardians, gods
Hapū	Subtribe, extended family group
Harakeke	Flax, *Phormium tenax*
Hui	Meetings
Iwi	Tribe
Kaitiaki	Guard, guardian, caretaker, manager, trustee
Kaitiakitanga	Guardianship
Kānuka	*Kunzea ericoides*
Karengo	Edible seaweed (*Bangiaceae* sp.)
Kauri	*Agathis australis*
Kaumātua	Elder
Kererū	New Zealand Wood Pigeon (*Hemiphaga novaseelandiae*)
Kaupapa Māori	Māori methodology
Mana	Respect, authority, control, power, status, spiritual power

Mana whenua	Local Māori with territorial rights and cultural authority over land
Mānuka	*Leptospermum scoparium*
Māori	Word used to describe the Indigenous people of New Zealand
Mātauranga	Information, knowledge, education
Mātauranga Māori	Māori knowledge
Mauri	Life principle
Ngahere	Forest
Oneone	Earth, soil
Pōhutukawa	*M. excelsa* or the New Zealand Christmas Tree
Rahui	Exclusion, ban, quarantine
Ramarama	*Lophomyrtus bullata*
Rangitahua	Raoul Island, part of Kermadec Island group 1000 km north of New Zealand
Rātā	*Metrosideros* spp., particularly *M. excelsa*, the pōhutukawa or New Zealand Christmas Tree
Rohutu	*Lophomyrtus obcordata*
Tangata whenua	Local people, aborigine, native
Tāne Mahuta	God (guardian spirit) of the forest and name of largest Kauri tree in New Zealand
Tangata Whenua Roopu	Māori community group(s)
Taonga	Treasured, sacred property
Te Roroa	A Māori tribe from the region between the Kaipara Harbour and the Hokianga Harbour in Northland, New Zealand
Te Tira Whakamātaki	The Māori Biosecurity Network (translates as the vigilant or watchful ones)
Tītī	Muttonbird
Treaty of Waitangi	Treaty signed between the representatives of the British Crown and Māori tribal leaders in 1840.
Wai	Water

Whakapapa	Genealogy, cultural identity, ancestry (extends to species assemblages within a holistic ecosystem paradigm)
Whānau	Immediate family, also used by Māori to describe individual Māori

Acknowledgements Nga mihi and thanks to Lincoln University for sponsoring this chapter's open access status.

Notes

1. Previously the Kauri Dieback Joint Agency Response, until 2015 when the name of Programme was changed to first the Kauri Dieback Management Programme to reflect the Ministry for Primary Industries (MPI) declaring PTA an unwanted organism (UO) in 2008 under New Zealand's Biosecurity Act (1993), thus initiating a central government response to assess future management options. In 2009, a five-year long-term management programme was implemented. MPI's priority from 2009 to 2014 was to manage Kauri Dieback as opposed to eradicating Kauri Dieback.
2. Auckland Regional Council, Waikato Regional Council, Northland Regional Council, Ministry of Agriculture and Fisheries, Department of Conservation.

References

Beauchamp, A. J. (2013). *The detection of* Phytophthora *Taxon "Agathis" in the second round of surveillance sampling—With discussion of the implications for kauri dieback management of all surveillance activity*. Client report: Joint Agency Kauri Dieback Response. http://www.kauridieback.co.nz/media/34150/surveillance%202%20final%20report%20pdf.pdf (Accessed March 21, 2016).

Beauchamp, T., & Waipara, N. (2014). Surveillance and management of kauri dieback in New Zealand. In *Proceedings of the 7th Meeting of the International Union of Forest Research Organizations (IUFRO), Working Party S07-02-09*, Phytophthora in Forests and Natural Ecosystems (p. 142). November 10–14, Esquel, Argentina.

Beever, R. E., Harman, H., Waipara, N., Paynter, Q., Barker, G., & Burns, B. (2007). *Native flora biosecurity impact assessment* (Landcare Research Contract Report LC0607/196). For MAF Biosecurity New Zealand, Volume 1 (103 pp), Volume 2 (202 pp).

Beever, R. E., Waipara, N. W., Ramsfield, T. D., Dick, M. A., & Horner, I. J. (2009). Kauri (*Agathis australis*) under threat from *Phytophthora*? In *Proceedings of the Fourth Meeting of the International Union of Forest Research Organizations (IUFRO) Working Party S07-02-09*. Phytophthoras in Forests and Natural Ecosystems, August 26–31, Monterey, CA, USA.

Bellgard, S. E., Pennycook, S. R., Weir, B. S., Ho, W., & Waipara, N. W. (2016). *Phytophthora agathidicida*. *Forest Phytophthoras, 6*(1). https://doi.org/10.5399/osu/fp.5.1.3748.

Bishop, R. (1999). *Kaupapa Māori Research: An indigenous approach to creating knowledge*. Paper presented at the Māori Psychology: Research and Practice, Hamilton.

Camara-Leret, R., Paniagua-Zambrana, N., Balslev, H., Barfod, A., Copete, J. C., & Macia, M. J. (2014). Ecological community traits and traditional knowledge shape palm ecosystem services in northwestern South America. *Forest Ecology and Management, 334*, 28–42. https://doi.org/10.1016/j.foreco.2014.08.019.

Chetham, J., & Shortland, T. (2013). *Kauri cultural health indicators: Monitoring framework*. Unpublished report prepared by Repo Consultancy Ltd for the Kauri Dieback Programme.

Coffin, A., van Eyndhoven E., & Beever, R. E. (2009). *Native flora impact assessment* (MAF Biosecurity New Zealand Technical Paper No. 2009/32). ISBN 978-0-478-35185-9 (31 pp).

Cunningham, C. (1998). A framework for addressing Maori knowledge in research, science and technology. *Te Oru Rangahau Maori Research and Development Conference*, Massey University.

de Freitas, C. T., Shepard, G. H., Jr., & Piedade, M. T. F. (2015). The floating forest: Traditional knowledge and use of Matupá vegetation islands by riverine peoples of the Central Amazon. *PLoS One, 10*(4), e0122542. https://doi.org/10.1371/journal.pone.0122542.

Department of Conservation. (2017). *Biosecurity Alert Myrtle Rust, Austropuccinia psidii*. Wellington: Department of Conservation.

Gadgil, P. D. (1974). *Phytophthora heveae*, a pathogen of kauri. *New Zealand Journal of Forestry Science, 4*, 59–63.

Harris, G. (2006). *Te Paraiti: The 1905–1906 potato blight epidemic in New Zealand and its effects on Māori communities.* Retrieved from https://repository.openpolytechnic.ac.nz/handle/11072/1212.

Hill, L., Stanley, R., Hammon, C., & Waipara, N. (2017). *Kauri Dieback Report 2017: An investigation into the distribution of Kauri Dieback, and implications for its future management, within the Waitakere.* Ranges Regional Park, Auckland Council Technical Report. Version 2: Update June 2017 (40 pp).

Horner, I. J., Hough, E. G., & Zydenbos, S. M. (2014). Pathogenicity of four *Phytophthora* species on kauri: In vitro and glasshouse trials. *New Zealand Plant Protection, 67,* 54–59.

Lyver, P. O., Taputu, T. M., Kutia, S. T., & Tahi, B. (2008). Tuhoe Tuawhenua matauranga of *kereru* (*Hemiphaga novaseelandiae novaseelandiae*) in Te Urewera. *New Zealand Journal of Ecology, 32*(1), 7–17.

Lyver, P. O., Jones, C. J., & Doherty, J. (2009). Flavor or forethought: Tuhoe traditional management strategies for the conservation of kereru (*Hemiphaga novaeseelandiae novaeseelandiae*) in New Zealand. *Ecology and Society, 14*(1), 40.

Mark-Shadbolt, M. (2017a). *Biosecurity (Report to Natural Resource Iwi Leaders Group).* Lincoln: Te Tira Whakamataki.

Mark-Shadbolt, M. (2017b). Te Tira Whakamātaki: Māori biosecurity network. *Crazy & Ambitious Conference.* https://www.youtube.com/watch?v=T3HgJ2s8OLU.

McCallum, R. E., & Carr, D. J. (2012). Identification and use of plant material for the manufacture of New Zealand indigenous woven objects. *Ethnobotany Research & Applications, 10,* 185–198.

McGeoch, M. A., Butchart, S. H. M., Spear, D., Marais, E., Kleynhans, E. J., Symes, A., et al. (2010). Global indicators of biological invasion: Species numbers, biodiversity impact and policy responses. *Diversity and Distributions, 16,* 95–108. http://doi.org/10.1111/j.1472-4642.2009.00633.x.

McSweeny, J. (2017). *MPI failed to engage over Myrtle Rust—Bay of Plenty iwi.* http://www.radionz.co.nz/news/national/337062/mpi-failed-to-engage-over-myrtle-rust-bay-of-plenty-iwi.

Mead, H. M. (2003). *Tikanga Māori: Living by Māori values.* Wellington: Huia.

Ministry for Primary Industries. (n.d.). *Kia Toitu He Kauri/Keep Kauri Standing.* https://www.kauridieback.co.nz/media/1393/kauri-diebackstrategy-2014-final-web.pdf.

Ministry for Primary Industries. (2017). *Myrtle Rust*. Retrieved October 19, 2017, from http://www.mpi.govt.nz/protection-and-response/finding-and-reporting-pests-and-diseases/pest-and-disease-search?article=1484.

Moller, H., Fletcher, D., Johnson, P., Bell, B., Flack, D., Bragg, C., et al. (2009). Changes in sooty shearwater (*Puffinus griseus*) abundance and harvesting on the Rakiura Titi Islands. *New Zealand Journal of Zoology, 36*(3), 325–341. https://doi.org/10.1080/03014220909510158.

New Zealand Treasury. (2012). *New Zealand economic and financial overview 2012*. Wellington: Treasury.

New Zealand's Biological Heritage. (2017). *Myrtle Rust reporter* [Press release]. Retrieved January 21, 2018, from http://www.biologicalheritage.nz/news/research-stories/myrtle-rust-reporter.

New Zealand's Biological Heritage National Science Challenge. (n.d.-a). *Myrtle Rust Update #1*. http://www.biologicalheritage.nz/programmes/maori-biosecurity-network/myrtle-rust/update-1.

New Zealand's Biological Heritage National Science Challenge. (n.d.-b). *Myrtle Rust Update #2*. http://www.biologicalheritage.nz/programmes/maori-biosecurity-network/myrtle-rust/update-2.

Nuttall, P., Ngakuru W., & Marsden, M. (2010). Te Roroa effects assessment for Kauri Dieback disease—(*Phytophthora* taxon Agathis—PTA). Report prepared for Te Roroa and the Kauri Dieback Joint Agency Response by Wakawhenua.

O'Connell-Milne, S. A., & Hepburn, C. D. (2015). A harvest method informed by traditional knowledge maximises yield and regeneration post-harvest for karengo (*Bangiaceae*). *Journal of Applied Phycology, 27*, 447–454.

Piddington, R. (1960). Action anthropology. *Journal of the Polynesian Society, 69*(3), 199–213.

Pihama, L., Cram, F., & Walker, S. (2002). Creating methodological space: A literature review of Kaupapa Māori research. *Canadian Journal of Native Education, 26*(1), 30–42.

Prussing, E., & Newbury, E. (2016). Neoliberalism and indigenous knowledge: Māori health research and the cultural politics of New Zealand's "National Science Challenges". *Social Science & Medicine, 150*, 57–66. https://doi.org/10.1016/j.socscimed.2015.12.012.

Robinson, C. J., Maclean, K., Hill, R., Bock, E., & Rist, P. (2016). Participatory mapping to negotiate indigenous knowledge used to assess environmental risk. *Sustainability Science, 11*(1), 115–126.

Shortland, T. (2011). *Cultural indicators for Kauri Ngahere*. A report prepared for the Tangata Whenua Roopu. Kauri Dieback Joint Agency Response, Whangarei, New Zealand.

Singh, R. K., Srivastava, R. C., Pandey, C. B., & Singh, A. (2015). Tribal institutions and conservation of the bioculturally valuable 'tasat' (*Arenga obtusifolia*) tree in the eastern Himalaya. *Journal of Environmental Planning and Management, 58*(1), 69–90. https://doi.org/10.1080/09640568.2013.847821.

Smith, L. T. (1999). *Decolonizing methodologies: Research and indigenous peoples*. Dunedin: University of Otago Press.

Souto, T., Deichmann, J. L., Nunez, C., & Alonso, A. (2014). Classifying conservation targets based on the origin of motivation: Implications over the success of community-based conservation projects. *Biodiversity and Conservation, 23*(5), 1331–1337. https://doi.org/10.1007/s10531-014-0659-9.

Te Tira Whakamātaki. (2017). *Myrtle Rust position statement*. Lincoln: Maori Biosecurity Network/Te Turi Whakamataki.

Teulon, D. A. J., Alipia, T. T., Ropata, H. T., Green, J. M., Viljanen-Rollinson, S. L. H., Cromey, M. G., et al. (2015). The threat of Myrtle Rust to Māori taonga plant species in New Zealand. *New Zealand Plant Protection, 68*, 66–75.

Uchida, J., Zhong, S., & Killgore, E. (2006). First report of a Rust disease on Ohia caused by *Puccinia psidii* in Hawaii. *Plant Disease, 90*(4), 524. https://doi.org/10.1094/PD-90-0524C.

Waipara, N. W., Hill, S., Hill, L. M. W., Hough, E. G., & Horner, I. J. (2013). Surveillance methods to determine tree health, distribution of Kauri Dieback disease and associated pathogens. *New Zealand Plant Protection, 66*, 235–241.

Walker, D., Ataria, J., Hughey, K., & Lambert S. (2013). Developing a culturally-based environmental monitoring and assessment tool for New Zealand Indigenous forests. *19th International Symposium on Society and Resource Management*. June 4–8, Estes Park, CO, USA.

Walter, R. K., & Hamilton, R. J. (2014). A cultural landscape approach to community-based conservation in Solomon Islands. *Ecology and Society, 19*(4), 41. https://doi.org/10.5751/ES-06646-190441.

Weir, B. S., Paderes, E. P., Anand, N., Uchida, J. Y., Pennycook, S. R., Bellgard, S. E., & Beever, R. E. (2015). A taxonomic revision of *Phytophthora* Clade 5 including two new species, *Phytophthora agathidicida* and *P. cocois*. *Phytotaxa, 205*(1), 21–38.

Whyte, W. F. (1981). *Street corner society: The social structure of an Italian slum*. Chicago and London: University of Chicago.

Wilson, S. (2009). *Kauri Dieback (Phytophthora taxon Agathis) Joint Agency Response: Tāngata Whenua Hui. Summary Report* (Unpublished), Maximise Consultancy (66 pp).

World Bank Group. (2016). *World Bank Group Forest Action Plan F16-20*. Washington, DC. Retrieved from http://documents.worldbank.org/curated/en/240231467291388831/Forest-action-plan-FY16-20.

Open Access This chapter is licensed under the terms of the Creative Commons Attribution 4.0 International License (http://creativecommons.org/licenses/by/4.0/), which permits use, sharing, adaptation, distribution and reproduction in any medium or format, as long as you give appropriate credit to the original author(s) and the source, provide a link to the Creative Commons license and indicate if changes were made.

The images or other third party material in this chapter are included in the chapter's Creative Commons license, unless indicated otherwise in a credit line to the material. If material is not included in the chapter's Creative Commons license and your intended use is not permitted by statutory regulation or exceeds the permitted use, you will need to obtain permission directly from the copyright holder.

6

User-Generated Content: What Can the Forest Health Sector Learn?

John Fellenor, Julie Barnett and Glyn Jones

1   Introduction

Forests are complex and integrated socio-ecological systems (Folke 2007). Given the varied nature of the associated stakeholder investment, societal expectation and environmental dynamics, this presents many challenges for their management (Kelly et al. 2012). Over recent decades, the human dimensions of these systems have rapidly evolved, reflecting growing concern with environmental degradation and facilitated by the emergence of information communication technologies (ICTs), especially user-generated content (UGC): publicly created and readily accessible online material. Such technologies are transforming human interaction and reshaping our experience of self and community.

J. Fellenor (✉) · J. Barnett
Department of Psychology, University of Bath, Bath, UK

G. Jones
Fera Science Ltd., National Agri-Food Innovation Campus, York, UK

For forest health sector (FHS) managers and stakeholders, this is an ontological issue: if our presuppositions about the world are changing, then our knowledge of that world, and how to manage it, should reflect these changes.

This chapter brings to the fore these ontological issues by reflecting on the relationship between UGC and forest health. We outline how concerns central to the human dimensions of tree health have evolved, providing an overview of the uses and potentials of UGC, before drawing on a rapid evidence review of literature to consider UGC in relation to forest health. Finally, we reflect on some broader questions from an ontological perspective, situating the use of UGC in broader debates about how digital technologies shape our relationship with the environment and ourselves.

2 The Changing Profile of Forest Management

Successful forest management requires consideration of a range of human dimensions, often manifesting as conflicting stakeholder perceptions, values and behaviours (Kearney and Bradley 1998). Due to the variability of culture, in which contrasting relational logics make up the non-human realm (Kohn 2013), forest management itself reflects different endogenous knowledge and beliefs (Pretzsch 2005; Sauget 1994). For hundreds of years, forests have been regarded as a commodity requiring the regulation of use, access and control (Michon et al. 2013). The notion of a 'moral forest', which incorporates environmental concerns, including climate change mitigation, has begun to be recognised. More recently, the 'recreational forest' has been constructed by and for an increasingly urban population. Serving the perceived therapeutic benefits of being in 'nature' (Hartig et al. 2014; Nilsson et al. 2011), the recreational forest represents lifestyle choices intersecting with ecotourism. In developing countries especially, forests provide income and employment via designation as national parks or conservation areas, attracting tourists, volunteers and international funding

(Bhuiyan et al. 2011). However, in ontological terms, the 'moral' and 'recreational' forest implies a separation between human and nature. Hence, the manner in which we internalise these notions, to articulate aspects of self-experience and our action in the world, suggests a deeper relation still, where humans and forests as material entities occupy a common ontology (González-Ruibal et al. 2011). This emphasises the intrinsic value of 'nature' and the importance of an 'ecocentric' view (Washington et al. 2017). Nonetheless, forest management has to account for myriad perspectives on human–forest relations. 'Forest health' is a contested term, with normative implications that one ecological state or goal is better than another (Sulak and Huntsinger 2012). The advent of the Internet, in reshaping both social and organisational concerns, affords further possibilities and challenges for forest management and how we envisage human–forest ontologies.

3   The Internet and User-Generated Content

The Internet enhances information sharing. It also presents new demands on forest managers because of a more 'present' public and set of stakeholders. Whilst often used synonymously, it is important to clarify how the term 'Internet' relates to 'World Wide Web' ('www'), 'social media', 'Web 2.0' and UGC. The Internet is essentially a network of networking-technology devices whilst 'www' is a space comprised of interlinked information, accessed via the hardware of the Internet. Web 2.0 describes websites enabling content generated by, and for, many interlinked users. Social media is an area of Web 2.0 that affords real-time creation and mediation of content and information sharing (Obar and Wildman 2015).

Public sector organisations are turning to social media to engage with different audiences and disseminate information (Panagiotopoulos and Bowen 2015). Social media, as a means of recruiting a broader audience, is entangled with a rapidly changing political economy of environmental conservation (Büscher 2013). For example, the decision-making processes of tourists, e.g. where to go and what to do, are increasingly influenced by social media shaping the expectations, perceptions, behaviours and the meanings they infer with regard to what nature is and what engagement with it entails (Cheng et al. 2017; Xiang and Gretzel 2010).

UGC is the common thread that connects social media and other aspects of Web 2.0, comprising content such as blogs (diary style text), wikis (collaboratively modifiable content), discussion forums, audio files and images. All social media platforms utilise UGC but not all UGC is limited to social media platforms (Luca 2015). UGC draws attention to a wider range of user, website and content (Kaplan and Haenlein 2010), i.e. data generated in different contexts for different purposes (Krumm et al. 2008). UGC impacts both economic and social processes (Luca 2015). As these processes are foundational dimensions of forest health management, it is incumbent on FHS organisations to have a clear grasp of what UGC is, what it does and what it might do.

UGC can be categorised thus:

- *Social networking*: e.g. Facebook, Twitter and Weibo. Interaction between public or closed communities to share information including images, videos and memes.
- *Social news aggregation*: e.g. Reddit, Buzzfeed and Digg. Selection, aggregation and up-voting of news items.
- *Image and video hosting*: e.g. Flickr, YouTube and Dailymotion. Used for uploading and sharing visual content.
- *Information, discussion, pattern search and learning*: e.g. Wikipedia and Google Trends. Researching and discovering search trends and producing/sharing resources.
- *Product search and discovery services*: e.g. TripAdvisor and Expedia. Enables product search and evaluation; provides personalised recommendations based on browsing history and interests.
- *Social commerce*: e.g. Amazon, eBay and Etsy. Sites or mobile apps supporting social interaction and user contributions to facilitate online retail.
- *Crowdsourcing*: e.g. Crowdrise, Kickstarter and IndieGoGo. Sites enabling individuals or groups to solicit funds, services or ideas.

Some of these platforms are immediately relevant to the FHS. Others may be equally important and yet remain unconsidered. YouTube, Facebook and Twitter are used by the Forestry Commission and the Department for Environment, Food and Rural Affairs in the UK to release information about various issues. These organisations are beginning to look at information and pattern recognition sites such as Google Trends to determine patterns in media coverage and consumption of FHS information. Social commerce sites are also relevant. For example, despite their own strict guidelines, eBay potentially provides a means for circumventing import regulations by users who sell plant material that may carry pests or pathogens.

Ontologically, UGC, in terms of the digital traces people create, blurs the distinction between the individual as a human presence and as a 'digital artefact' (Hogan 2010). The digital artefact mediates how people participate in the world and with each other. Whereas humans act in real time, the digital artefact is an accumulation of past interactions and performances and, as such, is a representation of an historical presence. However, both are often afforded the same ontological status—i.e. both are perceived as 'real' (Reed 2005). Reflecting on this issue enables a useful perspective on the role that UGC can play in forest health.

4   Rapid Evidence Review

Rapid evidence review is a quick and efficient way of synthesising the most relevant conceptual and empirical evidence pertaining to an issue or topic and meets the needs of stakeholders and policy makers working in, and responding to, rapidly evolving and dynamic socio-material environments (Thomas et al. 2013). With respect to how UGC intersects with the FHS we carried out a rapid evidence review of academic literature. We aimed to explore the ways in which UGC is utilised and to reflect on the issues associated with UGC, to enable a deeper appreciation of the benefits and challenges with regard to its potential. Our primary concern was a focus on the social and organisational processes underpinning the use of UGC and how these implicate different stakeholders and publics.

This approach was premised on the notion that literature around UGC use would be representative of domains, including communication, management, plant pathology and governance. We scrutinised ScienceDirect, Web of Science and Scopus as these databases provide access to eclectic material covering the range of scientific, technological and social scientific disciplines. Items included full-length articles published in refereed academic journals, conference articles, conference proceedings and theses.[1] An item was initially included if it addressed any conceptual or empirical aspect of UGC in the field of tree health, forestry or related areas such as environment, ecotourism and land management. Items without a primarily social or organisational focus and from the fields of mathematics, physics and computer science were excluded. Following Levy and Ellis (2006), we used forward and backward searching of items explicitly focussed on the use of UGC in the FHS and, if appropriate, added further items to the corpus. Eighty-six papers were reviewed in depth. For each paper, we established a categorisation of its premise, e.g. 'describes the application of a smart technology' or 'discusses technological challenges with respect to data mining'. We organised these categories into overarching areas capturing their most salient features (see Table 1). Whilst thirty-seven studies related to UGC and forest health, the majority related to the forest sector in general. Hence, to explore the role and potential of UGC, we draw broadly on those studies most applicable *and* transferable to forest health.

5   Findings and Discussion

The number of items where a specific type or aspect of UGC and its relation to forest health was the central concern of a paper was relatively limited. UGC was more often mentioned in passing or as a generic concluding comment, e.g. 'management practices can be enhanced by using social media'. Studies from different countries were represented. There was a marked increase in the volume of articles from 2015 onward.

6 User-Generated Content: What Can the Forest Health Sector Learn?

Table 1 Organisation of categories into overarching areas

Overarching area	Sub-categories	Number of papers in category
Management and communication	• How the web facilitates participation and communication • Developing management practices • Recommendations for social media use • Public engagement with tree initiatives • The role of the public and stakeholders as part of a collaborative process	29
Linking data and linking stakeholders	• Use of web portals to streamline data access • Models of unified databases to aid standardisation • Information sharing and interoperability • Use of Geographic information systems (GIS) as a means of enhancing data connectivity. • Remote sensing and plant tagging to aid micro- and macro-management of the environment	24
Citizen science and crowdsourcing	• Smartphone technology enhancing data collection • Citizen science; examples, strengths and weaknesses • Crowdsourcing applications	17
Monitoring invasive alien species (IAS), data mining and horizon scanning	• Horizon scanning for new outbreaks of pests/pathogens • Social media data mining, automatic classification of data, data scraping • Tracking of pests and pathogens • Technological challenges	16

5.1 Management and Communication

Perspectives on UGC in relation to forest management in general were prolific in the reviewed papers, but less so with regard to a focus on forest health.

Kelly et al. (2012) evaluated the adaptive management[2] process and an interactive website as one of several methods facilitating public participation in the Sierra Nevada Adaptive Management Project (SNAMP), a multi-agency initiative utilising adaptive management to examine the effects of fuel treatments.[3] Google Analytics data was collected to assess who was visiting the website and to infer temporal trends in user focus. This data was combined with a survey of members of the public involved with SNAMP, to explore their views on management, and a content analysis of UGC on the website's interactive discussion board. Peaks in web activity coincided with key public participation and outreach events. The discussion board received low use in relation to other methods such as public meetings and outreach. Sixty per cent of posts to the discussion board came from researchers and forty per cent from the public. It was concluded that whilst the web played a key role in the adaptive management process via dissemination of information to the public, 'the SNAMP public are not the typical content providers found in the online community literature' (ibid., p. 7), i.e. the discussion board facilitated researchers rather than the public.

These findings, alongside the evolving nature of UGC, indicate the need for an evolving and adaptive management approach to forest health. Management should account for the local and particular nature of forests and benefits from the input of local people and stakeholders, all essential components of adaptive approaches (Messier et al. 2015). Although Web 2.0 technologies enhance the scope for stakeholder groups to participate in discussions, they are not in themselves sufficient. They may provide one foundation for the collaborative decision-making and feedback required of adaptive management but are not a replacement for personal contact, direct communication or the 'mutual learning' that occurs through approaches such as participatory workshops or co-creation exercises (Kelly et al. 2012). Lei and

Kelly (2015) explored adaptive management, as a means of fostering collaboration between stakeholders, by comparing content analysis with an automated mapping algorithm[4] to identify patterns in public meeting notes made available for sharing on the SNAMP website. Analysis revealed that meetings largely focussed on key aspects of the project, such as the science involved and that, across time, discussion topics evolved; earlier discussions focussed on project logistics, whereas discussions about issues such as tree health were more persistent.

UGC incorporates the perspectives of individuals and groups. As part of an open flow of information, this enables policy makers and communities to become aware of how management is perceived and how different stakeholders are implicated. An open flow of information enables stakeholders to respond to their local forests and environment in times of crisis in a sensitive manner, and local communities can become more proactive in their own governance (Chandler 2015). Open information also facilitates an awareness of differences in endogenous beliefs. Finally, FHS managers are becoming aware of a meta-level of engagement with stakeholders in that UGC lends itself to the techniques of Big Data[5] analysis; stakeholder analysis is crucial for effective collaborative resource management. However, shifting the onus towards analysing what people do and say, in terms of UGC, requires that an equal amount of time and resources need to be applied to developing or adapting conceptual systems that can handle the vast amounts of UGC data available.

Communication also implicated UGC, e.g. the existing use or recommendation of social media to communicate with stakeholders, and a focus on the evolving ecology of communication within a broader organisational context. Stakeholders include those seeking to commodify forest products, as well as those concerned with their conservation (Gazal et al. 2016; Montague et al. 2016). Studies assessed social media use in the US forest products industry and social media adoption at the organisational level within business-to-business contexts. Twitter and Facebook are being used to facilitate communication and advertise and market products or services, implying the adoption of marketing models for management, rather than conservation or biosecurity. However, a lack of awareness, for example, of the issues involved in the movement of wood products hampers responding to the unintended consequences of such movement (Marzano et al. 2015), such as the spread of pests and pathogens.

These studies highlight the tension between different stakeholders and imperatives, making explicit the need for stakeholder engagement in all aspects of communication and consultation in decision-making processes. From the perspective of UGC, a simplistic stance towards how social media affects communication tended to be adopted, essentially a linear model of information dissemination, such that a message posted on social media reaches an easily specified audience and is attended to accordingly, akin to the 'hypodermic needle' model.[6] In this regard, Fellenor et al. (2017) and Hearn et al. (2014) draw attention to the deeper issues around UGC and communication.

Fellenor et al. (2017) harvested tweets mentioning ash dieback (ADB) disease during peak media attention to the issue during late 2012. The most prominent tweeters were people or organisations already affiliated to forest or environmental issues. Individual or group affiliations, interests and identities framed small groups of users engaged with ADB. Hence, engagement tended to reflect an existing concern, such as horse riders tweeting about cleaning horses' hooves to prevent spread, i.e. interactions. This contrasts with the perception that there is a homogeneous public waiting to be communicated with in a linear and unproblematic manner. Hearn et al. (2014) used communicative ecology theory to describe innovations in urban food systems according to their technical, discursive and social components, suggesting that social media combines with existing communication strategies to enhance the ability of organisations to achieve their goals. In relation to the ecology of communication, social media is part of a broader ontological shift where people can be connected in real time to the outcomes of their behaviour and practices.

5.2   Linking Data, Linking Stakeholders

Digital forestry (DF) is the systematic procurement and analysis of digital information to support sustainable forests and integrates all aspects of forestry information across different spatio-temporal scales (Zhao et al. 2005). At the heart of DF is the perceived need for open and accessible data to enhance communication and information dissemination. DF, alongside traditional methods, utilises digital technologies including remote sensors, GIS, GPS, visualisation software and computer modelling to collect and integrate vast amounts of data, with the aim of optimising forest management. If forest health management has to achieve multiple, complex and sometimes conflicting purposes, the tools and technology required have to be similarly complex and integrated (Tang et al. 2009). However, such technologies are often designed without integration in mind. Nonetheless, the emphasis is on enhancing the interoperability of systems by promoting connections between stakeholders using or producing digital technologies (Reynolds and Shao 2006). From this perspective, UGC becomes part of a much broader system which datafies[7] the interrelations between forests and people, an ontological shift in itself. The more these interrelations become datafied, the more transparent and readable the causal relations and contingencies which bring them together (Chandler 2015). Hence, whilst it is important to understand the uses and types of UGC at a pragmatic level, equal if not greater consideration has to be given to how the social, technological and organisational dimensions are entangled.

Six studies explored the quality and accessibility of information available to stakeholders, and the possibility of unifying systems and data. Despite initiatives seeking to harmonise the types of currently distributed information available about forests, data is often incomplete or incompatible due to the lack of interoperability of technological systems at both a global scale and local scale (McInerney et al. 2012). These authors developed a portal to provide access to forest-resources data, as well as providing the analytical capacity for monitoring and assessing forest change. The portal, ideally, integrates data from formal monitoring and from users employing technology as part of GIS, to serve initiatives concerned with forest health as well as the societal and ecological benefits of forests. Web-based social networks and users are themselves data sources representing a huge and heterogeneous repository of geo-referenced data that provide insights into the social impact of forest and other environmental phenomena, and complements 'standard' scientific data. Hence, UGC can be exploited by harvesting content from social media platforms and integrating it into a web portal for retrieval and scrutiny. Whilst both publics and experts need to be able to access distributed information of all types, the value of this information increases when it is integrated with an overarching modelling or geographic information infrastructure.

Google Fusion Tables (GFT) is a web-based data management and publishing application designed to allow non-specialist users to host, collaborate, manage and publish data (Shen 2012). They provide a common interface for different stakeholders, from individuals to government organisations, to upload, access and utilise a range of UGC. Bowie et al. (2014) utilised Google Maps and Fusion Tables to develop an interactive means of mapping and communicating the presence and ecological benefits of urban trees on a Toronto university campus. A secondary aim was to assess the efficacy of GFT for spatial data management. Data collected and integrated into the GFT included tree species, location, canopy cover, air pollution and climate data. GFT enables data storage that eschews the complexity of large formal databases in favour of systems that are easier to implement and interrogate for a variety of purposes. 'Scientific' aspects of data can be combined with UGC, e.g. the human dimensions of how people interact with trees, formal observation records and wiki-type collaboration. This study is implicitly from the perceptive of people and UGC integrated into overarching systems where different stakeholders and different data are intrinsically linked. UGC was also salient in studies exploring its integration into broader networks of various data types and technologies, literally connecting trees into a digital network. Qian et al. (2015), for example, integrated remote sensing and tree chipping with farming information, such as temperature and pesticides applied to trees, and collected via smart phone technology to assess the capacity to micromanage an orchard. Pushing the notion of connection even further, Luvisi and Lorenzini (2014) allude to the 'Internet of Things' (IoT) (Kopetz 2011),[8] suggesting that Web 2.0 technology will eventually facilitate 'wired, shared, digital, user friendly and rationalized [smart cities]' (ibid., 630). The premise that characterises the IoT, implicit in the literature around UGC and the forest sector, is the notion of interoperability via uniform access to data. Moreover, the IoT represents a new market for emerging ICTs and the tacit belief that objects, including trees, can be micro-managed for improved economic benefit.

Central to forest health is information flows around networks invoking different types of actor. Trees become part of this flow once technologies that wire them into the network, such as tagging devices, are introduced. The ontological issue is visible in that how actors are invoked, or rather what they are invoked as or for, depends on the perspective adopted. For the timber trade, trees are a commodity, for conservationists they need protecting, for computer scientists they are data to be incorporated into systems and models. Along with people, trees are enrolled into networks and treated as information. Hence, UGC is situated as part of a sociotechnical-material context, where the 'material' is trees, people and technological devices. Ontologically, this may be beneficial for forest health as long as all actors, humans included, are afforded a necessary and equal status. It is important not to lose sight of these dimensions because these are somewhat effaced by a simplistic reading of terms such as 'user' and 'UGC'.

5.3 Citizen Science and Crowdsourcing

Citizen science (CS) creates a nexus between policy, science, education and the public that, in conjunction with ICTs, pushes the boundaries of ecological research (Newman et al. 2012). Given that economic and political constraints are coincident with forest ecosystem services under increased pressure, forest managers have to constantly generate and evaluate cost-efficient means of monitoring forests and reaching and educating the public (Daume et al. 2014). Despite extensive literature around citizen science and crowdsourcing, our review revealed a paucity of literature where UGC was the central focus in relation to forest health. Instead, literature tended towards assessing the reliability of citizen science data in relation to expert data. For example, an exploration of the opportunities afforded by short-term hypothesis-led citizen science to quantify the relationship between the amount of damage to the leaves of the horse-chestnut tree, *Aesculus hippocastanum*, with the length of time that the horse-chestnut leaf mining moth, *Cameraria ohridella*, had been present (Pocock and Evans 2014). This study employed smart phone technology to test the concordance between participant-scored assessments of photographs of leaf damage with those from experts. Results indicated high concordance between scores, suggesting that citizen science data were accurate. Hence, the wide availability and existing uptake of technology facilitating UGC, such as smart phones, provide a cost-effective way of engaging the public with relatively little cost. However, UGC technology, especially smart phone apps and capacity, constantly evolves. If UGC is to be used as data, a commensurate, dynamic and evolving methodology that optimises its potential is necessary (Hawthorne et al. 2015).

Adriaens et al. (2015) reviewed two specific smart phone apps, 'That's Invasive!' and 'KORINA', for recording invasive alien species (IAS)[9] in North Western Europe, addressing the issues of data integration, openness, quality and interoperability. The challenges presented by these apps included omitted observer details, missing data due to server errors and image-resolution problems. KORINA had a low uptake and whilst this may reflect a short study period, it may also reflect a low degree of smart phone use amongst conservationists and/or low population density of the study area. Organisational attributes such as an organisation's culture are also a factor which can impede adaptive solutions (Dunning 2017; Lei and Kelly 2015). The non-governmental organisation responsible for managing the existing monitoring system was reluctant to promote apps to volunteers that did not already link to an existing system and perceived the new apps as either useless additions which would fragment recording, or as a competitor which might undermine the existing system (Adriaens et al. 2015). In terms of smart phone technology, whilst literature tended to focus on the technological challenges as well as the opportunities, organisational issues also need to be considered.

With regard to crowdsourcing,[10] Rallapalli et al. (2015) used gamification[11] to devise a Facebook game called Fraxinus to enable non-scientists to contribute to the genomics study of the ADB pathogen, *Hymenoscyphus pseudoalbidus*. DNA sequence alignment is resource intensive and can also be error prone. Human pattern recognition skills can improve such alignments; the game involved players aiming to produce the best alignment. In fifteen per cent of cases, computational alignments of genetic sequences were improved but most players engaged with the game in a transient manner, with the majority of the work performed by a small number of dedicated players. Findings such as this are important because they lead to further issues that need to be addressed, such as characterising the demographic engaging with such initiatives. Moreover, whilst individuals appear to be willing to share information using tools provided by Web 2.0, ensuring ongoing engagement from volunteers, especially those that require active, offline engagement, remains an issue to be addressed (Díaz et al. 2012).

As a collaborative outcome and because UGC is usually analysed in terms of large data sets, responsibility for a particular data point is often unknown. This can lead to concerns about data quality and is a factor behind the mistrust of citizen science in some quarters of the scientific community (Butt et al. 2013; López-Aparicio et al. 2017). Moreover, recognition of the potential of citizen science as a data source is also detracted from by a mistrust of UGC, especially UGC generated opportunistically; both UGC and CS data tend to be opportunistic (Daume 2015; Daume and Galaz 2016). Whilst social media is most often unstructured (e.g. tweet content can be presented in various ways), data generated for a specific purpose (e.g. using a dedicated phone app) comes with a structured format that makes curation somewhat easier. The aspects of CS UGC data that appear most challenging are the lack of complete and accurate geolocation data alongside the lack of accurate taxonomic detail.

A subset of UGC is the use of web tools to voluntarily create and disseminate geographic data, i.e. volunteered geographic information (VGI). VGI is considered as an empowering and democratising new form of citizen science (Foster and Dunham 2015) but may also reinforce the 'digital divide': the notion that disparities exist in access to and use of communication technologies because of differences in ethnicity, gender, class and socio-economic factors. In relation to communicative ecology, if ICTs change the nature of how organisations operate then we have to pay attention to the ideational, systemic and social aspects of these changes. Hence, what comes to the surface in terms of citizen science is a human dimension relating directly to the context of production of UGC that is not so much about the status of data but rather about a deeper layer which comprises these social and ideational and ontological aspects.

5.4 Monitoring Invasive Alien Species (IAS), Data Mining and Horizon Scanning

Policy makers need to be able to understand how emerging issues might affect current *and* future policy and practice. Hence, horizon scanning[12] has become a dominant activity across many policy domains, especially those relating to the environment. Having prospective information about IAS and the threats they pose to our forests means that actions can be carried out to reduce the likelihood of their ingress (Jones et al. 2017). This is more beneficial and cost-effective than trying to manage IAS once they have arrived.

The premise behind the use of UGC in relation to horizon scanning for IAS is that people use social media to discuss various aspects of their daily lives and this may include references to IAS, which FHS managers can utilise. Social media can be mined to discern where novel pests and pathogens are being talked about, monitor the proliferation and geographic range of pests and pathogens and predict future trends. Whilst platforms such as Instagram and Twitter lend themselves to content useful for flagging up potential IAS threats, or providing high-quality images (Daume and Galaz 2016), ninety per cent of Instagram users are under age thirty-five and the greatest proportion of users live in urban areas. Moreover, geolocation metadata is often absent from UGC but is crucial for event detection and building models and maps of IAS spread. If FHS managers use social media to reach an audience, they have to know who and where their audience is and how to leverage UGC content. FHS managers should be aware that the questions they need to ask about the ubiquity of social media and its potential in relation to IAS reflect a meta-level of enquiry into the social, ideological and particular technical affordances of the data and platforms in question.

Three papers in our review reflected on the generic conceptual challenges with regard to the presentation of social media data and the nature of metadata. Daume (2016) analysed a corpus of tweets with direct or descriptive references to IAS, sampled across a three-year period. Three target IAS, oak processionary moth (*Thaumetopoea processionea*), emerald ash borer (*Agrilus planipennis*) and Eastern Grey squirrel (*Sciurus carolinensis*), were followed and the sample assessed for information completeness and relevance. If tweets are merely descriptive and with no accompanying metadata or links, they are difficult to verify as relevant and accurate. Moreover, the sheer volume and structural features of data present practical challenges to using it (Brooker et al. 2016). Whilst there may be useful instances of data relating to sightings of IAS in new locations, the effort required to extract what amounts to a fraction of the overall data is considerable. However, social media can act as a real-time data source and provide early warnings for ecosystem shifts. Social media may be of use to IAS managers because it provides a communication channel with which to explore public perceptions and to garner public support or to provide information. These insights highlight that the social and organisational dimensions around UGC are entangled with scientific and pragmatic concerns. As traditional search methods often look at historical information, and in order to consider more current and less structured information, tools that can search social media are useful because of their up-to-date, real-time capacities.

The aggregation of large volumes of content is accompanied by the risk of losing important information. Actors that have a stronger affinity to social media may for example 'drown out' minority stakeholders or specific issues. The ease of information propagation, e.g. a 'like' on Facebook or a 'retweet', may not be a true reflection of the importance of certain issues. Nonetheless, Daume et al. (2014) suggest that aggregated social media content (ASMC) could be correlated with spatial and temporal patterns obtained through existing forest monitoring networks. ASMC may also generate information not covered by forest monitoring such as observations in private gardens, revealing new geographic areas that warrant closer inspection. Hence, ASMC represents a cost-effective and real-time data source.

Challenges remain with regard to how traditional data management practices may obstruct a rapid response to IAS, given that both horizon scanning and monitoring UGC involve the need to access and disseminate up-to-date and eclectic data. Whilst monitoring UGC may identify IAS, it does not necessarily prevent an incursion. According to Groom et al. (2015), IAS science struggles to meet the growing demand for IAS data. This partly reflects policy makers having to keep up to date with a rapidly changing digital environment, the risk of out-of-date information and developing policy frameworks that enable the use and sharing of data. Beyond this, there remains the question of the threshold at which the information obtained about an IAS identified from UGC would lead to action. In horizon scanning, issue selection is based on estimates of the likelihood and impacts of a risk in relation to a specific aspect of society or the environment (Van Rij 2010). However, risks may arise for a variety of reasons and these interact with the horizon scanning process itself. The challenges associated with utilising UGC for horizon scanning and identifying IAS are not just technological but also conceptual and organisational (Groom et al. 2015). Different types of expert and stakeholder knowledge need to be integrated into the process. UGC contributes to evidence thus forming a basis for decision-making but its content is not only a source of domain information but also reflects the societal context of its production. UGC as data can never be value-free. It is therefore important to develop an appreciation of the social, organisational, ontological and epistemological issues involved.

5.5 UGC, Forests and Our Sense of Self: Ontological Questions

How people understand forests and trees and how they attribute meaning and engage with them reflects their broader relationships and wider sociocultural influences and beliefs (Doody et al. 2014). Hence, those responsible for forest health management should consider how people construct their sense of self[13] in relation to their particular social, geographic and economic relationship with forests and trees (Cantrill and Senecah 2001). The question of ontology, of whether the Internet and UGC fundamentally change peoples' relationship with themselves and the world, is as important as questions about the pragmatics of using UGC. Turkle (2011), for instance, suggests that in our present era we have learned to see ourselves as 'plugged in technobodies' where our political and economic lives are articulated through a language of machine intelligence and distributed, networked and emergent organisation. This coincides with the erosion of traditional forms of community and institutions, and the emergence of a self, predicated on notions of multiplicity, heterogeneity and fragmentation. For those interested in and managing the FHS, this translates into a need to understand the ontological underpinnings of why and how people participate in activities contributing to the care of forests and trees. Understanding the endogenous knowledge of communities, their particular relationship with forests, is crucial. If a dispersed and online general public is less likely to engage with an issue than a localised, motivated and active community (Massung et al. 2013), then a problem for managers seeking to communicate and utilise 'plugged in' 'communities' is how to overcome this inertia.

Implicit in the reviewed literature was a conflation of the categories of the person in the human sense and as a digital artefact. This results in mis-conceptualising who the object of communication, e.g. the audience, is. When we act towards the artefact rather than the person, there is a tendency to idealise what can be achieved. In some sense, an 'idealised citizen' has been tacitly imagined as this object: an individual interpolated in such a way that they are responsive to how the government and other organisations want them to act. This conceptualisation fails to problematise the complexities of subjectivity, and that the virtually mediated environments which extend into many aspects of our lives and which result in plural identities are complex and increasingly predicated, for example, on consumption as a mandatory practice (Şerban 2016).

6 Conclusion

Our primary concern in this chapter was to focus on the social and organisational processes underpinning the use of UGC in the FHS. UGC needs to be understood not only in terms of what it facilitates (i.e. linking stakeholders and data) but how this facility exists as part of the changing face of the human dimensions of forest health.

Organisations, researchers and workers in the FHS, interested in UGC, need to pay equal attention to social psychological processes as much as they do the mechanics and technical aspects of utilising ICTs and developing technological infrastructures.

Processes of commodification, transformed into a perception of forests in terms of the recreational and therapeutic benefits they afford, sit uncomfortably with the belief that forests need to be looked after on their own terms and be available free from human exploitation. Communication and organisational change in relation to the FHS needs to reflect both action *and* research that can be identified from across different levels, including the technological, the discursive (ideologies and beliefs underpinning the content of communication) and the social (the different stakeholder groups involved and their relationships). If ICTs and UGC radically change the nature of how organisations operate, then we have to pay attention to the ideational, systemic and social aspects of these changes. Understanding why and how people engage with UGC rests on a set of complex relationships that belie the notion of homogeneous audiences, unitary selves, straightforward communication and ideal citizens. Alongside research and development which focusses on implementing UGC and social media in the FHS, we feel that equal, if not primary, consideration needs to be afforded to how UGC and social media change our perception of ourselves and the world in the first place.

Notes

1. Following exploratory work, final search terms were applied to article abstract, title and key words. Terms capture the manner in which ICTs and UGC are usually represented: ('user generated content' OR 'social media' OR 'web 2.0' OR 'smart phone') AND ('forest' OR 'tree health'). Year selection was 2012–2017.
2. Adaptive management approaches acknowledge address forest systems as complex and adaptive and eschew traditional top-down management in favour of innovation, collaboration, learning and action in the face of incomplete and uncertain scientific knowledge (Lawrence 2017; Westgate et al. 2013).

3. Thinning, and removing trees and underbrush to mitigate fire risk.
4. Specifically self-organising maps; see Kohonen (2013).
5. 'Big Data' denotes massive amounts of structured and unstructured data that cannot be analysed using traditional techniques. The challenges involved in making sense of such data include issues of storage, curation and creating utilities that can harvest and process it accordingly. Different disciplines have different ideas on what Big Data is and what it can be used for.
6. An outdated model of communications based on the premise that an intended message is directly received and wholly accepted by the receiver.
7. 'Datafication' denotes the transformation of our social lives into online quantified data, enabling real-time surveillance and predictive analysis.
8. The envisaged convergence of technologies including wireless communication, real-time analytic capacity, machine learning, remote sensing and embedded systems.
9. Organisms with a tendency to spread to a degree that causes damage to other species, the human economy and health.
10. A definition of crowdsourcing is individuals or organisations using contributions from Internet users to obtain services. Hence, some crowdsourcing projects will also be CS projects but some will not.
11. 'Gamification' is a motivational technique using game elements, such as point scoring, competition and questing in a non-gaming context.
12. The practice of seeking, gathering and analysing information about emerging threats so that policy makers can develop a resilient, long-term plan of action more able to cope with uncertainty.
13. 'Self' is an extensive and complicated concept. For our purposes, it can be thought of as a materially situated yet inward directed awareness, providing for a sense of continuity and consistency of experience across time and place.

References

Adriaens, T., Sutton-Croft, M., Owen, K., Brosens, D., van Valkenburg, J., Kilbey, D., et al. (2015). Trying to engage the crowd in recording invasive alien species in Europe: Experiences from two smartphone applications in northwest Europe. *Management of Biological Invasions, 6*(2), 215–225.

Bhuiyan, M. A. H., Siwar, C., Ismail, S. M., & Islam, R. (2011). Ecotourism development in recreational forest areas. *American Journal of Applied Sciences, 8*(11), 1116.

Bowie, G. D., Millward, A. A., & Bhagat, N. N. (2014). Interactive mapping of urban tree benefits using Google Fusion Tables and API technologies. *Urban Forestry & Urban Greening, 13*(4), 742–755.

Brooker, P., Barnett, J., Cribbin, T., & Sharma, S. (2016). Have we even solved the first 'big data challenge?' Practical issues concerning data collection and visual representation for social media analytics. In *Digital methods for social science* (pp. 34–50). Basingstoke, UK: Springer.

Büscher, B. (2013). Nature 2.0. *Geoforum, 44,* 3.

Butt, N., Slade, E., Thompson, J., Malhi, Y., & Riutta, T. (2013). Quantifying the sampling error in tree census measurements by volunteers and its effect on carbon stock estimates. *Ecological Applications, 23*(4), 936–943.

Cantrill, J. G., & Senecah, S. L. (2001). Using the 'sense of self-in-place' construct in the context of environmental policy-making and landscape planning. *Environmental Science & Policy, 4*(4), 185–203.

Chandler, D. (2015). A world without causation: Big data and the coming of age of posthumanism. *Millennium, 43*(3), 833–851.

Cheng, M., Wong, I. A., Wearing, S., & McDonald, M. (2017). Ecotourism social media initiatives in China. *Journal of Sustainable Tourism, 25*(3), 416–432.

Daume, S. (2015). *Social media mining as an opportunistic citizen science model in ecological monitoring: A case study using invasive alien species in forest ecosystems.* (Ph.D., University of Göttingen).

Daume, S. (2016). Mining Twitter to monitor invasive alien species—An analytical framework and sample information topologies. *Ecological Informatics, 31,* 70–82.

Daume, S., Albert, M., & von Gadow, K. (2014). Forest monitoring and social media—Complementary data sources for ecosystem surveillance? *Forest Ecology and Management, 316,* 9–20.

Daume, S., & Galaz, V. (2016). "Anyone know what species this is?"—Twitter conversations as embryonic citizen science communities. *PLoS One, 11*(3), e0151387.

Díaz, L., Granell, C., Huerta, J., & Gould, M. (2012). Web 2.0 Broker: A standards-based service for spatio-temporal search of crowd-sourced information. *Applied Geography, 35*(1), 448–459.

Doody, B. J., Perkins, H. C., Sullivan, J. J., Meurk, C. D., & Stewart, G. H. (2014). Performing weeds: Gardening, plant agencies and urban plant conservation. *Geoforum, 56,* 124–136.

Dunning, K. H. (2017). Missing the trees for the forest? Bottom-up policy implementation and adaptive management in the US natural resource bureaucracy. *Journal of Environmental Planning and Management, 60*(6), 1036–1055.

Fellenor, J., Barnett, J., Potter, C., Urquhart, J., Mumford, J., & Quine, C. (2017). The social amplification of risk on Twitter: The case of ash dieback disease in the United Kingdom. *Journal of Risk Research,* 1–21. https://doi.org/10.1080/13669877.2017.1281339.

Folke, C. (2007). Social–ecological systems and adaptive governance of the commons. *Ecological Research, 22*(1), 14–15.

Foster, A., & Dunham, I. M. (2015). Volunteered geographic information, urban forests, & environmental justice. *Computers, Environment and Urban Systems, 53,* 65–75.

Gazal, K., Montague, I., Poudel, R., & Wiedenbeck, J. (2016). Forest products industry in a digital age: Factors affecting social media adoption. *Forest Products Journal, 66*(5), 343–353.

González-Ruibal, A., Hernando, A., & Politis, G. (2011). Ontology of the self and material culture: Arrow-making among the Awá hunter–gatherers (Brazil). *Journal of Anthropological Archaeology, 30*(1), 1–16.

Groom, Q. J., Desmet, P., Vanderhoeven, S., & Adriaens, T. (2015). The importance of open data for invasive alien species research, policy and management. *Management of Biological Invasions, 6*(2), 119–125.

Hartig, T., Mitchell, R., De Vries, S., & Frumkin, H. (2014). Nature and health. *Annual Review of Public Health, 35,* 207–228.

Hawthorne, T., Elmore, V., Strong, A., Bennett-Martin, P., Finnie, J., Parkman, J., et al. (2015). Mapping non-native invasive species and accessibility in an urban forest: A case study of participatory mapping and citizen science in Atlanta, Georgia. *Applied Geography, 56,* 187–198.

Hearn, G., Collie, N., Lyle, P., Choi, J. H.-J., & Foth, M. (2014). Using communicative ecology theory to scope the emerging role of social media in the evolution of urban food systems. *Futures, 62,* 202–212.

Hogan, B. (2010). The presentation of self in the age of social media: Distinguishing performances and exhibitions online. *Bulletin of Science, Technology & Society, 30*(6), 377–386.

Jones, G., Hugo, S., & Agstner, B. (2017). *Horizon scanning: Supplementary report to the Future Proofing Plant Health (FPPH) project.*

Kaplan, A. M., & Haenlein, M. (2010). Users of the world, unite! The challenges and opportunities of social media. *Business Horizons, 53*(1), 59–68.

Kearney, A. R., & Bradley, G. (1998). Human dimensions of forest management: An empirical study of stakeholder perspectives. *Urban Ecosystems, 2*(1), 5–16.

Kelly, M., Ferranto, S., Lei, S., Ueda, K.-I., & Huntsinger, L. (2012). Expanding the table: The web as a tool for participatory adaptive management in California forests. *Journal of Environmental Management, 109*, 1–11.

Kohn, E. (2013). *How forests think: Toward an anthropology beyond the human*. Berkeley and Los Angeles, CA, USA: University of California Press.

Kohonen, T. (2013). Essentials of the self-organizing map. *Neural Networks, 37*, 52–65.

Kopetz, H. (2011). *Real-time systems: Design principles for distributed embedded applications*. Boston, MA, USA: Springer Science & Business Media.

Krumm, J., Davies, N., & Narayanaswami, C. (2008). User-generated content. *IEEE Pervasive Computing, 7*(4), 10–11.

Lawrence, A. (2017). Adapting through practice: Silviculture, innovation and forest governance for the age of extreme uncertainty. *Forest Policy and Economics, 79*, 50–60.

Lei, S., & Kelly, M. (2015). Evaluating collaborative adaptive management in Sierra Nevada Forests by exploring public meeting dialogues using self-organizing maps. *Society & Natural Resources, 28*(8), 873–890.

Levy, Y., & Ellis, T. J. (2006). A systems approach to conduct an effective literature review in support of information systems research. *Informing Science, 9*, 181–212.

López-Aparicio, S., Vogt, M., Schneider, P., Kahila-Tani, M., & Broberg, A. (2017). Public participation GIS for improving wood burning emissions from residential heating and urban environmental management. *Journal of Environmental Management, 191*, 179–188.

Luca, M. (2015). User-generated content and social media. In S. Anderson, J. Waldfogel, & D. Stromberg (Eds.), *Handbook of media economics* (pp. 563–592). Oxford: North-Holland.

Luvisi, A., & Lorenzini, G. (2014). RFID-plants in the smart city: Applications and outlook for urban green management. *Urban Forestry & Urban Greening, 13*(4), 630–637.

Marzano, M., Dandy, N., Bayliss, H. R., Porth, E., & Potter, C. (2015). Part of the solution? Stakeholder awareness, information and engagement in tree health issues. *Biological Invasions, 17*(7), 1961–1977.

Massung, E., Coyle, D., Cater, K. F., Jay, M., & Preist, C. (2013). *Using crowdsourcing to support pro-environmental community activism*. Paper presented at the Proceedings of the SIGCHI Conference on Human Factors in Computing Systems.

McInerney, D., Bastin, L., Diaz, L., Figueiredo, C., Barredo, J. I., & Ayanz, J. S.-M. (2012). Developing a forest data portal to support multi-scale decision making. *IEEE Journal of Selected Topics in Applied Earth Observations and Remote Sensing, 5*(6), 1692–1699.

Messier, C., Puettmann, K., Chazdon, R., Andersson, K., Angers, V., Brotons, L., et al. (2015). From management to stewardship: Viewing forests as complex adaptive systems in an uncertain world. *Conservation Letters, 8*(5), 368–377.

Michon, G., Nasi, R., & Balent, G. (2013). Public policies and management of rural forests: Lasting alliance or fool's dialogue? *Ecology and Society, 18*(1), 30.

Montague, I., Gazal, K. A., Wiedenbeck, J., & Shepherd, J.-G. (2016). Forest products industry in a digital age: A look at e-commerce and social media. *Forest Products Journal, 66*(1), 49–57.

Newman, G., Wiggins, A., Crall, A., Graham, E., Newman, S., & Crowston, K. (2012). The future of citizen science: Emerging technologies and shifting paradigms. *Frontiers in Ecology and the Environment, 10*(6), 298–304.

Nilsson, K., Sangster, M., & Konijnendijk, C. C. (2011). Forests, trees and human health and well-being: Introduction. In K. Nilsson, M. Sangster, C. Gallis, T. Hartig, S. de Vries, K. Seeland, et al. (Eds.), *Forests, trees and human health* (pp. 1–19). Dordrecht: Springer.

Obar, J. A., & Wildman, S. S. (2015). Social media definition and the governance challenge: An introduction to the special issue.

Panagiotopoulos, P., & Bowen, F. (2015). *Conceptualising the digital public in government crowdsourcing: Social media and the imagined audience*. Paper presented at the International Conference on Electronic Government.

Pocock, M. J., & Evans, D. M. (2014). The success of the horse-chestnut leaf-miner, *Cameraria ohridella*, in the UK revealed with hypothesis-led citizen science. *PLoS One, 9*(1), e86226.

Pretzsch, J. (2005). Forest related rural livelihood strategies in national and global development. *Forests, Trees and Livelihoods, 15*(2), 115–127.

Qian, J.-P., Yang, X.-T., Wu, X.-M., Xing, B., Wu, B.-G., & Li, M. (2015). Farm and environment information bidirectional acquisition system with individual tree identification using smartphones for orchard precision management. *Computers and Electronics in Agriculture, 116*, 101–108.

Rallapalli, G., Saunders, D. G., Yoshida, K., Edwards, A., Lugo, C. A., Collin, S., et al. (2015). Lessons from Fraxinus, a crowd-sourced citizen science game in genomics. *Elife, 4,* e07460.

Reed, A. (2005). 'My blog is me': Texts and persons in UK online journal culture (and anthropology). *Ethnos, 70*(2), 220–242.

Reynolds, K. M., & Shao, G. (2006). From data to sustainable forests. In G. Shao & K. M. Reynolds (Eds.), *Computer applications in sustainable forest management: Including perspectives on collaboration and integration* (Vol. 11). Dordrecht: Springer Science & Business Media.

Sauget, N. (1994). Of land, woods and men: Farmers talk about the land, the evolution of woodland areas and the landscape. *Landscape Issues, 11*(1), 52–58.

Şerban, O. (2016). A process identity: The aesthetics of the technoself. Governing networking societies. *Balkan Journal of Philosophy, 8*(2), 165–174.

Shen, W. (2012). Introducing New Fusion Tables API. Retrieved from http://googleresearch.blogspot.ca/search/label/Fusion%20Tables.

Sulak, A., & Huntsinger, L. (2012). Perceptions of forest health among stakeholders in an adaptive management project in the Sierra Nevada of California. *Journal of Forestry, 110*(6), 312–317.

Tang, L., Shao, G., & Dai, L. (2009). Roles of digital technology in China's sustainable forestry development. *International Journal of Sustainable Development and World Ecology, 16*(2), 94–101.

Thomas, J., Newman, M., & Oliver, S. (2013). Rapid evidence assessments of research to inform social policy: Taking stock and moving forward. *Evidence & Policy: A Journal of Research, Debate and Practice, 9*(1), 5–27.

Turkle, S. (2011). *Life on the screen: Identity in the age of the Internet*. New York: Simon & Schuster.

Van Rij, V. (2010). Joint horizon scanning: Identifying common strategic choices and questions for knowledge. *Science and Public Policy, 37*(1), 7–18.

Washington, H., Taylor, B., Kopnina, H., Cryer, P., & Piccolo, J. (2017). Why ecocentrism is the key pathway to sustainability. *The Ecological Citizen, 1*(1), 35–41.

Westgate, M. J., Likens, G. E., & Lindenmayer, D. B. (2013). Adaptive management of biological systems: A review. *Biological Conservation, 158,* 128–139.

Xiang, Z., & Gretzel, U. (2010). Role of social media in online travel information search. *Tourism Management, 31*(2), 179–188.

Zhao, G., Shao, G., Reynolds, K. M., Wimberly, M. C., Warner, T., Moser, J. W., et al. (2005). Digital forestry: A white paper. *Journal of Forestry, 103*(1), 47–50.

7

The Social Amplification of Tree Health Risks: The Case of Ash Dieback Disease in the UK

Julie Urquhart, Julie Barnett, John Fellenor, John Mumford, Clive Potter and Christopher P. Quine

1 Introduction

Risk experts have long observed that newly emerging diseases generate complex and sometimes contradictory interactions between attempts by governments to manage disease outbreaks, media coverage of those events and the diverse risk perceptions of stakeholders and publics. The difficulty for policy makers is that the technical risk assessment tools and methodologies they rely on to set priorities, recommend and justify preventative actions and target scarce resources may not always be

---

J. Urquhart (✉)
Countryside & Community Research Institute,
University of Gloucestershire, Oxstalls Campus,
Gloucester, UK

J. Barnett · J. Fellenor
Department of Psychology, University of Bath,
Bath, UK

© The Author(s) 2018
J. Urquhart et al. (eds.), *The Human Dimensions of Forest and Tree Health*,
https://doi.org/10.1007/978-3-319-76956-1_7

well attuned to often rapidly evolving public risk understandings and the social and cultural processes which shape these. In the case of pest and disease threats to trees, woods and forests, the identification of ash dieback in the UK in 2012 elevated tree health from an issue predominantly of expert and high-level stakeholder concern to a major focus of public scrutiny and media attention over a period of just a few weeks, bringing in its train widespread criticism of the government's ability to ensure effective biosecurity in the live plant trade (Urquhart et al. 2017a; Mumford 2013). The resulting social intensification of public risk concern, if sustained, seemed likely to have profound implications for the way tree pest and disease threats would need to be handled and communicated by the government, its agencies and stakeholders. It posed reputational risks for government if a more risk-aware and critical public perceived disease prevention efforts to be 'too little, too late', control programmes poorly designed and risk communications confused and inconsistent.

Clearly, if government and stakeholder efforts to safeguard tree health in the UK are to be effective, it is essential that policy makers and risk managers have a better understanding of how both experts and publics view future risks to tree health. Evidence-based research is, therefore, needed to analyse the emerging nature of public risk concerns and to suggest ways in which policy makers and risk managers can better engage with these based on an understanding of formative processes and underlying values. We need to know which publics are affected by or engaged with tree health risks. We also need to know how their respective understandings of risk develop over the

---

J. Mumford · C. Potter
Centre for Environmental Policy, Imperial College London,
South Kensington Campus, London, UK

C. P. Quine
Forest Research, Northern Research Station,
Roslin, Midlothian, UK

course of outbreaks through exposure to official risk communications, public debate and/or personal experience. Further work is then needed to characterise the implications of this for public engagement, risk communication, priorities for action and risk analysis more broadly. A particular concern here is how uncertainty should be captured and characterised within policy and public databases, such as the UK Plant Health Risk Register.

Stakeholder and public engagement and participation are integral to the process of environmental policy-making in order to help formulate the problem and enable more effective decision-making (e.g. Gormley et al. 2011; COA 2013). However, we know from previous work in the human and animal health fields that public risk understandings do not develop in isolation but are influenced by cultural associations, social interactions, personal experience, assessments of institutional competence and the historical benchmarking of previous disease risk events (Lewis and Tyshenko 2009; Selbon et al. 2005). A useful way to conceptualise these interacting influences is provided by the Social Amplification of Risk Framework (SARF), developed in the late 1980s in order to integrate technical analyses of risk with the social, cultural and individual factors influencing how publics experience it (Kasperson et al. 1988). The SARF emphasises the socially constructed nature of all risk perceptions and lays stress on the dynamic processes through which risk is communicated and interpreted by many different social agents. It draws attention to the complex nature of risk perceptions and understandings and as such may offer scope for constructive dialogue between risk assessors, risk communicators, policy makers and publics.

This chapter draws on social research undertaken as part of the UNPICK (Understanding public risk in relation to tree health) research project (2015–2017), designed to investigate how UK publics perceive, understand and make sense of the growing threats to tree health from invasive pests and diseases. The risks posed by tree pests and pathogens have been widely recognised in expert circles, but the degree to which this awareness is shared by publics and some stakeholders is still unclear.

There is a potential conflict between government attempts to manage the risks, media coverage about their importance and likely impact and the different ways in which various publics and stakeholders make sense of the threats. A key aim of the project was to explore the interrelationships of media representation, expert assessments and public perceptions of tree pest and disease outbreaks in an integrated way using the SARF as an analytical lens. The research adopted a variety of social science approaches, including interviews with policy makers, managers and scientists involved in making decisions about how to deal with ash dieback; content analysis of traditional and social media related to the outbreak; an online national survey of public attitudes to tree health; Q Methodology interviews with members of the public in areas affected by ash dieback; and an analysis of helpline contacts.

In this chapter, we focus on the ash dieback outbreak in the UK to exemplify how SARF can help us to understand how risk issues associated with an outbreak may be 'intensified' or 'attenuated', the knock-on effects of these processes and how discrepancies between 'expert' and public assessment of the risk may arise. The chapter proceeds with, firstly, an outline of the SARF, followed by an explanation of the methods adopted in the study. This is followed by a synthesis of the findings from the various methods adopted by the project, and a discussion of the implications of the study. Detailed results from each method are beyond the scope of this chapter, and readers are directed to the published outputs of the project for a more in-depth presentation of the findings from this work (Urquhart et al. 2017a, b, under review-a, under review-b; Fellenor et al. 2017, under review-a, under review-b).

2 Social Amplification of Risk Framework (SARF)

SARF was first introduced in 1988 by Kasperson, Renn, Slovic and colleagues (Kasperson et al. 1988) in response to a perceived need for a broader understanding of risk and how it is perceived by different social actors. In its original conception, the framework was presented as an overarching approach designed to integrate the 'technical' assessment of risk alongside the 'social or perceptual' analysis of hazards (Renn et al. 1992; Kasperson 1992). The primary rationale was to try to understand why some risks or events assessed by experts as not significant sometimes elicit strong public concerns and result in substantial impacts upon society and economy (e.g. the bovine spongiform encephalitis (BSE) outbreak in the 1990s), while others, deemed by experts to pose a significant risk by experts (e.g. smoking), are associated with a more graduated or even 'attenuated' response from publics and society (Kasperson 2012a).

SARF recognises that responses to risk are not only determined by exposure to the physical impacts (or harms) caused by a hazard event itself, but are also shaped by interactions between the transfer of information about hazard events and the responses of individuals and social groupings to these 'risk signals'. Critically, because responses are mediated through a variety of psychological, social, institutional and cultural processes, the result can be to intensify or attenuate individual and collective perceptions of risk and shape risk behaviour. This is defined by the authors as 'social amplification' (Kasperson et al. 1988; Renn et al. 1992; Renn 1991) (see Fig. 1).

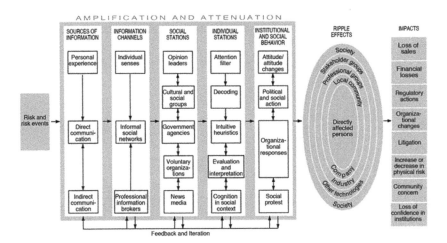

Fig. 1 The social amplification of risk framework (from Kasperson 2012a)

The framework borrows the metaphor of amplification from classical communications theory (Lasswell 1948; Shannon and Weaver 1949) to analyse how social agents generate and translate 'risk signals' (Bakir 2005). Risk signals are both transmitted and processed by individuals and social entities called 'amplification stations', with social amplification most likely to occur when risks are serious and the situation is fraught with uncertainties (Kasperson 2012a). These agents of amplification may include scientists, risk management institutions, the media, activists, peer groups, social networks and public agencies. One of the key insights of the framework is that amplified risk perceptions can lead to behavioural responses that, in turn, result in secondary impacts described by Kasperson and colleagues as 'ripple effects' (Kasperson et al. 1988, p. 181) (Fig. 1). These ripples are the secondary and tertiary impacts that may extend far beyond (geographically, temporally and socially) the direct harms of the hazard event and include: enduring changed mental perceptions and sensitivities; economic impacts for particular sectors and throughout the economy, increased pressure for policy reform; changes in the physical nature of the hazard (feedback mechanisms); and repercussions for other technologies and activities (for instance, by changing public willingness to accept potentially hazardous technologies) (Kasperson et al. 1988; Renn et al. 1992; Kasperson and Kasperson 1996).

A key part of the communication process is that risks and risk events are portrayed through various risk signals (i.e. images, signs and text involved in the transfer of information about the risk) which interact with a range of psychological, social, institutional or cultural processes in ways that intensify or attenuate perceptions of the risk and its manageability (Kasperson 2012b). SARF, therefore, suggests that alongside consideration of the risk signal it is important to understand the social response mechanisms through which information about the event is interpreted (Burns et al. 1993). How the public responds to the risk signal is tempered by factors such as the perceived seriousness of the 'risk event' and by what the event signifies. Understanding these processes requires an appreciation of the role played by the heuristics, mental models and short cuts people use to make sense of, and evaluate, complex risk information, alongside levels of trust and the potential for stigmatisation.

In her use of social representation theory, Moscovici (1984) examined how individuals or groups may compare a new or emerging risk to a previous risk event via the linked mental processes of 'anchoring' and 'objectification'. Anchoring involves comparing the unfamiliar to existing knowledge and enables new information to be interpreted in terms of existing beliefs and memories of previous hazards. Objectification refers to the heuristic devices that people use to transform unfamiliar and abstract notions into concrete common-sense realities. Heuristic mechanisms are influenced by the extent to which the public perceives a risk to be catastrophic, deadly and uncontrollable (dread risks) and the extent to which the risk is poorly understood, unknown to those exposed and has delayed effects (unknown risk) (Slovic 1987). For example, when the media attributes specific storms or floods to climate change they are objectifying an abstract phenomenon (Höijer 2010). This can often involve the use of images, metaphors, tropes or symbols. The importance of various 'availability heuristics'—the mental short cuts to judgement that people use to assess risks—has been widely studied. Kuhar et al. (2009), for instance, found that those respondents who had personally observed (and drawn conclusions about) 'red tides' affecting the Florida coast had much higher awareness of the health risks of eating seafood than those only exposed to official health advisories.

Further, the nature of social and political groups influences the responses of its members and represents an ideological interpretation of risk (Kasperson et al. 1988). Renn et al.'s (1992) concept of 'social stations of amplification', for instance, recognises that individuals act as members of larger social units and cultural groups that co-determine the social processing of risk (Kasperson 2012a). Thus, individuals may perceive risk through the lens of values of the organisation or group to which they belong and its cultural biases (Dietz and Stern 1996).

A key element here is the degree to which there is trust in the institutions responsible for managing and communicating about the risk. Burns et al. (1993) concluded that when an event is perceived as improperly managed, there are high levels of uncertainty about the risk, or that future risk is great, the public are likely to perceive a greater threat. In this context, there are reputational risks for the government if risks are inadequately communicated and a more critical public perceives risk managers as incompetent.

The following section sets out how the SARF can help to inform understanding of public attention to tree health issues and outlines the methods that were adopted to explore the ash dieback case study.

3 Methodology and Methods

The methods adopted in this study represent a layered, sequential analysis of the assessment, communication and public understanding of tree health risks by (i) offering a critical analysis of how ash dieback has been framed by scientists, policy makers and risk managers over time; (ii) exploring how communications about ash dieback from these expert sources have been deliberated on and interpreted via an increasingly complex set of traditional and social media channels, and how the public, as a form of 'citizen media', may act as a 'social amplification station'; (iii) examining how various publics perceive, understand and act on the risks associated with ash dieback; and (iv) integrating the three streams of work through the SARF. Our contention here is that there is not one 'risk' waiting to be identified, but that different actors will construct their own socio-spatial perceptions of risk. These may change over time as information, knowledge and direct experience of the outbreak develops.

Drawing on SARF, Fig. 2 illustrates how the framework was applied as an analytical tool to integrate the assessment of expert, policy maker

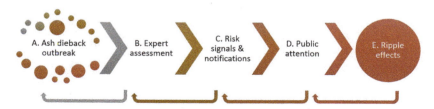

Fig. 2 Conceptualisation of response to ash dieback outbreak using the SARF

and public responses to the ash dieback outbreak. Firstly, for any given tree pest or disease outbreak, experts and risk managers will assess the nature of and degree of risk involved. Notifications about the outbreak may be released by government agencies or others responsible for outbreak management, based on expert assessments of the risk. These notifications may be picked up by the news media, who in turn will translate the risk signals and present their own interpretation of the outbreak. Wider publics and stakeholders respond to these risk signals through a range of social, psychological and cultural filters to construct their own perception of the risk. This in turn leads to 'ripple effects', or changed behaviours or ways of thinking about tree health issues. SARF also recognises that risk perception is rarely a linear process and feedback processes occur which further influence how publics' and other actors' perceive the risk over time. For instance, policy makers and risk managers may adapt their management or communication strategy in response to the public and media response to an outbreak (see, for instance, Tomlinson 2016).

3.1 The Ash Dieback Outbreak

Ash dieback is a disease caused by the fungal pathogen *Hymenoscyphus fraxineus*. It affects many species of ash, but in particular the common ash (*Fraxinus excelsior*) and narrow-leaved ash (*Fraxinus angustifolia*) (Kowalski 2006; FR 2012). The disease causes leaf loss, bark lesions and dieback of the crown and usually results in tree death over a period of years. In Europe, the disease was first identified in Poland in 1992 (Kowalski 2006) and is now widespread across the continent. It was discovered in the UK in 2012 at a tree nursery in Buckinghamshire on ash saplings that had been imported from the Netherlands, but it is also believed that spores of the pathogen may have blown in from continental Europe (Heuch 2014). Ash dieback has been identified across the UK, but its impact is currently the greatest in eastern regions, such as East Anglia and Kent, where both young and mature trees in woodlands and the wider landscape are visibly affected by the disease.

3.2 Investigating Scientists, Policy makers and High-Level Stakeholders as Risk Amplification Stations

The idea that expert judgements about risk may be subject to social processes and contestation just as much as expressions of public concern frames the first stage of the analysis. Firstly, a documentary analysis was undertaken to review academic, policy and grey literature to outline the technical risk assessment process and the official management response to ash dieback (see Fig. 3). The second stage was semi-structured interviews with a range of experts, including scientists, policy makers and key stakeholders (e.g. NGOs, nursery sector and foresters) (Fig. 3). A total of 21 individuals were interviewed between March and November 2015. Interviews lasted between 45 and 90 minutes and were audio recorded and transcribed verbatim. A thematic analysis was undertaken on the transcripts, involving both manual and digital (Nvivo 12.0) coding, in order to identify the sources of information that respondents' drew on to form their perceptions, the affective and cognitive filters through which the outbreak was viewed, and the role of their interactions with others in shaping those perceptions. For a detailed overview of the method adopted, see Urquhart et al. (2017a).

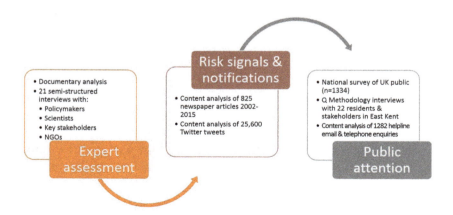

Fig. 3 Methods adopted to explore the interactions between expert assessment, media attention and public concern about ash dieback in the UK

3.3 Assessing the Impact of Traditional Media Coverage and Social Media Feedback

The second stage of analysis involved examining how the risk framings and communications about ash dieback from scientists and various biosecurity professionals have been filtered and interpreted in traditional and social media (Fig. 3). First, an analysis of traditional media was undertaken. British newspaper articles from 2002 to 2015 were analysed using LexisNexis to assess how the media described events associated with ash dieback and the extent to which previous tree health issues or other risk events were referenced in relation to the ash dieback outbreak (Fellenor et al. under review-a).

An important additional layer of analysis was to examine how the social media coverage of ash dieback developed over the course of the early stages of the outbreak. Analysis focused on the social media platform Twitter to consider the social amplification of risk in relation to ash dieback disease. An empirical analysis was made on 25,600 tweets to see what people said about ash dieback on Twitter, who was talking about it and how they talked about it (see Fellenor et al. 2017 for a full account of analytical approach).

3.4 Understanding the Drivers of Public Attention to Tree Health Risks

This stage of the research involved three levels of analysis (Fig. 3). Firstly, a nationally representative survey was undertaken in April 2016 to assess broad public awareness and concern about tree health issues, as well as willingness to adopt biosecure behaviours (see Urquhart et al. 2017b). The questionnaire was deployed by a professional panel survey company (http://www.respondi.com) using an online survey tool, resulting in 1334 responses suitable for analysis. Questions in the survey sought to elicit respondents' awareness of tree health risks, their concern and interest in these issues and their willingness to adopt biosecure behaviours. Cross-tabulations, factor analysis and ordinal logistic regression modelling were used to identify variables likely to influence respondents' awareness and concern. A further analysis compared the results of this survey with a prior survey undertaken in 2013 to investigate change over time in public attention to tree health risks.

Secondly, interviews were undertaken with a sample of 22 residents and stakeholders in East Kent, using Q Methodology. A full explanation of Q Methodology and how it was applied is provided in Urquhart et al. (under review-a). In short, it involved asking respondents to sort a series of 44 statements relating to attitudes and beliefs about ash dieback and tree health more broadly according to the extent the statements aligned to their personal views. The resulting 'Q sorts' were factor analysed to identify clusters of respondents with similar points of view.

Thirdly, we investigated direct expressions of concern from observing publics by examining a database of 1282 email and telephone enquiries to Forest Research's Disease and Diagnostics Advisory Service (FRAS) over the last 5 years (Fellenor et al. under review-b). This allowed us to track the nature of public attention to ash dieback in a naturally emerging data set, as opposed to being elicited via a research survey. The data set was analysed using Textometrica,[1] a free online tool for visualising and exploring short texts. See Fellenor et al. (under review-b) for a full account of the analysis method.

In order to integrate the empirical findings from across the different data sets, the research team met for a series of group analysis sessions in which the data were considered as a whole using the SARF. These were further presented and deliberated on at a workshop with high-level policy makers across relevant government departments in October 2017 to validate the findings and further integrate and synthesise the results across the various streams of work.

Reflecting our aim of describing perceptions of tree health risks through a SARF lens, the following sections discuss the processes identified in Fig. 2 in the light of the empirical findings. We provide insight into the socially constructed nature of experts' and policy makers' risk assessments, evidence of social amplification (or not) in both traditional and social media, a spatially and temporally nuanced exploration of public attention to tree pest outbreaks, and the interaction between experts, policy makers, media and publics to create a dynamic, evolving and complex tree health 'riskscape'.

4   The Objective Expert?

The original framing of SARF as a communication-reception process implies that expert risk assessment, and any communication and signalling of risk that results, constitutes the 'real' or benchmark risk against which the public's 'perceived' risk is either amplified or attenuated (Merkelsen 2011). There is an implicit assumption that expert risk perceptions are based on objective technical assessments. This conceptualisation is empirically problematic when there are high levels of scientific uncertainty and where experts may disagree about the nature of the risk they are trying to communicate, as in the tree health case (Busby et al. 2009; Busby and Onggo 2012; Pidgeon and Barnett 2013). It further downplays the extent to which experts may themselves socially construct risk on the basis of shared worldviews, subjective beliefs and institutional affiliations (Duckett et al. 2015; Urquhart et al. 2017a).

The analysis of the data from the interviews with scientists, policymakers, practitioners and high-level stakeholders suggests that expert risk perceptions are heterogeneous and dynamic, and they draw on a wide range of evidence to construct their understanding of the risks posed by a tree pest or disease outbreak. Along with official notifications and technical risk assessments, they also rely on their own experience, anecdotal evidence, interactions with stakeholders and media accounts. Heuristic devices used by our respondents included a reference to past outbreaks in order to explain or contextualise their perceptions about the current risk. For instance, Dutch elm disease was drawn on to justify their own framing of the risks posed by ash dieback, as expressed by one tree nursery owner:

> There's reckoned to be 60 million ash trees in the country … so it far outweighs the cataclysm that was Dutch elm disease, in my view.

It was also cited as they tried to make sense of why ash dieback was taken up by the media, with one scientist respondent suggesting:

> I think it is actually probably because of Dutch elm disease, whenever there's anything that affects trees in this country, I think the 'Great British Public' are, you know, nature lovers.

Similarly, the government's aborted sell-off of England's public forest estate[2] was used to contextualise the government's response to the disease, as described by a representative of a landowners' association:

> I think it kind of all goes back to they [the Government] found themselves just incredibly vulnerable after the disaster of trying to sell off the public forest estate. They just did not expect that kind of response. ... It galvanised quite a lot of influential public opinion ... and I think they were just very nervous of anything to do with trees and woods, and here was a disease.

In many instances, respondents indicated high levels of concern in the early stages of outbreaks when there is often limited scientific evidence, a lack of clarity on management responsibilities or regulatory mechanisms, making effective management and control very difficult to plan, justify and implement. The issue of uncertainty poses one of the greatest challenges facing policy makers in making objective risk assessments for tree health outbreaks. For many tree pests and diseases, there is uncertainty about the likelihood of introduction and spread but also about the effectiveness of any attempts to control, manage or contain an outbreak once it is underway. Inevitably, under conditions of uncertainty, policy makers and decision makers may feel particularly exposed to risks to their reputation. Indeed, in the ash dieback case, much of the initial government response to the outbreak arguably reflected concerns about reputational risks related to intense media scrutiny during the early stages of the outbreak in 2012, as one government policy maker indicated:

> Right from the word go, officials at number ten were involved in the policy and media handling of what the government's response was going to be. So, there was strong pressure right from the very top for the government to be seen to be doing something about this.

Tree health managers, regulators and policy makers may therefore respond both to the hazard event itself ('A' on Fig. 2) but also to what they perceive as public concern ('D' on Fig. 2). Our analysis suggests that where there are concerns over uncertainty and reputational risk, decision makers are particularly likely to be sensitive to what they believe the public is thinking and often see messages disseminated in the media ('C' on Fig. 2) as a proxy for public concern. One policy maker suggested that 'In my view, the main driver was the media, and then the government response to the media. It didn't have as much to do with the science or the practicalities of it at all'.

Risk managers may therefore attribute risk perceptions to wider publics and other stakeholders in their efforts to ensure the social acceptability of any interventions. Indeed, the analysis suggests the response to institutional or reputational risks in public bodies is often driven by how risk managers and policy makers assume the public feel about a particular pest or disease rather than on the basis of any empirical evidence of public concern. This highlights a need for a better understanding of public perception of risk as well as recognition of the importance of reputational drivers for government action. An understanding of levels of public knowledge, what prompts their interest and attention and how they access information about pests and diseases, would help in designing risk communication strategies. It would also help risk managers address both institutional risks and societal risks associated with tree pest and disease outbreaks.

The findings from the analysis of the interview transcripts concur with Busby and Onggo (2012) and implies that experts are social actors just as much as publics, interacting, observing and being influenced by others' judgements in different settings. In this dynamic interaction, cultural context likely influences actors (e.g. policy makers, publics, institutions and media) and is used to frame risk debates, as outlined by Renn (2003): 'All actors participating in the communication process transform each message in accordance with their previous understanding of the issue, their application of values, worldviews, and personal or organizational norms, as well as their own strategic intentions and goals' (p. 377). Different individuals and groups will thus assess risk differently because they attach systematically different values to what is being harmed and may view the consequences of that harm differently (Jackson et al. 2006). Thus, rather than seeing divergences between expert and lay views as evidence of amplification, social risk amplification may best be understood as an attribution or judgement that one individual or group of individuals make of the risk assessments or judgements of another or others.

5 The Media as a 'Social Station of Amplification' for Tree Pest Outbreaks

Our analysis of the traditional media coverage of ash dieback revealed that early reporting featured risk signals such as 'killer', 'disease' and 'spread', highlighting the spread of the disease across Europe and blaming the government for preventing its incursion into the UK (Fellenor et al. under review-a). As SARF notes, risk events are rarely seen in isolation, and the media attention referenced previous tree health outbreaks such as Dutch elm disease in the 1970s and more recent outbreaks such as Oak processionary moth (*Thaumetopoea processionea*), *Phytophthora ramorum* and Horse-chestnut leaf miner (*Cameraria ohridella*). It further warned of potential new invaders not yet present in the UK, but on the watch list of future risks, such as Emerald ash borer (*Agrilus planipennis*) and Xylella (*Xylella fastidiosa*).

According to SARF, traditional media (newspapers, radio and television) are ascribed a 'pivotal role as a "station" relaying "signals" and constructing public representations of risk' (Murdock et al. 2003, p. 156). The role that news media play as 'risk articulators' has always been given prominence in studies of risk communication and awareness within a social amplification framework. However, early critics took issue with the linear representation within SARF of media reporting of risk events as merely information transmission, positing instead a much more interactive involvement by journalists and media editors as they react to the storylines that their initial reporting may have set in motion. Furthermore, the media may also seek to 'shape' risk perceptions through adopting particular positions or stances in order to promote a particular agenda. A number of scholars have looked at how key actors use the media (Rayner 1988; Petts et al. 2001; Bakir 2005), such as institutions and lobby groups seeking to influence media coverage in order to convey a particular message or draw attention to their own interests and agendas. Indeed, our expert interviews (Sect. 4) suggested that a number of environmental NGOs and industry groups used the early media attention on ash dieback as an opportunity to raise tree health on the political agenda by actively amplifying the risks in their briefings to journalists. A representative of a landowners' association said:

We very quickly decided that this was an opportunity for us to raise the whole profile of tree health within government circles. So we were very happy to brief the press and make it as big a story as possible, and as threatening.

Less well studied has been how social media may influence, often very rapidly, public views on hazard events. As far as we are aware, there has been no consideration of social media and SARF, although there are a small number of studies of social media and risk perception (e.g. Gaspar et al. 2014). With increasing use of a range of platforms, such as social networking sites, blogs, online video, text messages and portable digital devices (Smith 2010), publics are becoming more actively involved than ever before in shaping risk stories (Veil et al. 2011). By posting firsthand accounts and images of emerging hazard events, the public operates in effect as a 'citizen media' platform and as a 'social amplification station'. In addition, social media presents an important communication tool for risk communicators for both disseminating risk information and engaging in dialogue with the public in order to best manage the risk issue.

Analysis of Twitter showed several waves of interest in tree health, suggesting that a majority of information tweeted was resending (retweeting) what was already available in official notifications or traditional media. Moreover, assessing the tweets for particular synonyms for risk revealed that they largely reflected what was said in specific traditional media stories, which were then repeated on Twitter, rather than as original content created by users. Given the limited character count available for tweets, fragments of the original media stories were transported to the Twitter platform, reflecting how certain features of media messages are emphasised and amplified. Our analysis revealed tweets pertaining to initial concerns with its 'spread' and the 'fight' against the disease. Later, these themes fell in prominence and themes of 'blame', and then finally, 'too late', were most common. A further observation was how information is tailored in line with group identities and individual interests. For example, information on tree health can piggyback onto other interests circulating on Twitter. For instance, for users with a primary interest in countryside recreation, tree pests may be of interest in the context of whether it may or may not diminish their recreational experience. Thus, Twitter users may have an active role in re-presenting risk to a wider audience, but the intention is often to reshape the risk within their own worldview or in relation to core interests. For some, this involves a call for official action, a response to their personal sense of responsibility to help or may be seen as just another example of natural events.

6 Are the Public Concerned About Tree Health Risks?

As outlined above, assumptions are often made by policy makers and risk managers about how publics view risk issues, often on the basis of media coverage of the risk event concerned. But to what extent does this align with actual public opinion? The first point to stress, perhaps obviously, is that public opinion about tree pests and diseases is not homogenous, as demonstrated in our national survey and the East Kent case study. Different individuals have different views about the seriousness of tree pest outbreaks and their likely impacts, and many are unaware of tree health issues (21% of respondents had never heard of the issue, and a further 57% indicated they knew very little about it).

Around one in three respondents indicated they were either extremely concerned or very concerned about tree health issues, and around half indicated a willingness to adopt biosecure behaviours, such as avoiding bringing plants and wood products to the UK, buying from trusted locally grown sources and cleaning footwear and bike tyres. Members of environmental organisations and those who feel a strong sense of identity with a place (home, village, park, etc.) are likely to have higher awareness and levels of concern about tree pests and diseases. Further, those who visit woodlands regularly are likely to be more aware than non-visitors, and gardeners are more likely to be concerned than non-gardeners. Women, older respondents, those with a strong sense of affinity with a place, members of environmental organisations, woodland visitors and gardeners were most likely to express a willingness to adopt biosecure behaviours.

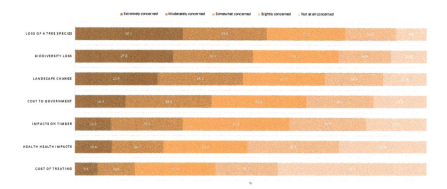

Fig. 4 Stated concerns about impacts of tree pests and diseases

The national survey results suggest that the public's various concerns about tree health are rooted in wider interests, such as access to the countryside, aesthetic values, recreation and gardening. Concerns about the ecological and landscape impacts of tree diseases appear to be greater than economic concerns (such as the cost of treating or removing diseased trees) or human health impacts (see Fig. 4).

A comparison with results from a survey conducted in 2013 showed a decline over the three-year period in awareness, concern and willingness to take actions that prevent tree health problems occurring. The 2013 survey was undertaken shortly after the period of intense media scrutiny on the ash dieback outbreak when it was identified in the autumn of 2012 (Fuller et al. 2016). This may explain the higher level of awareness and concern at that time, but as no baseline of public perceptions prior to the ash dieback outbreak exists, it is difficult to be clear whether the interest in 2013 represented a peak in attention at the time. Although our study suggests that individuals with higher levels of knowledge about invasive tree pests and diseases are more likely to be concerned about the issue, it also suggests that a primary source of information for awareness is the media. The most frequently cited source of information about tree pests and diseases was traditional media such as TV, newspapers and radio. Thus, the way the issue is framed in media accounts is likely to influence public opinion, at

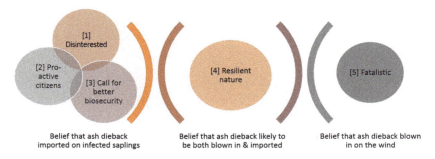

Fig. 5 Narratives associated with beliefs about pathways of introduction for ash dieback

least in the initial phases of an outbreak when it is relatively unknown, perhaps skewing more longer-term attention to tree health issues.

In the Q Methodology analysis conducted in East Kent, a diverse set of five narratives on public perceptions about ash dieback (*Hymenoscyphus fraxineus*) emerged (Fig. 5), typified as disinterested (a lack of concern or interest in tree); proactive citizens (locally aware and active); call for better biosecurity (concerned about preventing future outbreaks); resilient nature (belief that nature is resilient and, with help from science, will cope); and fatalistic (pessimistic about future tree health) (Urquhart et al. under review-a). Opinions varied greatly between the narratives on what, if anything, should be done about tree health and who should be blamed for tree pest and disease outbreaks. A key factor in shaping public attitudes was people's beliefs about how the disease arrived in the UK and if anyone was to blame (Fig. 5). Attitudes also reflected broader worldviews about the vulnerability or resilience of nature and cultural perspectives, independent of the actual events around ash dieback.

While the survey and the East Kent case study represent a research intervention that involves eliciting data from respondents, we also undertook an analysis of naturally occurring data in the form of emails and calls to Forestry Commission and Defra's helpline during the early phase of the ash dieback outbreak (Fig. 6). Interestingly, the analysis indicated that the helpline contacts generally had few media references and did not relate to ash dieback in a way that was typical of the media

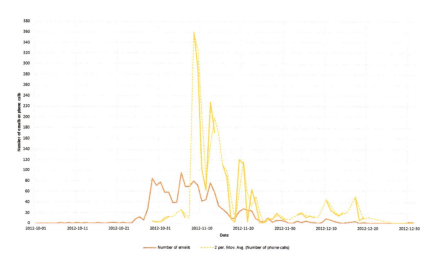

Fig. 6 Emails and calls to Defra and the Forestry Commission helpline during the ash dieback crisis in 2012

coverage. The surge in emails and calls appears mainly to reflect an interest in obtaining more information and offering to help (e.g. to give the location of infected trees), rather than showing panic or concern. Most significantly, the content of emails reflected a rational and reasonable public response to ash dieback and not one which might have otherwise have reflected an 'irrational' public. Correspondents and callers generally wanted to help, for example by reporting a case.

7   Conclusions

Viewing a tree health outbreak through the lens of SARF allows us to consider the interactions between experts, policy makers, publics and the media in the construction of tree health risks. By exploring the dynamic interrelationships between these different actors and the social, psychological and cultural processes through which they determine risk, we have provided a more nuanced understanding of tree health risks that can inform risk communication strategies. We suggest that such strategies need to be sensitive to different cultural perspectives on public risk perceptions and that notifications that merely present scientific data, without consideration of how calls for behaviour change, for instance, may threaten underlying cultural values and beliefs are thus unlikely to succeed (Urquhart et al. under review-a).

This section sets out a number of implications that emerged from the integrated analysis presented in this chapter. Firstly, it is important to recognise that there is no single public to address on tree health, nor any simple way to capture the degree of attention, interest or concern shown by these publics. Typically, there are many different publics, with varying degrees of concern about a given issue. This makes measurement using conventional survey methods difficult. Specific worldviews, experiences and interests of different publics can reinforce positive, relevant and personalised responses aimed at managing tree health issues.

Secondly, tree health events or outbreaks are not seen in isolation but are assessed by both publics and experts in the light of earlier experiences and events. In anticipation or response to a 'tree health event' or issue, the event should be seen in broader historical, social and political terms, not just through the biology and ecology of the threat in question. Further, risk assessment has traditionally focused on the environmental and economic consequences of potential pests and diseases. The assessments should be broadened, and problem definitions of tree health issues should incorporate wider dimensions relevant to the public, such as how specific groups and their needs or interests will be affected.

Thirdly, there may be a gap (or mismatch) between communication undertaken in the early stages of an outbreak and long-term communications required to bring about changes in behaviour. There could be benefits in linking the short- and long-term communications more directly. This will need to take into account consistency between tree health advice and other messages, such as to enjoy nature or visit the countryside (e.g. by ensuring that increased use is mindful of biosecurity). The distinction perceived between traditional and social media communication campaigns may be underestimating the flow between these media. Understanding how traditional and social media influence each other and how this interaction shapes the potential to communicate and amplify positive messages and responses will help to improve tree health management.

Empirically, the study reported in this chapter contributes to our understanding of what drives public risk concerns and how far this is differentiated across groups with different exposures to tree pests and diseases. It provides an analysis of the public and media response to the ash dieback outbreak through an integrated analysis of the historical, social and risk communication influences at work. Furthermore, the research has generated important insights into the ways individuals are encountering tree pests and diseases in different settings and the extent to which they are able to relate the associated risks to their own actions and behaviours. Using SARF as an analytical tool allowed us to consider the interactions between expert and policy risk assessment, media attention and public opinion. Rather than a linear process of expert assessment informing policy decisions, leading to notifications that are amplified in the media and absorbed by the public, our analysis revealed a dynamic relationship whereby policy and expert risk assessments are reassessed in the light of media and public scrutiny. Meanwhile, media and public attention will evolve in response to the degree to which they perceive the government as handling outbreaks in an appropriate manner. SARF also allows us to consider the 'ripple effects' from a risk event ('E' in Fig. 2). In the ash dieback case, as well as the biological, ecological and landscape impacts of widespread decline of ash, there were significant institutional ripple effects. The government's response represents a step change in policy attention to tree health issues more broadly, with biosecurity and tree health being higher on the political agenda, additional funding and resources being made available for scientific research and improvement to contingency planning and coordination of working across government departments and their partners.

In summary, the empirical evidence generated by this project contributes to the policy evidence base by specifically addressing expert and policy risk perceptions alongside media and public attention. Analysing these different data sets through the lens of SARF allowed us to not only delineate the nature of public concern, but also better understand how policy makers and risk managers may attribute 'concern' to the public by responding to media coverage of an outbreak. Finally, in a policy domain (tree health) previously dominated by operational risk analyses, the work contributes to a broader framing of disease risks, building social science capacity while integrating technical and social perspectives. The need for further work that seeks to develop a better understanding of the underlying cultural determinants of tree health risk perceptions is crucial if societal expectations are to be managed and behavioural change encouraged as new and emerging tree pest and disease outbreaks arise.

Acknowledgements The research reported in this chapter was produced as part of the UNPICK (Understanding public risk in relation to tree health) project funded jointly by a grant from BBSRC, Defra, ESRC, the Forestry Commission, NERC and the Scottish Government, under the Tree Health and Plant Biosecurity Initiative (Grant Number BB/L012308/1). It draws on material published in the peer-reviewed outputs of the project and a policy briefing (Potter et al. 2018).

Notes

1. Available on http://textometrica.humlab.umu.se/.
2. In October 2010, the government announced plans to sell off parts of the public forest estate in England. However, after intense media coverage and public criticism of the decision, the government rescinded the decision to dispose of the estate and instead set up a new independent public body to hold the nation's forests in trust for future generations.

References

Bakir, V. (2005). Greenpeace v. Shell: Media exploitation and the Social Amplification of Risk Framework (SARF). *Journal of Risk Research, 8*(7), 679–691. https://doi.org/10.1080/13669870500166898.

Burns, W. J., Slovic, P., Kasperson, R. E., Kasperson, J. X., Renn, O., & Emani, S. (1993). Incorporating structural models into research on the social amplification of risk: Implications for theory construction and decision making. *Risk Analysis, 13*(6), 611–623. http://doi.org/10.1111/j.1539-6924.1993.tb01323.x.

Busby, J. S., & Onggo, S. (2012). Managing the social amplification of risk: A similation of interacting actors. *Journal of the Operational Research Society*, 1–16. https://doi.org/10.1057/jors.2012.80.

Busby, J. S., Alcock, R. E., & MacGillivray, B. H. (2009). Interrupting the social amplification of risk process: A case study in collective emissions reduction. *Environmental Science & Policy, 12*(3), 297–308. https://doi.org/10.1016/j.envsci.2008.12.001.

COA. (2013). Risk Analysis Framework 2013. Edited by Office of the Gene Technology Regulator Department of Health and Ageing. Canberra: Commonwealth of Australia.

Dietz, T., & Stern, P. (Eds.). (1996). *Understanding risk*. Washington, DC: National Academy Press.

Duckett, D., Wynne, B., Christley, R. M., Heathwaite, A. L., Mort, M., Austin, Z., et al. (2015). Can policy be risk-based? The cultural theory of risk and the case of livestock disease containment. *Sociologia Ruralis, 55*(4), 379–399. https://doi.org/10.1111/soru.12064.

Fellenor, J., Barnett, J., Potter, C., Urquhart, J., Mumford, J., & Quine, C. P. (under review-a). Ash dieback and other tree pests and pathogens: Dispersed risk events and the social amplification of risk framework. *Journal of Risk Research*.

Fellenor, J., Barnett, J., Potter, C., Urquhart, J., Mumford, J., Quine, C. P., & Raum, S. (under review-b). 'I'd like to report a suspicious looking tree': Public concern, public attention and the nature of reporting about ash dieback in the UK. *Public Understanding of Science*.

Fellenor, J., Barnett, J., Potter, C., Urquhart, J., Mumford, J., & Quine, C. P. (2017). The social amplification of risk on twitter: The case of ash dieback disease. *Journal of Risk Research*, 1–21. https://doi.org/10.1080/13669877.2017.1281339.

FR. (2012). Rapid assessment of the need for a detailed Pest Risk Analysis for *Chalara fraxinea*. Forest Research.

Fuller, L., Marzano, M., Peace, A., Quine, C. P., & Dandy, N. (2016). Public acceptance of tree health management: Results of a national survey in the UK. *Environmental Science & Policy, 59*(May), 18–25. http://doi.org/10.1016/j.envsci.2016.02.007.

Gaspar, R., Gorjão, S., Seibt, B., Lima, L., Barnett, J., Moss, A., et al. (2014). Tweeting during food crises: A psychosocial analysis of threat coping expressions in Spain, during the 2011 European EHEC outbreak. *International Journal of Human-Computer Studies, 72*, 239–254. https://doi.org/10.1016/j.ijhcs.2013.10.001.

Gormley, A., Pollard, S., & Rocks, S. (2011). *Green leaves III: Guidelines for environmental risk assessment and management.* Cranfield: Cranfield University.

Heuch, J. (2014). What lessons need to be learnt from the outbreak of Ash Dieback Disease, *Chalara fraxinea* in the United Kingdom? *Arboricultural Journal: The International Journal of Urban Forestry, 36*(1), 32–44. https://doi.org/10.1080/03071375.2014.913361.

Höijer, B. (2010). Emotional anchoring and objectification in the media reporting on climate change. *Public Understanding of Science, 19*(6), 717–731. https://doi.org/10.1177/0963662509348863.

Jackson, J., Allum, N., & Gaskell, G. (2006). Bridging levels of analysis in risk perception research: The case of the fear of crime. *Forum: Qualitative Social Research, 7*(1, Art. 20), 1–26.

Kasperson, R. E. (1992). The social amplification of risk—Progress in developing an integrative framework. In S. Krimsky & D. Golding (Eds.), *Social theories of risk.* Westport, CT: Praeger.

Kasperson, R. E. (2012a). A perspective on the social amplification of risk. *The Bridge, 42*(3), 23–27.

Kasperson, R. E. (2012b). The social amplification of risk and low level radiation. *Bulletin of the Atomic Scientists, 68*(3), 59–66. https://doi.org/10.1177/0096340212444871.

Kasperson, R. E., & Kasperson, J. X. (1996). The social amplification and attenuation of risk. *Annals of the American Academy of Political and Social Sciences, 545,* 95–105.

Kasperson, R. E., Renn, O., Slovic, P., Brown, H. S., Emel, J., Goble, R., et al. (1988). The social amplification of risk: A conceptual framework. *Risk Analysis, 8*(2), 177–187. http://doi.org/10.1111/j.1539-6924.1988.tb01168.x.

Kowalski, T. (2006). *Chalara fraxinea* sp. nov. associated with dieback of ash (*Fraxinus excelsior*) in Poland. *Forest Pathology, 36*(4), 264–270. http://doi.org/10.1111/j.1439-0329.2006.00453.x.

Kuhar, S. E., Nierenberg, K., Kirkpatrick, B., & Tobin, G. A. (2009). Public perceptions of Florida red tide risks. *Risk Analysis, 29*(7), 964–969. https://doi.org/10.1111/j.1539-6924.2009.01228.x.

Lasswell, H. (1948). The structure and function of communication in society. In L. Byrson (Ed.), *The communication of ideas.* New York: Institute for Religious and Social Studies.

Lewis, R. E., & Tyshenko, M. G. (2009). The impact of social amplification and attenuation of risk and the public reaction to mad cow disease in Canada. *Risk Analysis, 29*(5), 714–728. https://doi.org/10.1111/j.1539-6924.2008.01188.x.

Merkelsen, H. (2011). Institutionalized ignorance as a precondition for rational risk expertise. *Risk Analysis, 31*(7), 1083–1094. http://doi.org/10.1111/j.1539-6924.2010.01576.x.

Moscovici, S. (1984). The phenomenon of social representations. In R. M. Farr & S. Moscovici (Eds.), *Social representations*. Cambridge: Cambridge University Press.

Mumford, J. D. (2013). Biosecurity management practices: Determining and delivering a response. In A. Dobson, K. Barker, & S. Taylor (Eds.), *Biosecurity: The socio-politics of invasive species and infectious diseases*. Abingdon: Routledge.

Murdock, G., Petts, J., & Horlick-Jones, T. (2003). After amplification: Rethinking the role of the media in risk communication. In N. Pidgeon, R. E. Kasperson, & P. Slovic (Eds.), *The social amplification of risk* (pp. 156–178). Cambridge: Cambridge University Press.

Petts, J., Horlick-Jones, T., & Murdock, G. (2001). Social amplification of risk: The media and the public. Contract Research Report 326/2001 for the Health & Safety Executive.

Pidgeon, N., & Barnett, J. (2013). *Chalara and the social amplification of risk*. Report to Defra.

Potter, C., Urquhart, J., Mumford, J., Barnett, J., Fellenor, J., & Quine, C. P. (2018). UNPICK policy briefing note. Edited by Imperial College London.

Rayner, S. (1988). Muddling through metaphors to maturity: A commentary on Kasperson et al., the social amplification of risk. *Risk Analysis, 8*(2), 201–204. http://doi.org/10.1111/j.1539-6924.1988.tb01172.x.

Renn, O. (1991). Risk communication and the social amplification of risk. In R. E. Kasperson & P. J. M. Stallen (Eds.), *Communicating risks to the public* (pp. 287–324). Dordrecht: Springer. https://doi.org/10.1007/978-94-009-1952-5.

Renn, O. (2003). Social amplification of risk in participation: Two case studies. In N. Pidgeon, R. E. Kasperson, & P. Slovic (Eds.), *The social amplification of risk* (pp. 374–401). Cambridge: Cambridge University Press.

Renn, O., Burns, W. J., Kasperson, J. X., Kasperson, R. E., & Slovic, P. (1992). The social amplification of risk: Theoretical foundations and empirical applications. *Journal of Social Issues, 48*(4), 137–160. http://doi.org/10.1111/j.1540-4560.1992.tb01949.x.

Selbon, M., Raude, J., Fischler, C., & Flahault, A. (2005). Risk perception of the 'mad cow disease' in France: Determinants and consequences. *Risk Analysis, 25*(4), 813–826. http://doi.org/10.1111/j.1539-6924.2005.00634.x.

Shannon, C. E., & Weaver, W. (1949). *The mathematical theory of communication*. Urbana: University of Illinois Press.

Slovic, P. (1987). Perception of risk. *Science, 236,* 280–285.

Smith, A. (2010). *Government online: The internet gives citizens new paths to government services and information*. Washington, DC: Pew Internet & American Life Project.

Tomlinson, I. (2016). The discovery of ash dieback in the UK: The making of a focusing event. *Environmental Politics, 25*(4), 709–728. https://doi.org/10.1080/09644016.2015.1118790.

Urquhart, J., Potter, C., Barnett, J., Fellenor, J., Mumford, J., & Quine, C. P. (under review-a). Risk communication and the subjective differences in the public perceptions of ash dieback: A Q methodology study. *Land Use Policy*.

Urquhart, J., Potter, C., Barnett, J., Fellenor, J., Mumford, J., & Quine, C.P. (under review-b). Managing the institutional risks of tree pest and disease outbreaks in Britain: The case of ash dieback. *Forest Policy and Economics*.

Urquhart, J., Potter, C., Barnett, J., Fellenor, J., Mumford, J., & Quine, C. P. (2017a, November). Expert risk perceptions and the social amplification of risk: A case study in invasive tree pests and diseases. *Environmental Science & Policy, 77,* 172–178. http://doi.org/10.1016/j.envsci.2017.08.020.

Urquhart, J., Potter, C., Barnett, J., Fellenor, J., Mumford, J., Quine, C. P., & Bayliss, H. (2017b). Awareness, concern and willingness to adopt biosecure behaviours: Public perceptions of invasive tree pests and pathogens in the UK. *Biological Invasions, 19*(9). https://doi.org/10.1007/s10530-017-1467-4.

Veil, S. R., Buehner, T., & Palenchar, M. J. (2011). A work-in-process literature review: Incorporating social media in risk and crisis communication. *Journal of Contingencies and Crisis Management, 19*(2), 110–122. http://doi.org/10.1111/j.1468-5973.2011.00639.x.

8

Implementing Plant Health Regulations with Focus on Invasive Forest Pests and Pathogens: Examples from Swedish Forest Nurseries

E. Carina H. Keskitalo, Caroline Strömberg,
Maria Pettersson, Johanna Boberg, Maartje Klapwijk,
Jonàs Oliva Palau and Jan Stenlid

1 Introduction and Aim

International trade and climate change have resulted in a situation where Invasive Alien Species (IAS), including Invasive Pests and Pathogens (IPPs), constitute one of the major threats to biodiversity (e.g. COM 2011 244 final; Caffrey et al. 2014; Ricciardi 2006; O'Brien and Leichenko 2000), especially in forests (Holmes et al. 2009; cf. Manion and Griffin 2001; Stenlid and Oliva 2016). Invasive species

---

E. C. H. Keskitalo (✉)
Department of Geography and Economic History,
Umeå University, Umeå, Sweden

C. Strömberg · M. Pettersson
Department of Business Administration,
Technology and Social Sciences,
Luleå University of Technology, Luleå, Sweden

will likely come to have a growing impact in a situation of increasing global transport of goods and commodities and as the climate continues to change. Long-distance transportation of, for example, living plants to and from plant nurseries and retail stores means that plant pests (pest insects and pathogens) can be introduced, with limited possibility of identifying and tracing the pest until after they have been established and the damage has already occurred. As a result of climate change, invasive pests may be able to establish in areas that previously had unfavourable climates (Holmes et al. 2009; Brunel et al. 2013). Even species that are not considered pests in their native habitat might become pests in their new environment owing to different biotic and abiotic circumstances (Holmes et al. 2009; Brunel et al. 2013).

Trade of living plant material is one of the main pathways for species invasions (Santini et al. 2013; Liebhold et al. 2012). Species thought to have spread globally through the trade of living plant material are *Hymenoscyphus fraxineus*, the causal agent of ash dieback (Woodward and Boa 2013), several *Phytophthora* species, e.g. *P. ramorum* (Grünwald et al. 2012), and the Citrus longhorn beetle (*Anoplophora glabripennis*) (Haack et al. 2009). In the case of pathogens especially, but also for insects, the delayed observation of first symptoms/damage and the delayed identification of the causal agent have led to an unbridled spread of certain species across Europe. Movement of living plant material globally and within Europe makes nurseries an important part of the pathway of species invasions and could make them a hub for species invasions (Bergey et al. 2014).

It is therefore of the utmost importance that plant protection measures against potential invasive species be introduced. However, it has proven difficult to effectively implement such protective measures

---

J. Boberg · J. Oliva Palau · J. Stenlid
Department of Forest Mycology and Plant Pathology,
Swedish University of Agricultural Sciences, Uppsala, Sweden

M. Klapwijk
Department of Ecology,
Swedish University of Agricultural Sciences, Uppsala, Sweden

in practice (e.g. Liebhold et al. 2012). In the context of IPPs, several studies have highlighted the need for increasing application of different incentives and instruments to manage invasive species, including taxes, levies and quarantine measures, as well as raising public awareness (Klapwijk et al. 2016; Caffrey et al. 2014; Perrings et al. 2005, Shine et al. 2000; Stenlid et al. 2011). For instance, Smith et al. (2014, 1325) have argued, "[a] strong strategic legislative framework is essential for addressing the complex challenges of invasive alien species". However, as illustrated in Pettersson et al. (2016), it is not only the legislative and policy framework that constitutes the limitation, but also the possibilities for practical implementation on the ground. One good example of this issue is the case of nurseries that commercialize living plants, but also constitute a main pathway for introduction of pests and pathogens (Liebhold et al. 2012; Jung et al. 2016; Santini et al. 2013). Implementation considerations regarding plant health in this case include possibilities for monitoring, controlling and reporting from nurseries, as well as options to restrict spread via plant trade. The risks of spreading IPPs via plant trade must be understood within the plant trade system if crucial action (Liebhold et al. 2012) is to be imposed by the formal institutional framework.

In Sweden, where approximately 10% of the export value and 3% of GDP relate to forest products (e.g. Skogsindustrierna 2000, 2013), the risks associated with the plant trade may particularly be associated with forest nurseries, although ornamental nurseries pose a significant threat as well. Many ornamental nurseries import plants from large European distribution hubs (Dehnen-Schmutz et al. 2010), which means that plants may originate from anywhere in the world (due to the auctioning system). Although some nurseries grow their plants locally, these plants may also be contaminated by other plants while in transit. In ornamental nurseries, some plant species represent a larger risk for the spread of IPPs than others do. For example, plant pathogens of the genus *Phytophthora* are associated with species of *Rhododendron* (Lilja et al. 2011). Another practice that represents a risk is subcontracted growing, where seeds from Swedish tree provenances are sent to be planted and produce seedlings, for example in Germany, which are then sent back for out-planting in Sweden. Trees, like beech, that are purchased from continental Europe can also serve as potential hosts for new pest or pathogen species. Moreover, there is a risk especially linked to the spread of *Phytophthora* species, as these species are generally waterborne. Thus, recirculating drainage water could spread the pathogen to the entire nursery, and using surface water (e.g. from rivers) for irrigation could potentially introduce *Phytophthora* species present in the water. A theoretical risk here could be introducing *Phytophthora alni* (which specifically infects alder and is present in many rivers in Sweden [Redondo et al. 2015a]) into forest nurseries that produce alder seedlings.

Given these considerations of limitations both in the regulative framework and at the monitoring and detection level, the present study outlines the legislative framework for managing invasive pest and pathogen risks, at EU and Swedish levels. The study also specifically reviews the risks related to invasive species found in Swedish forest nurseries, asking: How is the system monitoring and detecting new invasive pest and pathogens?

The study focuses on plant health in Sweden with a special emphasis on risks to forests (cf. Skogsindustrierna 2000). Methodologically, the study of the regulative framework draws on a review of plant health legislation for the prevention of harmful/quarantine organisms at EU and Swedish levels; risks identified in this review are explained and qualified by previous literature as well as by plant pathologists' personal observational accounts of recent identification of species in Sweden. The study of monitoring and detection in Swedish forest nurseries draws on semi-structured interviews focused on forest plant nurseries and organizations involved in implementing the regulatory framework. Thus, persons from Swedish business organizations in forestry, as well as other forestry-related organizations, governmental authorities, and the largest companies in Sweden involved in production of plants in forest nurseries, were strategically selected, identified and contacted.[1] A total of seven interviews were conducted: five interviews covering nine different forest nurseries across four companies,[2] as well as interviews with the Federation of Swedish Family Forest Owners (Swe. *LRF Skogsägarna*), referred to as LRF and the Swedish Board of Agriculture, referred to as BoA. To allow company interviewees to speak freely about their considerations, they are not referred to by name but rather as "C1", "C2",

"C3" and "C4" (as in "Company 1", "Company 2", etc.). The two interviews made with C3 are referred to as "C3:1" and "C3:2". The interviews were conducted during the fall of 2014 and the fall of 2015. They lasted approximately 1 hour and 15 minutes, and were recorded and fully transcribed.[3] The interview guide, as well as the thematic coding of the interviews, focused on the forest nurseries' work concerning regulated plant pests, including monitoring and identification routines related to plant health and what potential concerns interviewees identified within the system.

2 Plant Health Regulation for the Prevention of Harmful/Quarantine Organisms at EU and Swedish Levels

The current EU plant health regime is a complex system that builds on the original intra-community trade, as well as imports of plant and plant products from non-EU member states (MacLeod et al. 2010). The aim is to protect the EU against harm caused by the introduction and spread of organisms injurious to plants and plant products through legal instruments including prohibition/banning and certification. The current approach is based on the listing of harmful organisms (defined as pests of plants or of plant products, which belong to the animal, fungal or plant kingdoms, or which are viruses, mycoplasmas or other pathogens) into different categories, from particularly harmful organisms whose introduction and spread must be banned by all member states, to the listing of plants and plant products that must be subject to a plant health inspection, including special rules for protected zones (Annex I–VI, Directive 2000/29/EC).

In the 2010 evaluation of the plant health regime, weaknesses in the current system were pointed out, especially in relation to the regimes' preventive capacity, which was not thought to be fit for purpose. The regime's response to the various and increasing issues in relation to protection against organisms injurious to plants and plant products had been primarily ad hoc solutions, rather than strategic adaptations to the development (European Commission 2010).

Against the backdrop of the outcome of that evaluation and in response to the increasing influx of harmful organisms caused by intensified globalization of trade, it was decided that the regime should be replaced. A proposal for a new regulation concerning protective measures against plant pests to replace the current regime was accepted in 2013 (COM 2013) and came into force in December 2016 (Regulation (EU) 2016/2031). The new regulation, which replaces seven existing EU Directives on harmful organism, will, however, not be fully applicable until December 2019 to allow for authorities and other actors to adjust to the new rules (Regulation (EU) 2016/2031, Art. 113).

Unlike the current plant health regime, the new regulation addresses all pests, which thus will be listed together but divided into three main categories following risk assessment. The main categories are:

- Union quarantine pests. These are pests that are not present in the EU or, if present, they are localized and subject to official control (Art. 3, 4 and Appendix I, 2 b). The introduction, movement, holding, multiplication or release of union quarantine pests is, as a main rule, prohibited in accordance with Art. 5 of the Regulation.
- Protected zone quarantine pests. Pests occurring in most parts of the EU, except in certain protected zones into which they cannot be allowed to spread (Art. 32).
- Union regulated non-quarantine pests. Pests that are widely present in the whole of the EU, but which, due to their impact on plant quality, may not occur on seeds or planting material (Art. 36).

In order to ensure an efficient use of the resources allocated to deal with pests within the EU, the regulation also introduces the concept "priority pests". Priority pests are union quarantine pests that are currently not established in the EU territory but whose introduction and spread are likely to have "most severe" consequences for the economy, the environment and/or society (Art. 6). Priority pests are subject to additional measures, including surveys, action plans for eradication, contingency plans, etc.

In addition to the implementing acts for each category, member states are given some discretion to adopt additional or stricter measures. In order to ensure effective action against pests that are not categorized as union quarantine pests, member states may take protective measures against the pests if they consider that the criteria for such pests are fulfilled. According to Article 29, member states are also "allowed to adopt more stringent eradication measures than required by Union legislation" on certain conditions and if this does not conflict with the free movement of, in this case, plants and plant products. The new regulation obliges anyone who is aware of the presence of a quarantine pest to notify the competent authorities; it encourages member states to conduct surveys for the presence of pests; and it sets out eradication measures, including area restrictions, as well as rules for the establishment of contingency and eradication plans. As an additional level of precaution, the regulation moreover opens up for the possibility to introduce temporary restrictions on imports as well as limiting the movement within the EU for certain high-risk commodities. Imports of these commodities must be preceded by a detailed risk assessment determining if the trading is acceptable and under what conditions (Shiffers 2017). This must be considered rather controversial, not least since it is generally not considered possible to prevent such risks more broadly, as it would constitute a trade limitation under the World Trade Organisation (WTO); trade in specific species cannot be prohibited unless identified and scientifically proven risks are in evidence (Pettersson et al. 2016).

The upshot of this framework is that much plant health work must focus on monitoring and detection and on establishing routines to hinder pests that, despite restrictions, enter through the large volumes of plant trade taking place (and through the established trade practices of nurseries). However, this system results in a situation whereby, despite the large array of known and unknown threats, monitoring and surveys (undertaken by the Swedish Board of Agriculture) are generally limited to pests and pathogens listed on the quarantine list. As a consequence, among the *Phytophthora* species considered important pathogens, only *P. ramorum* is regulated, meaning that when it is found measures will be taken to eradicate it. From this also follows that the surveys only aim at finding this particular species, while others will not necessarily be identified. Because many different *Phytophthora* species found in nurseries are of unknown origin, they are not subject to regulation, even though they could potentially cause damage.

The main consideration is, thus, the limitations on which species to be made subject to regulation. In Sweden, however, for specified plant pests, the rules to control or hinder spread in the Plant Protection Act and in Swedish Board of Agriculture regulations are substantial. It follows from the Act that the government or another authority may order property owners or users to take action to combat plant pests; decide on decontamination of facilities and objects, as well as decontamination or destruction or limited use of plants, plant products and packaging; prohibit sowing or planting; issue regulations on cultivation or harvesting; prohibit or impose conditions for the handling of plants, plant products, pests, soil, etc., including import, export or possession; and take or prescribe necessary measures also regarding private property. In addition, regulations on notification for handling of plant pests and decisions on sampling or examination of plants, plant products, soil, facilities, etc., may be issued to control the spread of plant pests and to verify the presence or absence of such pests. To this effect, regulations on health certificates may be issued. A health certificate is a document, accompanying a plant or a plant product that establishes its health status. The Swedish Board of Agriculture is responsible for both issuing health certificates (upon application) and annulling deficient certificates (p. 11 and 11 a, Ordinance [2006:817]). Examples of health certificates are plant passports[4] under EU law and labelling in accordance with ISPM 15 (Prop. 2012/13:174). Pursuant to the notification requirement, any suspicions that plants or plant crops have been infested with pests must be reported to the competent authority.[5] Swedish regulations (SJVFS 2004:53) also exist on heat treatment, kiln drying and marketing of sawn wood, wood packaging material, etc., under the Plant Protection Act.[6]

For these types of actions, some of the main limitations identified so far have mainly been that health certificates and plant passports are not yet mainstreamed across countries with regard to content and specifications of documents; different regulations exist in different countries for what these should contain and on what basis they are established

(Pettersson et al. 2016). In addition, limitations may exist with regard to the extent to which pests, some of which are non-regulated, may spread without being detected in a systematic way. At present, many of recent descriptions of invasive species have been made by trained university personnel after the pests have already spread in nature. Discoveries are often made by chance, as was the case with *Dothistroma septosporum* (J. Stenlid, pers. comm.), *Phytophthora alni* (C. H. Olsson, pers. comm.) and *Diplodia pinea* (J. Oliva, pers. comm.). In addition, several new un-reported invasive *Phytophthora* species, as well as new infection sites, were found in Sweden as part of surveys conducted within research projects (i.e. not directly intended for monitoring) (Redondo et al. 2015b; Cleary et al. 2016).

3   Interviewee Descriptions of Practical Monitoring and Detection in Nurseries

The interviewees C1, C2, C3 and C4 represent Swedish forest nurseries where Scots pine (*Pinus sylvestris*), lodgepole pine (*Pinus contorta*), Norway spruce (*Picea abies*) and in some cases larch (*Larix sibirica and L. xeurolepis*) are grown (interviews with C1, C2, C3:1 and C4).[7] Some of the nurseries purchase plants, but in small amounts only and mainly from other Swedish forest nurseries, although on occasion plants are purchased from other European countries by some of the studied nurseries (interviews with C1, C2 and C3:1). The seeds used by the nurseries generally come from Swedish seed orchards, although some of the interviewed nurseries also import seeds, e.g. from Central Europe, the Baltic states and Canada (interviews with C1, C2, C3:1 and C4). For growing substrate, the nurseries use peat from Sweden and in some cases from Finland (interviews with C1, C2, C3:1 and C4). Water for irrigation in the nurseries comes from rivers, surface water (such as streams or lakes) or groundwater (interviews with C1, C2, C3:1 and C4).

In line with the requirements set by the legislative framework, interviewees noted that intensive monitoring is the only way detection can be guaranteed and, thus, this is a key activity for managing plant health in nurseries. Monitoring is seen as the main measure to control infections that the nurseries themselves, given the current plant trade, could not prevent. Interviewees from nurseries presented several reasons for checking the plants and taking measures if problem occurs, for the sake of both customers and the companies (interviews with C1, C2 and C4). Checking the plants and taking measures if problems occur are necessary for developing good products that customers want to buy (interviews with C1, C2 and C4). Interviewees thus noted that plant health was among the central concerns to their companies: in the long run, if the plants the companies delivered were not healthy, the consequence would be a bad reputation and, thus, a competitive disadvantage (interviews with C1, C2 and C4). Furthermore, if plants are checked regularly, problems are discovered before the damage becomes too severe, which in the end saves money for the company (interviews with C1, C2, C4 and BoA). Another reason for checking the plants regularly is that if this is done, and problems occur in the forest after plantation, the nursery staff can go back and check whether it was due to something in the process through which this plant was developed (e.g. problems at certain points of development) or due to problems with plants purchased from specific locations (interviews with C1 and C4).

It was noted by several interviewees that the first line of action when a problem occurs in the nursery (e.g. detection of unhealthy plants) is to try to identify the cause (interviews with C1, C2, C3:1 and C4). Interviewees noted that if this could not be done by the nursery staff, the Swedish University of Agricultural Sciences (SLU) and/or the Forestry Research Institute of Sweden (Skogforsk) was often contacted (interviews with C1, C2, C3:1 and C4). SLU and Skogforsk may thus constitute important resources for the nurseries, potentially helping them by, e.g. clarifying the potential risk for damage, taking samples and identifying species (interviews with C1, C2, C3:1 and C4).

When the cause is identified, various measures, such as treatment with fungicides, are taken (interviews with C1, C2, C3:1 and C4). However, according to several of the interviewees, few chemicals are approved for forest nurseries, because it is too expensive for manufacturers to develop new chemicals specifically for forest nurseries or to register forest nurseries as users for existing chemicals used in regular nurseries and agriculture (interviews with C1, C3:1, C3:2, C4 and LRF). If a problem results in sick or dead plants, they are discarded, e.g. burnt or composted (interviews with C1, C2 and C4). The companies are not compensated for costs resulting from these problems or for any costs associated with the measures taken (interviews with C1, C2, C3:1 and C4).

Interviewees noted that some of the most typical problems in nurseries are caused by different fungal species (e.g. *Botrytis cinerea* and *Lophodermium seditiosum*), mosses (e.g. *Marchantiophyta*) and weeds (e.g. *Chamerion angustifolium*) (interviews with C1, C2, C3:1 and C4). The interviewees presented several causes for these particular problems. The nursery is perceived as a perfect environment for fungi due to the warm and humid conditions, and because there are many plants of the same species in one place (interviews with C1, C3:1, C3:2 and C4). Different pathways were described for spores of fungi as well as seeds of weeds to get into the forest nurseries. The plants are grown and delivered to different actors along the product development chain in boxes that are reused many times, which means that spores and seeds may follow, for instance, when the boxes are brought back to the nursery from the forest after delivery (interviews with C1 and C4). Preventative measures against this were regularly taken. To avoid risk of infection, the boxes are washed before they are reused (interviews with C1 and C4). However, spores and seeds also come from the area surrounding the nursery and also in the peat (interviews with C1 and C4). Another pathway described for fungi, weeds, etc., to get into the nursery is when plants are purchased from another forest nursery and brought into the nursery for storage, or when plants are moved between a company's own forest nurseries (interview with C1).

As a result, several interviewees raised the issue of the potential risks of introducing new plant pests when moving, but also particularly when importing, plants. Here, specific risks mentioned were the *Phytophthora* species (interviews with C1, C2, C3:1, LRF and BoA). Some interviewees mentioned their concern that new kinds of species would be established in the nurseries as well as in the forest due to climate changes (e.g. *Phytophthora* species, the pinewood nematode [*Bursaphelenchus xylophilus*] and the Mountain pine beetle [*Dendroctonus ponderosae*]) (interviews with C2, C3:1 and LRF). As a means of avoiding these types of considerations and problems, the growing plants are checked by the nursery staff every day (interviews with C1, C2, C3:2 and C4). Samples of the growing plants are continuously sent for closer inspection by experts (in laboratories owned by the companies themselves and/or Skogforsk) (interviews with C1, C2, C3:2 and C4). Before the plants are delivered to the customer, quality controls are made and the plants are checked (interviews with C1, C2, C3:2 and C4).

However, limitations existed in that plants purchased from another nursery are usually not checked by the nursery staff, but only opened and checked by the customer after delivery (interviews with C1, C2 and C3:2). Furthermore, plants are naturally checked by the customer when planting, but there is no guarantee that they will be thoroughly examined (interviews with C1 and C2). However, weeds in the nursery are cleaned out (often manually) to avoid their spreading and taking over, and flowers, brushwood, etc., in the surrounding areas are cleared (interviews with C1 and C4). When importing plants, the risks of introducing new plant pests (e.g. *Phytophthora* species) are discussed within the company and with the Board of Agriculture and the Swedish Forest Agency (interview with C3:1).

Interviewees primarily exchange information with parties that could support detection if unknown material was found, and to a lesser extent with governmental agencies and mainly when setting up import systems. When asked whether governmental authorities (such as the Board of Agriculture or the Swedish Forest Agency) are informed when problems occur, interviewees noted that there is no such requirement, and that authorities are not involved in most instances (interviews with C1, C3:2 and C4). However, the Board of Agriculture and the Swedish Forest Agency do visit the forest nurseries on a regular basis, inspecting documentation, traceability of seeds, plant passports and the occurrence of weeds in the nursery (interviews with C1, C3:1, C3:2 and C4). The interviewee at the Board of Agriculture noted that the Swedish government is currently considering implementing systems to create further incentives for actors in the plant production chain to take action regarding plant pest risks (interview with BoA); to support the implementation of developing requirements at the EU level, these could include, for instance, an improved risk management function or plant health support unit, and further development of crisis management plans for different species (cf. Pettersson et al. 2016).

4   Discussion and Conclusion

The present study has examined the regulatory framework for plant health and the possibilities it offers for forest nurseries to manage plant health risks, primarily through legislative review and interviews in organizations responsible for implementing legislation and practitioners carrying out legislation, i.e. nurseries. The interviews conducted in the nurseries themselves clarified how monitoring and detection are performed in Sweden, in line with EU and Swedish legislation. The main considerations reported by the interviewees included difficulties due to the lack of specific and affordable chemicals for treatment, in particular in the case of forest nurseries, as well as reliance on similar monitoring having been carried out at other levels and areas of the system, for instance plants purchased from other nurseries. As a result, risks stem from the potential of introducing new plant pests when moving plants. Because plants purchased from other nurseries are not checked at the purchasing nursery, but only by the customer upon delivery, or more likely at planting, it was noted that there is no guarantee they will be thoroughly checked.

Similar to earlier research, the study illustrates that much of the focus is placed on the different parties and networks involved in the plant trade monitoring and detecting risks. This is both because the present WTO framework in effect means that only transport of proven high-risk organisms can be prevented (Pettersson et al. 2016; Klapwijk et al. 2016) and because the plant trade framework, as it currently operates, inherently results in these kinds of risks since it involves large volumes of plants being routinely moved between nurseries and purchasers across country lines. Thus, under the existing WTO framework, this type of trade in plants is supported rather than impeded, with resultant risks being a natural consequence of the way the system operates. The new EU plant health regulation, however, might imply a change in this regard with its preventive rules, established criteria for risk assessments and possibilities to prohibit both imports and movement within the EU.

In addition, the interviews also revealed limited integration between agencies and nurseries, as nurseries in these cases were seldom in contact with the agencies. Interviewees also noted that they were only at limited occasions in contact with potential detection support functions that they referred to at Skogforsk or SLU. As earlier studies have shown that there presently exist limitations in collaboration surrounding plant health within the agency system as well (Pettersson et al. 2016), it is possible that improvement in integration between agencies, detection support functions and nurseries could support improved control, monitoring and detection at these levels. The development of a new plant health support unit (cf. Pettersson et al. 2016) could, for instance, serve as a coordinating instrument. Integration might also be facilitated by new requirements at the EU level indicating the need for more coordinated plant health approaches and potentially possibilities to check a larger number of species (not only those on a quarantine list).

The present study would seem to indicate to some extent the still relatively early stage of development of coordination—at international, EU and national levels—concerning plant risks in trade, which has also been illustrated in other studies (cf. Pettersson et al. 2016; Klapwijk et al. 2016). The study also illustrates the risks inherent within a wide-ranging plant trade system as supported under international trade law, where, despite multiple control mechanisms, a single infection that slips through can potentially spread across a wide area of territory. Given climate change and increasing globalization, as some of the interviewees noted, risks do exist both for domestic pests (increased establishment of fungi, mosses and weeds) and for increasing establishment of IPPs.

Notes

1. One company did not reply and the business organization, the Swedish Forest Industries Federation, and two other organizations contacted did not consider themselves relevant to be interviewed due to limited involvement with this specific issue.
2. Of these interviews, two were group interviews and one was a telephone interview, all of which on the request of the interviewees to either have additional staff present or undertake the interview by telephone.
3. The exception was the shorter phone interview that lasted about 15 minutes and was not possible to record due to technical difficulties.
4. For further information, see SJVFS (2010:13, p. 7 and Appendix 7).
5. The Plant Protection Act is not applicable to insect ravages in forests. In case of such outbreaks, regulations under the Forestry Act apply.
6. In addition, also Swedish forest legislation allows for regulations to prevent or impose conditions on the use of forest reproductive material of indigenous or foreign origin in the establishment of new forest stands if warranted from a silvicultural point of view (p. 7, Para. 1, Forestry Act). This means that forest material from outside the EU may not be introduced in Sweden without permit. A permit may in turn only be granted if the admission is in compliance with Directive 1999/105/EC (s. 10–10a), and invasive tree species may not be used as forest reproductive material in mountainous forestland (with high natural and cultural values). The trading of forest material within the EU is also subject to control with regard to invasive species. Certain types of wood require a plant passport to ensure that the wood is free from plant pests. In addition, foreign tree species (in general) may only be used as forest reproductive material in exceptional cases, although it is generally allowed to grow *Pinus contorta* in certain parts of the country (SKSFS 1993:2, 2010:2).
7. All of the nurseries in the interview study also noted that they are FSC and PEFC certified.

References

Bergey, E. A., Figueroa, L. L., Mather, C. M., Martin, R. J., Ray, E. J., Kurien, J. T., et al. (2014). Trading in snails: Plant nurseries as transport hubs for non-native species. *Biological Invasions, 16*(7), 1441–1451.

Brunel, S., Fernández-Galiano, E., Genovesi, P., Heywood, V. H., Kueffer, C., & Richardson, D. M. (2013). Invasive alien species: A growing but neglected threat. In *Late lessons from early warnings: Science, precaution, innovation* (pp. 518–540). Copenhagen: European Environmental Agency.

Caffrey, J. M., Baars, J.-E., Barbour, J. H., Boets, P., Boon, P., Davenport, K., et al. (2014). Tackling invasive alien species in Europe: The top 20 issues. *Management of Biological Invasions, 5*(1), 1–20.

Cleary, M., Ghasemkhani, M., Blomquist, M., & Witzell, J. (2016). First report of *Phytophthora gonapodyides* causing stem canker on European beech (*Fagus sylvatica*) in Southern Sweden. *Plant Disease*, PDIS-04-16-0468-PDN.

COM (2011) 244 final. *Our life insurance, our natural capital: An EU biodiversity strategy to 2020.*

COM (2013) 267 final. *Proposal for a Regulation of the European Parliament and of the Council on protective measures against pests of plants.*

Council Directive 1999/105/EC of 22 December 1999 on the marketing of forest reproductive material.

Council Directive 2000/29/EC of 8 May 2000 on protective measures against the introduction into the Community of organisms harmful to plants or plant products and against their spread within the Community.

Dehnen-Schmutz, K., Holdenrieder, O., Jeger, M. J., & Pautasso, M. (2010). Structural change in the international horticultural industry: Some implications for plant health. *Scientia Horticulturae, 125,* 1–15.

European Commission. (2010). *Evaluation of the community plant health regime*. Final Report. https://ec.europa.eu/food/sites/food/files/plant/docs/ph_biosec_rules_final_report_eval_en.pdf.

Government Bill. Prop. 2012/13:174. Ändringar i växtskyddslagen.

Grünwald, N. J., Garbelotto, M., Goss, E. M., Heungens, K., & Prospero, S. (2012). Emergence of the sudden oak death pathogen *Phytophthora ramorum*. *Trends in Microbiology, 20,* 131–138.

Haack, R. A., Hérard, F., Sun, J., & Turgeon, J. J. (2009). Managing invasive populations of Asian longhorned beetle and citrus longhorned beetle: A worldwide perspective. *Annual Review of Entomology, 55*(1), 521.

Holmes, T. P., Aukema, J. E., Von Holle, B., Liebhold, A., & Sills, E. (2009). Economic impacts of invasive species in forests: Past, present, and future. *Annals of the New York Academy of Science, 1162,* 18–38.

Jung, T., Orlikowski, L., Henricot, B., Abad-Campos, P., Aday, A. G., Aguín Casal, O., et al. (2016). Widespread *Phytophthora infestations* in European nurseries put forest, semi-natural and horticultural ecosystems at high risk of Phytophthora diseases. *Forest Pathology, 46*(2), 134–163.

Klapwijk, M. J., Hopkins, A. J. M., Eriksson, L., Pettersson, M., Schroeder, M., Lindelöw, Å., et al. (2016). Reducing the risk of invasive forest pests and pathogens: Combining legislation, targeted management and public awareness. *Ambio, 45,* 223–234.

Liebhold, A. M., Brockerhoff, E. G., Garrett, L. J., Parke, J. L., & Britton, K. O. (2012). Live plant imports: The major pathway for forest insect and pathogen invasions of the US. *Frontiers in Ecology and the Environment, 10,* 135–143.

Lilja, A., Rytkonen, A., Hantula, J., Muller, M., Parikka, P., & Kurkela, T. (2011). Introduced pathogens found on ornamentals, strawberry and trees in Finland over the past 20 years. *Agricultural and Food Science, 20,* 74–85.

MacLeod, A., Pautasso, M., Jeger, M. J., & Haines-Young, R. (2010). Evolution of the international regulation of plant pest and challenges for future plant health. *Food Security, 2,* 49–70.

Manion, P. D., & Griffin, D. H. (2001). Large landscape scale analysis of tree death in the Adirondack Park, New York. *Forest Science, 47,* 542–549.

O'Brien, K. L., & Leichenko, R. M. (2000). Double exposure: Assessing the impacts of climate change within the context of economic globalization. *Global Environmental Change, 10*(3), 221–232.

Ordinance (2006:817) on plant protection.

Perrings, C., Dehnen-Schmutz, K., Touza, J., & Williamson, M. (2005). How to manage biological invasions under globalization. *Trends in Ecology & Evolution, 20*(5), 212–215.

Pettersson, M., Strömberg, C., & Keskitalo, E. C. H. (2016). Possibility to implement invasive species control in Swedish forests. *Ambio, 45,* 214–222.

Redondo M. A., Boberg J., Olsson C. H., & Oliva J. (2015a). Winter conditions correlate with *Phytophthora alni* subspecies distribution in southern Sweden. *Phytopathol, 105*(9), 1191–1197.

Redondo, M. Á., Boberg, J., Stenlid, J., & Oliva, J. (2015b). First report of *Phytophthora pseudosyringae* causing basal cankers on horse chestnut in Sweden. *Plant Disease, 100,* 1024.

Regulation (EU) 2016/2031 of the European Parliament and of the Council of 26 October 2016 on protective measures against pests of plants, amending Regulations (EU) No. 228/2013, (EU) No. 652/2014 and (EU) No. 1143/2014 of the European Parliament and of the Council and repealing Council Directives 69/464/EEC, 74/647/EEC, 93/85/EEC, 98/57/EC, 2000/29/EC, 2006/91/EC and 2007/33/EC.

Ricciardi, A. (2006). Are modern biological invasions an unprecedented form of global change? *Conservation Biology, 21*(2), 329–336.

Santini, A., Ghelardini, L., De Pace, C., Desprez-Loustau, M. L., Capretti, P., Chandelier, A., et al. (2013). Biogeographical patterns and determinants of invasion by forest pathogens in Europe. *New Phytologist, 197,* 238–250.

Shiffers, B. (2017). New European Union plant health regime: A more stringent regulation that could impact trade from developing countries in the near future (Guest editorial). *Tunisian Journal of Plant Protection, 12*(1), 1–3. Available at https://orbi.uliege.be/bitstream/2268/212920/1/Schiffers.TunisianJPlantProtect.2017.pdf.

Shine, C., Williams, N., & Gündling, L. (2000). *A guide to designing legal and institutional frameworks on alien invasive species* (No. 40). IUCN.

SJVFS 2004:53. Regulations of the Swedish Board of Agriculture.

Skogsindustrierna. (2000). *Europe needs the forest industry.* Stockholm: Skogsindustrierna/Swedish Forest Industries Federation.

Skogsindustrierna. (2013). *Facts and figures.* Stockholm: Skogsindustrierna/Swedish Forest Industries Federation, Stockholm.

SKSFS 1993:2. Regulations of the Swedish Forest Agency.

SKSFS 2010:2. Regulations of the Swedish Forest Agency.

Smith, A. L., Bazely, D. R., & Yan, N. (2014). Are legislative frameworks in Canada and Ontario up to the task of addressing invasive alien species? *Biological Invasions, 16*(7), 1325–1344.

ST 8795 2016 REV 2 - 2013/0141 (OLP). Position of the Council at first reading with a view to the adoption of a Regulation of the European Parliament and of the Council on protective measures against pests of plants, amending Regulations (EU) No. 228/2013, (EU) No. 652/2014 and (EU) No. 1143/2014 of the European Parliament and of the Council and repealing Council Directives 69/464/EEC, 74/647/EEC, 93/85/EEC, 98/57/EC, 2000/29/EC, 2006/91/EC and 2007/33/EC—Adopted by the Council on 18 July 2016.

Stenlid, J., & Oliva, J. (2016). Phenotypic interactions between tree hosts and invasive forest pathogens in the light of globalization and climate change. *Philosophical Transactions of the Royal Society, B 371,* 20150455. https://doi.org/10.1098/rstb.2015.0455.

Stenlid, J., Oliva, J., Boberg, J. B., & Hopkins, A. J. M. (2011). Emerging diseases in European forest ecosystems and responses in society. *Forests, 2,* 486–504.

Woodward, S., & Boa, E. (2013). Ash dieback in the UK: A wake-up call. *Molecular Plant Pathology, 14,* 856–860.

9

The Economic Analysis of Plant Health and the Needs of Policy Makers

Glyn Jones

1 Introduction

The increasing threat to UK forests, woodlands and trees from invasive pests and disease (Freer-Smith and Webber 2017) has resulted in the issue of protecting plant health becoming a major policy area for Government. Rapidly expanding trade (both in terms of numbers of products and volumes), dynamic pathways and changing trends (e.g. demand for large trees) have provided multiple (and changing) opportunities for new pests and disease to enter and establish (dependent on multiple introductions and where they arrive) (e.g. Dehnen-Schmutz et al. 2010; Bradley 2012; Liebhhold et al. 2012; Santini et al. 2013). It has been estimated that 26,000 plant species have been introduced into the UK compared to a native flora of 1600 (Crawley et al. 1996). This provides a conveyor belt pathway for non-native pests and disease.

G. Jones (✉)
Fera Science Ltd., National Agri-Food Innovation Campus, York, UK

This increasing frequency of pest/disease outbreaks therefore poses a threat to biodiversity and the diverse ecosystem services and benefits provided by trees and woodlands to society, for example provision of raw materials, ornamental resources, cultural services as recreation and aesthetic values and regulating services such as carbon sequestration and water purification (Binner et al. 2017 list around 200 published references across these services). In some cases, outbreaks can lead to the destruction of large areas of natural and/or commercial plantings, impacting on ecosystem functions and causing significant economic losses. Moreover, there are public health implications of tree pests and pathogens. Trees and greenspace have been shown to have strong links to human health and well-being through multiple pathways (Wolf et al. 2015). Trees are important filters of harmful pollutants, and green spaces have been shown to promote physical activity with positive implications for reduced blood pressure and stress (Park et al. 2010), reduced obesity (Bell et al. 2008) and improved general health (Maas et al. 2006).

The consequences of the "conveyor belt" are increasing pressure upon public-sector capacity and budgets for plant health management. In turn, this has focused attention on improving the allocation of public resources to reduce the risks and consequences of pest/disease outbreaks occurring in the future. This includes how best to manage (eradicate, contain or do nothing) newly established populations in order to reduce social damages associated with the impacts of plant pests and diseases to ecosystem services. Policy can address such issues to some degree with ex-ante contingency plans. Other policy decisions relate to resources for surveillance as well as incentivising land managers to report outbreaks.

Whilst the field of economics can provide valuable insights for the management of plant health issues, it often overlooks a critical need from a policy-making perspective. Cook et al. (2017) summarise this perspective particularly well. They acknowledge that decision-makers are time pressured and that methods to inform action must be expedient. Decision-makers seldom have the luxury of long-term research. Often, they are tasked with producing a response within hours, days or weeks. Such rapid action is necessary given the small window of opportunity for eradication. Further, the context of an outbreak constantly changes due to external pressures.

This chapter provides a brief overview of economic approaches to plant health and describes the tools that can be applied rapidly by policy makers to estimate the costs and benefits associated with public and private responses to plant pests and disease. The chapter concludes with a short discussion about how the economic analysis of plant health interacts with policy.

2 The Public/Private Nature of Plant Health

Lansink (2011) provides a clear description of the economic aspects of biosecurity and the relative roles of public and private sectors. Plant health is affected by the actions of all those actors involved in plant production and transport as well as other sectors that can provide a pathway for pests and disease. "Buyers" of plant health are plant producers, importers and consumers (of food, ornamentals as well as the ecosystem services provided by woodlands, parks, etc.). Plant health is provided by those involved in the supply chain. However, market failure in the provision of plant health means that the state has to step in to provide and support higher levels of biosecurity and plant health. The market failure results from 2 sources:

- The public good nature of biosecurity
- Asymmetric information between buyers and sellers

To this can be added the concept of filterable externalities whereby one person's protective actions reduce or *filter* the undesirable events experienced by others (Shogren and Crocker 1991), i.e. there are positive external impacts from biosecurity.

Together these will result in plant health or biosecurity being undersupplied by the market on its own without public intervention. The public good nature derives from two distinguishing characteristics: the "consumption" of plant health does not reduce the availability to others (non-rival), i.e. if a neighbouring plant nursery adopts more biosecure practices, my nursey also benefits with a lower probability of a pest incursion, but this does not reduce any benefit available to others, and no one can be excluded from the "consumption" (non-excludable), i.e. my neighbouring biosecure nursery cannot prevent my nursery from benefitting.

Lansink (2011) suggests that the public good element of plant health can have different appearances:

- Producers who pursue high-risk activities do not account for the risks they impose on others, whether they are other actors in the supply chain or to society more generally. Potential losses can be large with, for example, the loss of export markets or significant elements of the landscape and ecosystem services
- Plant producers generally differ in terms of phytosanitary standards. This can be due to individual choices made by growers or reflect the regulatory standing of particular parts of the industry
- Perrings et al. (2000) have noted that plant health status may depend upon the weakest performer, e.g. where export markets are only accessible if the production area is free from a pest or disease, but this status can be lost due to a single importer with lower standards.

This weakest link characteristic has been noted before the problem was framed in public good terms. Charles Fernalds' address to the Association of Economic Entomologists in 1896 appealed for legislation to prevent spread through trade, and these laws should apply to neighbouring countries as there was no entomological dividing line.[1]

Information as to the health of plant material clearly differs between sellers and buyers. Asymmetric information in this context relates to information about the phytosanitary quality of the product that is generally available for the sellers but less so for the buyers. In a modern trading system, this may not actually be the case since some "sellers" may only have the product on site for a very short period or indeed, not at all. Whilst certification can cross this bridge to some degree, the issue of asymptomatic diseases always makes tracing the original source of infestation/infection particularly difficult.

Public-sector responses are under pressure and not just from budgetary pressures. The increasing volumes of trade in plants and plant products, the internationalisation of firms and new and evolving pathways (e.g. online marketplaces) stretch public resources more thinly. With this backdrop, Lansink also cites the increasing call from society on private-sector responsibility as leading to the introduction by the private sector of various forms of self-regulation but that this still remains under-utilised. He concludes by saying that whilst efforts should be made to encourage such self-regulation, it could be substantially improved if public-sector policies would provide benefits to farms or firms complying with the regulations following self-regulation. It is worth noting the early stages of development of industry generated schemes in the UK that reflect both the impact of recent arrival, spread and establishment of ash dieback (*Hymenoscyphus fraxineus*) and the impending threat of *Xylella fastidiosa* and the corresponding regulatory environment that would severely restrict affected sites and those close to affected sites (personal communication with the Horticultural Trades Association and the Woodland Trust).

There is an important differentiation within plant health that relates to the time path of the pest/disease. The impacts from pests in general can follow different time paths, and it has been suggested that those typical for pests of the natural environment follow a trajectory that makes planning interventions problematical (Waage et al. 2005). This is illustrated in Fig. 1, which shows the potential shapes of expected impact over time. Expected Impact A (EI_A) shows a linear impact trajectory that might be typical of a pest of an agricultural commodity. EI_B shows a rapid early impact that levels out and is used to illustrate the potential impact of an animal disease. The third impact timeline, EI_C, is seen as indicative for pests of trees and reflects the longevity of the host and often cryptic nature of the pest disease. The figure further illustrates the importance of being able to detect pests and disease that follow EI_C at very low prevalence levels before the damages rise exponentially. Furthermore, we need methods to recognise when we are on the upper trajectory of this curve, and it is thus no longer cost-effective or practical to attempt eradication

Thus, the temporal dimension of the size and impact of an outbreak is crucial (Epanchin-Niell and Liebhold 2015). Selecting the best time to act to "limit" the impact of an outbreak is a key management

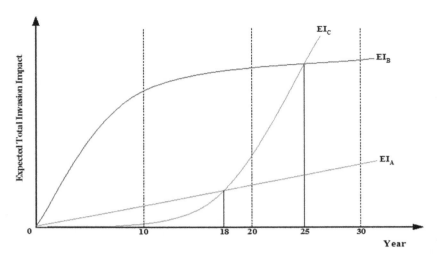

Fig. 1 Expected impacts of invasive species over time

decision. In Fig. 1, EI_C would be a lower priority than EI_A or EI_B based upon analyses of the level of damages (and neglecting the rate at which they increase) and taking a short-term perspective lower than 25 years (when EI_C becomes greater than EI_B).

A number of papers have considered a real options approach to consider when is the optimal time to intervene—whether policy makers should "wait and see" and apply responses in the future rather than immediately (Ndeffo Mbah et al. 2010; Sims and Finoff 2013; Sims et al. 2016). The motivation for this is the view that a policy response is often delayed in order to gather more information about the pest or the invaded area. Broadly, they conclude that it is optimal to wait and see for highly predictable outbreaks and to act early for the opposite. Sims et al. (2016) include an analysis of impact irreversibility (economic and ecological) which can affect the low uncertainty/wait and see response, but not the high uncertainty/act early case.

With respect to UK policy, the decision-maker has limited response options and needs to be able to respond immediately in the case of statutory pests. Further, most of these pests and diseases would be characterised by high levels of uncertainty with respect to the scale and location of the pests, the rate of spread, impact on the hosts and, in the case of trees, the location of the hosts. Thus, in such cases, the mandated early response matches that suggested by the real options approach. Where lists for mandatory action exist, the insights from the real options approach should be directed at how pests and disease get onto the list and/or the range of possible responses available to decision-makers should an outbreak occur.

Pests of plants can have impacts on a wide range of goods and services. Some of these impacts are priced in markets (e.g. lost crop production), whereas others affect the natural world where the impacts are not directly (or even indirectly) priced in markets (e.g. reduced carbon sequestration or negative impacts on landscapes). The former set of impacts has traditionally been assessed with the discipline of agricultural/forestry economics and the latter in environmental economics.

Quantifying the impact of plant pests remains fundamentally difficult (Parker et al. 1999). The value of marketable goods can be expressed, but predicting changes in future values remain challenging (Baker et al. 2005). Non-marketable goods can be given values by utilising a range of environmental valuation techniques. There are, however, a number of issues with respect to the validity and accuracy of valuations from such methods (e.g. Parks and Gowdy 2013).

Gowdy (2007) suggests that the usefulness of applying methods for valuing non-market impacts depends on the following factors:

- Time. It can be applied when the effects of a particular event stabilise after a certain period. If effects are not stable and highly uncertain, monetising methods are not recommended.
- Scale. Such methods can be applied when effects concern a relative small clearly defined area. When impacts cover a larger area, the calculation of the impacts is much more complicated, both from a theoretical (aggregating impacts, longer time span, increasing uncertainty) and a practical point of view (involvement of large populations).

Nevertheless, the UK Treasury Green Book for Appraisal and Evaluation in Central Government (HM Treasury 2003) states that for Social Cost-Benefit Analysis, the valuation of non-market impacts is a challenging but essential element and "should be attempted wherever feasible" (p. 57). It further states that "The full value of goods such as …. environmental assets cannot simply be inferred from market prices" (p. 57), but that such important social impacts should not be neglected. It acknowledges that whilst the approaches can be complex, they are equally as important as market impacts.

Epanchin-Niell (2017) provides an excellent summary of approaches for determining cost-effective resource allocation. The approaches are described in Table 1, with comments added as to their potential for use in a time-constrained UK policy context. The first two are more relevant for UK policy making. The basic concepts of benefit–cost analysis are understood by policy makers, and the tool is applied both for ex-ante assessments of potential policy options but also ex-post in the event of an outbreak to assess the range of management responses.

As mentioned in the introduction, decision-makers are generally operating under severe time constraints. They are almost always non-economists, but they do have access to economic analysis. Given this dynamic, any economic analysis needs to be clearly understandable and be capable of explaining how issues of concern to decision-makers have been incorporated. Such concerns may relate to legal obligations or the public acceptability of options under consideration. These can be more clearly incorporated within the benefit–cost framework in a way more easily understood and accepted by non-economist decision-makers. Optimal control methods require high levels of mathematical understanding to appreciate their outputs and could be viewed as "black box" and outputs difficult to accept if the intuition behind them is lacking. They may support the broad direction of policy but are less likely to be utilised during outbreak events.

Reflecting that economic methods need to address the requirements of plant health decision-makers, the EU 7th Framework project PRATIQUE (2010) provided an overview of economic methods for the assessment of impacts due to pest and disease of plants. It outlined how to assess which economic methods were appropriate given a range of considerations for any analysis and outlined some key considerations to be taken into account when selecting the least complex method to

Table 1 Approaches to identify cost-effective responses (adapted from Epanchin-Niell 2017)

Approach	Use	Method	Potential for use in UK policy context
Benefit–cost analysis	Determine whether a project is a cost-effective investment	Determine whether benefits > costs	Currently undertaken. Generally ex-ante for policy options and ex-post for responses to pest outbreaks
Return on investment analysis	Prioritise allocation of budget across a set of projects (e.g. rank projects based on cost-effectiveness)	Select projects in decreasing order of benefit-to-cost ratio	Prioritisation is not applied in such a formal way but results from expert qualitative assessment of potential threats. A quantitative assessment such as RoI could improve resource allocation
Optimisation	Determine the efficient level of investment (i.e. maximise net benefits); design management to best achieve management objective	Apply optimal control, etc., within a bio-economic modelling framework	This approach is not used, and most decision-making is not focussed on searching for an optimal solution but on comparing practical options. However, recent government-funded research is considering how to develop a consistent approach to modelling impact using bio-economic models. Complex epidemiological models are often applied ex-ante and ex-post but rarely with an economic component
Optimal policy design	Determine optimal policy parameters (e.g. tax level, inspection rate) to alter private behaviour or decision-making to achieve management objectives	Apply dynamic optimisation, optimal control, etc., using a bio-economic model that accounts for private decision-making	Analysis of policy options in plant health tends to use benefit–cost assessments of policy options. Policy making does take account of private decisions, but not in the sense of

conduct an economic assessment of the impacts of plant pests and disease. It consists of three stages: (1) understanding the characteristics of the pest and receptor environments (hosts, habitats, industries, tourism, etc.), (2) defining the scope/scales of the economic assessment (impact range) and (3) choosing an appropriate assessment method, taking available resources into account. If the required scope does not match with available impact assessment methods, then the scope needs to be adjusted or the omissions properly recorded and clearly outlined.

They consider possible economic, environmental and social methods that can be applied, the impact range that they can cover, and assess the resources required for implementation in terms of data, skills and time.

It illustrates that the resource requirement for simple benefit–cost analyses are low for data, skills and time input which fits the view from Cook et al. (2017) of time-pressured decision-makers. This is particularly the case at the time of outbreaks but still leaves the possibility of using more sophisticated methods for longer-term questions of surveillance allocation and incentivisation mechanisms.

Plant pest and disease incursions are frequently associated with high levels of uncertainty requiring additional techniques to be able to aid decision-making as more information becomes available and uncertainty is reduced. This is very much the case with respect to pests and diseases that affect the natural environment.

There are methods to produce monetary estimates of the value society gets from ecosystems affected by pest and disease, and these could be applied for assessing environmental impacts in economic terms. However, they are not simple to apply, and there are a number of issues surrounding using the existing valuation evidence. Furthermore, application of those methods presupposes that environmental effects are known. PRATIQUE did not go particularly far in assessing environmental valuation methods as they were deemed too complex for official Pest Risk Assessment purposes and expert judgment was deemed the principal method for assessing environmental impacts. In the UK, there has been significant concern over the impact of tree pests and disease with the major rationale for public intervention being the impact of the social and environmental values provided by trees and woodlands. The debate about how best to include these values in decision-making is ongoing and discussed later.

2.1 Bio-Economic Models

One of the key determinants as to the potential success of a policy response is the rate of spread of a pest or disease. Thus, the response option needs to be assessed against a baseline impact model that incorporates the spread over time. There is a growing interest in using bio-economic models as a tool for policy analysis to better understand the relative effectiveness of management options particularly on natural resources and human welfare (e.g. Barbier and Bergeron 2001; Ruben et al. 2001). One of the potential benefits of these models is that one can get a better and more comprehensive indication of the feedback effects between human activity and spread. Bio-economic models have been applied to the sphere of invasive species and plant health for a number of years (e.g. Perrings et al. 2000; Cook et al. 2007). These models generally include spread models that include natural- and human-assisted spread and the impact on hosts as well as the effect of the management response. However, as Epanchin-Niell (2017) points out, such models require substantial empirical analysis to develop inputs and to parameterise and calibrate them to realistically represent systems.

It is recognised that studies investigating the impacts of invasive species require a substantial and continuous economic input at every stage, from risk assessment and prevention measures to long-term practical management (Finnoff et al. 2005). Bio-economic modelling integrates natural and human (economic) systems to express output in a monetised format. Changes to ecosystems, e.g. due to alien impacts, can alter human behaviour. Classic examples are provided by the plant pathogens *Phytophthora infestans* (potato blight) and *Hemileia vastatrix* (the cause of coffee rust) when, probably through trade, they spread from their native ranges. Such plant diseases can have significant economic and social impacts: *P. infestans* caused the Irish potato famine that resulted in the death of an estimated 1.5 million people and led to the emigration of approximately 1 million to America in the mid-nineteenth century (Donnelly 2001). Coffee leaf rust, caused by *H. vastatrix*, devastated the coffee industry in Sri Lanka that consequently became uneconomic and was eventually destroyed, to be replaced by tea in the

1890s (Kushalappa and Eskes 1989). Complexity is increased given that people can adapt to change—changing their behaviour or changing the environment. When people adapt, they alter the interaction with the natural system leading to further changes in the ecosystem, thus creating feedback loops and so interactions continue (Finnoff et al. 2005). Such feedback loops are a characteristic of bio-economic models.

Bio-economic modelling has a history of use in optimising resource management (e.g. Conrad and Clark 1995; Finnoff et al. 2010). Models can incorporate and quantify uncertainty and appraise the effectiveness of policy instruments (e.g. Sims et al. 2010) allowing the most economically profitable strategies that coincide with the most ecologically conservative policies to be identified. Caley et al. (2008) suggested that bio-economic modelling could be used to make more informed decisions during plant health risk assessments. Although regarded as challenging, the probabilities of introduction, establishment and economic, social and environmental impacts can simultaneously be evaluated by integrated, bio-economic modelling. A review of bio-economic models for the management of exotic species by Olson (2006) showed that the majority of bio-economic models focus on a single pest or single pathway. The aim of integrating economic and ecological factors is to get more precise estimates of the risk of invasive species on human and natural systems as well as account for interdependencies between economic and ecological factors (Finnoff et al. 2006). Such models are useful in predicting adoption of a policy, the impacts the policy will have and assessing the robustness of assumptions made through sensitivity analyses.

Given the difficulties in detecting the early stages of most infestation, and in particular before damage becomes visible, surveillance and monitoring measures, which are largely reliant on visual inspections and trapping, are generally only partially effective at reducing spread. In effect, the impact of the pest is a function of the spread rate which is driven by pest intrinsic growth (i.e. reproduction) and natural and anthropogenic dispersal rates and the density of host population (Aukema et al. 2008). The effectiveness of responses will slow down the spread to varying degrees and therefore stop or reduce the rate at which the damage and cost increase in time (Epanchin-Niell and Hastings 2010). In most cases, there is considerable uncertainty about the spread of pests and disease in time and the damage costs associated. Given the possible differences in geographic distribution and density of hosts, the rate of spread and consequently the magnitude of the impact are also uncertain.

In order to account for these uncertainties, stochastic bio-economic models have been used to predict impacts of pest and disease through simulated scenarios based on a range of key spread parameters. The stochastic simulation methods are commonly used in pest risk assessments and in particular the modelling of complex invasive infestations with uncertain ecological parameters and hence variability in impact (Cook et al. 2007; Kriticos et al. 2013). Use of Monte Carlo simulations is a preferred method in cases where the variability in growth and spread parameters is an important factor in determining the magnitude of eradication cost (Olson and Roy 2002; Epanchin-Niell and Hastings 2010). These methods allow users to compute numerical probabilities of the uncertain input parameters, such as likelihood of pest entry and establishment and/or rate of spread, thus allowing assessors to predict the range of possible impacts, and the most likely impact, over a specified probability distribution of input parameters (Epanchin-Niell et al. 2012).

> Cook et al. (2017) present a generalisable bioeconomic model that, they suggest, is needed to be able to present results to decision makers. They state:"Virtually all decision support people are time-pressured. To influence private or government action our methods of calculating benefits and costs must be expedient. Biosecurity staff seldom have the luxury of researching specific species in detail over months or years. Instead, they are usually asked to predict the economic, environmental and social impacts of threatening or newly-arrived species in areas they have not been observed in before; all within a matter of hours, days or (at best) weeks. Time is critical, particularly in the case of new species incursions, because the window of opportunity for successfully removing them is usually short. Moreover, the context to which a response effort is to be

made constantly changes due to external pressures, like political and economic cycles."

Cook et al. (2017) lay out the basis for a simple bio-economic model that can be deployed rapidly to predict economic impacts and thus reflect the needs of decision-makers. The scientific and economic concepts applied are relatively simple, but they are well documented and can form the initial foundation for decisions. Whilst their approach is practical and accessible, these authors recognise that it is just a snapshot in time and that decision support systems are being developed that can revolutionise the way information is presented to decision-makers such as those based upon interactive map-based technologies. They apply a generic biological model to predict impacts that incorporates the probability of each step of the invasion process (NB they do not advocate its use in every situation). This approach is justified based upon their experience that very little is known about these steps at the time decisions are made and that decision tools need to be explicit about the uncertainty at the same time as generating predictions with limited data.

The generic spread model is designed to capture the key ecological processes and grounded in ecological theory, to reliably simulate the invasion pathway using parameters either easily measured or estimated using knowledge of the biological classification and/or ecology of the pest concerned. Again, the amount of time and resource available is key in justifying this or other spread models. The model incorporates arrival, establishment (thus, it can consider pests yet to arrive in the location of interest), local population growth and spread which includes the potential for satellite populations to be generated (i.e. it combines short- and long-distance dispersal). The mechanism of spread is based upon diffusion models that incorporate the intrinsic rate of population increase and a diffusion coefficient. The pest grows over time following a logistic curve until the carrying capacity is reached. Satellites are generated at a given rate and will relate to natural phenomena (e.g. weather) or human behaviour (e.g. trade) to jump ahead of the initial outbreak.

This framework allows discussion between decision-makers around key attributes of the pest and the invaded system in the contexts that they work in—the risk within the entry pathway, entry points and therefore the potential for establishment, the effect of different spread rates in the new environment and so on. With a limited set of parameter estimates, they can at least make basic informed predictions of the possible outcomes of response options.

The economic impact focuses on the losses caused to plant products with market prices using a simple partial equilibrium model (i.e. the only part of the market affected is the pest host; substitute and complement products are unaffected). Parameter requirements are restricted to the gross value of production, costs of production, prices, yield impacts and demand elasticities (the sensitivity of demand to changes in other economic variables, such as the prices and consumer income). Cook et al. (2017) suggest that non-market impacts are addressed separately via methods such as multi-criteria analysis. They do not address the case prevalent in the UK, and likely in many other countries, when non-market estimates exist but infrequently in the precise context needed for plant health issues. For example, in the UK, there are estimates for the ecosystem service values provided by trees and woodland. However, these are mostly for the values associated with new woodland areas, whereas plant pests infer losses of services. It is well documented that losses loom larger than gains (Kahneman and Tversky 1979), and thus, such estimates would be underestimated. Further, since plant health is a relatively low-level environmental policy priority in many countries, the resources to fill this significant information gap are unlikely to be forthcoming. In the UK, it is difficult to assess impacts of tree pests and disease across different species since ecosystem service values associated with the will differ, perhaps significantly, but evidence for this is missing. Despite this, recent work in the UK (Sheremet et al. 2017) has shown that there is a willingness to support publicly funded tree disease programmes but that this is conditional on ownership and control measure.

3   Interactions with Policy

There is a strong economic argument for the public support of plant health policies, and this is now generally accepted at national-level decision-making. However, the degree to which it influences the efficient allocation of resources for plant health (or indeed the allocation of resources within the total plant health budget) is open to debate. Ward (2016) provided a view on the cost-effectiveness of the typical responses from those who manage plant pest and disease to a new outbreak based upon decades of managing public plant health inspection resources in the UK. Figure 2 shows the potential temporal lag from the pest entering and spreading through the landscape to stakeholder awareness and willingness to respond to manage the outbreak, i.e. the response is generally reactive rather than proactive. As the graph illustrates, this has consequences for cost-effectiveness (i.e. how much reduction in pest damage per plant health pound spent—the y-axis represents a "border", be it a national one or a business one, and thus, the spend could be public or private) of the available management options and therefore the potential for economic savings by investing in early action.

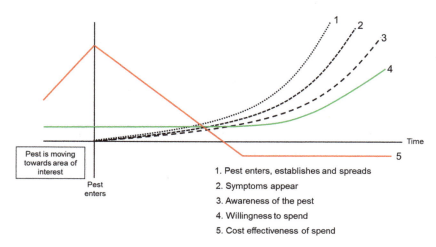

Fig. 2   Pest progression, willingness to respond and cost-effectiveness

The cost-effectiveness line shows the value linked to preventing the pest/disease arriving in the first place and implies a threshold for the spread beyond which management responses should cease and that adaptation measures ("learning to live with") are most appropriate (the point where the cost-effectiveness line dips below the horizontal axis). In reality, as outlined in the above sections, the cost-effectiveness curve is exceptionally difficult to predict given the state of knowledge regarding the context of the initial finding of the outbreak and the effectiveness of the management options available.

The previous section showed that there are limits to what conventional cost-benefit analyses can be expected to deliver. Cook et al. (2017) discuss the context of use of bio-economic models by those responsible for biosecurity and the short time frames that are typical of the decision-making process during outbreaks. It is not uncommon for almost all decision support people to be time-pressured, and they therefore require methods to estimate the benefits and costs of response options that can be delivered in a short time window. The decision-making period for those responsible for biosecurity matters is short, and they rarely have sufficient time to fully research the problem at hand before making an initial decision. During an outbreak, there is a need to rapidly predict the economic, environmental and social impacts. This could be from days to a few weeks, and an initial decision will be applied under conditions of significant uncertainty. This early decision-making is necessary since, in the case of new species incursions, the window of opportunity for successfully removing them is usually short. The resources available to the decision-makers also vary over time due to external pressures (such as political and economic cycles), meaning that priorities will inevitably change over such cycles.

A biosecurity decision-maker at a national level will likely have dedicated economic expertise available to produce cost-benefit assessments within the policy response frameworks. These economists can directly apply the methods described. They will input from a range of scientific disciplines to construct and run models and will need to describe the estimated impacts to policy leads. This raises the question as to the degree to which the policy leads are au fait with the economic modelling methods and the inherent shortcomings. Cook et al. (2017) suggest that "Public officials and community stakeholders charged with the responsibility of making these decisions are often naive about what science can and cannot say about complex systems. In these situations, policy makers tend to rely on a limited number of 'heuristic principles' (Kahneman and Knetsch 1992) to help them simplify the process of judgment". Such heuristics might give greater weight to particular locations or to economic impacts over environmental. To this end, it is important that such officials are aware of inherent shortcomings in order to account for the uncertainties within the decision-making process. This includes an awareness of how the decision might change as new information becomes available and, importantly, a willingness to change the decision in this light.

There are underlying economic factors behind the risk of outbreaks, rates and modes of spread, their impacts and the public and private strategies to manage plant pest and diseases (e.g. Perrings et al. 2000). These suggest that there are a number of critical factors in determining optimal (given the current level of information) responses. Some of these critical factors are:

(a) Prevalence when found: this is affected by efficacy (and effort) of the detection system, which includes the effect of asymptomatic characteristics of the pest/disease (Parnell et al. 2015). The prevalence could be beyond the point of no return, and thus, decision-makers could ignore eradication as a possibility at the outset and possibly even containment depending on other factors. Figure 1 shows the typical low level of impact in the earlier time period illustrating the problems of early detection before exponential growth of impact occurs. This may be the case for a publicly funded surveillance system, but, privately, some landowners may have limited incentives to report given the often severe regulatory requirements.

(b) Rate of spread of the pest/disease: this is a factor of the pest characteristics as well as the degree to which human activities contribute. At early stages of an incursion, this usually comes from the available literature and expert opinion.

(c) Impact per host: this can vary from yield/quality reduction to mortality effects or morbidity effects that lead to mortality through other means. Knowledge of the precise impacts of the pest/disease on ecosystem services is almost always imperfect.
(d) Value of host: this requires an understanding on how the impacts lead to changes in ecosystem services provided by the host and consequently on human welfare.
(e) Efficacy of control options (including size of the control area). Again, this is often uncertain with clear felling (in the case of tree pests and disease) often being the default option.

These factors (and others such as social acceptability and the regulatory context) all interact and play a key role in determining the economic efficiency of plant health interventions.

The decision-makers also engage with a wide range of stakeholders whose stakes may well put them in conflict with each other. Concerned industries encompass a wide range of sectors: agriculture, horticulture (food and ornamental crops), forestry, landscaping and management of parks and gardens (including local authorities)—resulting in a complex and heterogeneous set of commercial stakeholders. The stakeholder landscape is much more far-reaching and is dynamic, changing with time, geography and knowledge level. As a pest moves through different "stages" of an outbreak, different actors come into contact with it, have capabilities to identify or otherwise deal with it and take actions that affect its progress.

Thus, public decision-makers have a particularly complex and difficult task. Economics can assist with the framework for assessing options as long as there is clarity as to how the favoured policy response was derived and what the economic analysis includes and excludes. This also requires an understanding by all stakeholders that decisions are made under conditions of significant uncertainty that will never be fully removed.

Note

1. Fernald also suggested awareness of the public good nature of plant health:

   Last year the Chairman of the Board of Selectmen in a Massachusetts town refused to use any of the public money for the protection of trees along the streets from the canker worms, because the idea was 'agin natur'. This year that same man's apple trees are as bare of leaves as though a fire had run through his orchard, and therefore I am of the opinion that it will be 'agin natur' for that man to gather a crop of fruit from his trees this fall.

References

Aukema, B., Carroll, A., Zheng, Y., Zhu, J., Raffa, K., Moore, R., et al. (2008). Movement of outbreak populations of mountain pine beetle: Influences of spatiotemporal patterns and climate. *Ecography, 31*, 348–358.

Baker, R., Cannon, R., Bartlett, P., & Barker, I. (2005). Novel strategies for assessing and managing the risks posed by invasive alien species to global crop production and biodiversity. *Annals of Applied Biology, 146*, 177–191.

Barbier, B., & Bergeron, G. (2001). *Natural resource management in the hillsides of honduras. Bio-economic modeling at the micro-watershed level* (Research Report No. 123). Washington, DC: International Food Policy Research Institute.

Bell, S., Hamilton, V., Montarzino, A., Rothnie, H., Travlou, P., & Alves, S. (2008). Greenspace Scotland Research Report. Greenspace and quality of life: A critical literature review [Electronic Version]. *Greenspace Scotland; Transforming urban spaces; OPENspace; Sniffer.*

Binner, A., Smith, G., Bateman, I., Day, B., Agarwala, M., & Harwood, A. (2017). *Valuing the social and environmental contribution of woodlands and trees in England, Scotland and Wales.* Forestry Commission Research Report. https://www.forestry.gov.uk/pdf/FCRP027.pdf/$FILE/FCRP027.pdf. (Accessed December 13, 2017).

Bradley, A. B. (2012). Global change, global trade, and the next wave of plant invasions. *Frontiers in Ecology and the Environment, 10*(1), 20–28.

Caley, P., Groves, R., & Barker, R. (2008). Estimating the invasion success of introduced plants. *Diversity and Distributions, 14*(2), 196–203.

Conrad, J. M. & Clark, C. W. (1995). *Natural resource economics: Notes and problems*. Cambridge: Cambridge University Press.

Cook, D. C., Thomas, M. B., Cunningham, S. A., Anderson, D. L., & De Barro, P. J. (2007). Predicting the economic impact of an invasive species on an ecosystem service. *Ecological Applications, 17*, 1832–1840.

Cook, D. C., Wilby, A. & Fraser, R. W. (2017). *Improving plant biosecurity policy evaluation and prioritisation: The economic impacts of pests and diseases*. World Scientific Publishing Europe Ltd.

Crawley, M. J., Harvey, P. H., & Purvis, A. (1996). A comparative ecology of the native and alien floras of the British Isles. *Philosophical Transactions of the Royal Society of London B Biological Sciences, 351*, 1251.

Dehnen-Schmutz, K., MacLeod, A., Reed, P., & Mills, P. R. (2010). The role of regulatory mechanisms for control of plant diseases and food security—Case studies from potato production in Britain. *Food Security, 2*(3), 233–245.

Donnelly, J. S., Jr. (2001). *The great Irish potato famine*. Stroud: Sutton Publishing Limited. Xii 292. 07509 2632 5.

Epanchin-Neill, R. (2017). Economics of invasive species policy and management. *Biological Invasions, 19*(11), 3333–3354.

Epanchin-Neill, R. S., & Hastings, A. (2010). Controlling established invaders: Integrating economics and spread dynamics to determine optimal management. *Ecology Letters, 13*(4), 528–541.

Epanchin-Neill, R. S., & Liebhold, A. M. (2015). Benefits of invasion prevention: Effect of time lags, spread rates, and damage persistence. *Ecological Economics, 116*, 146–153.

Epanchin-Neill, R. S., Haight, R. G., Berec, L., Kean, J. M., & Liebhold, A. M. (2012). Optimal surveillance and eradication of invasive species in heterogeneous landscapes. *Ecology Letters, 15*, 803–812.

Fernald, C. H. (1896). The association of economic entomologists, address by the president: The evolution of economic entomology. *Science, New Series, 4*(94), 541–547.

Finnoff, D., Shogren, J. F., Leung, B., & Lodge, D. M. (2005). The importance of bioeconomic feedback in nonindigenous species management. *Ecological Economics, 52*(3), 367–381.

Finnoff, D., Settle, C., Shogren, J. F., & Tschirhart, J. (2006). Invasive species and the depth of bioeconomic integration. *Choices: The Magazine of Food, Farm & Resource, 21*(3), 147–151.

Finnoff, D., Lewis, M. A., & Potapov, A. B. (2010). Control and the optimal management of a spreading invader. *Resource and Energy Economics, 32*, 534–550.

Freer-Smith, P., & Webber, J. F. (2017). Tree pests and diseases: The threat to biodiversity and the delivery of ecosystem services. *Biodiversity and Conservation, 26*(13), 3167–3181.

Gowdy, J. (2007). Toward an experimental foundation for benefit-cost analysis. *Ecological Economics, 63*, 649–655.

Kahneman, D., & Knetsch, J. (1992). Valuing public goods: The purchase of moral satisfaction. *Journal of Environmental Economics and Management, 22*, 57–70.

Kahneman, D., & Tversky, A. (1979). Prospect theory: An analysis of decision under risk. *Econometrica, 47*(2), 263.

Kriticos, D., Venette, R., Koch, F., Rafoss, T., Van der Werf, W., & Worner, S. (2013). Invasive alien species in the food chain: Advancing risk assessment models to address climate change, economics and uncertainty. *NeoBiota, 18*, 1.

Kushalappa, A. C., & Eskes, A. B. (1989). Advances in coffee rust research. *Annual Review of Phytopathology, 27*, 503–531.

Lansink, A. O. (2011). Public and private roles in plant health management. *Food Policy, 36*(2), 166–170.

Liebhhold, A. M., Brockerhoff, E. G., Garrett, L. J., Parke, J. L., & Britton, K. O. (2012). Live plant imports: The major pathway for forest insect and pathogen invasions of the US. *Frontiers in Ecology and the Environment, 10*, 135–143.

Maas, J., Verheij, R. A., Groenewegen, P. P., de Vries, S., & Spreeuwenberg, P. (2006). Green space, urbanity, and health: How strong is the relation? *Journal of Epidemiology and Community Health, 60*(7), 587–592.

Ndeffo Mbah, M. L., Forster, G., Wesseler, J., & Gilligan, C. (2010). Economically optimal timing for crop disease control under uncertainty: An options approach. *Journal of the Royal Society, Interface/The Royal Society, 7*(51), 1421–1428.

Olsen, L. (2006). The economics of terrestrial invasive species: A review of the literature. *Agriculture and Resource Economics Review, 35*, 178–194.

Olson, L. J., & Roy, S. (2002). The economics of controlling a stochastic biological invasion. *American Journal of Agricultural Economics, 84*(5), 1311–1316.

Park, B. J., Tsunetsugu, Y., Kasetani, T., Kagawa, T., & Miyazaki, Y. (2010). The physiological effects of *Shinrin-yoku* (Taking in the forest atmosphere or forest bathing): Evidence from field experiments in 24 forests across Japan. *Environmental Health and Preventive Medicine, 15*, 18–26.

Parker, I. M., Simberloff, D., Lonsdale, W. M., Goodell, K., Wonham, M., Kareiva, P. M., et al. (1999). Impact: Toward a framework for understanding the ecological effects of invaders. *Biological Invasions, 1*, 3–19.

Parks, S., & Gowdy, J. (2013). What have economists learned about valuing nature? *A Review Essay. Ecosystem Services, 3*(2013), e1–e10.

Parnell, S., Gottwald, T. R., Cunniffe, N. J., Alonso Chavez, V., & van den Bosch, F. (2015). Early detection surveillance for an emerging plant pathogen: A rule of thumb to predict prevalence at first discovery. *Proceedings of the Royal Society of London B: Biological Sciences, 282*(1814). pii: 20151478. http://doi.org/10.1098/rspb.2015.1478.

Perrings, C., Williamson, M. H., & Dalmazzone, S. (2000). *The economics of biological invasions*. Cheltenham: Edward Elgar Publishing.

PRATIQUE. (2010). Review of impact assessment methods for pest risk analysis, PD No. D2.1. Author(s): Johan Bremmer, Tarek Soliman, Marc Kenis, Urs Schaffner, Monique Mourits, Wopke van der Werf, Alfons Oude Lansink.

Ruben, R., Kuyvenhoven, A., & Kruseman, G. (2001). Bioeconomic models and ecoregional development: Policy instruments for sustainable intensification. In D. R. Lee & C. B. Barrett (Eds.), *Tradeoffs or synergies? Agricultural intensification, economic development and the environment*. Cambridge, MA, USA: CABI Publishing.

Santini, A., Ghelardini, L., De Pace, C., Desprez-Loustau, M. L., Capretti, P., Chandelier, A., et al. (2013). Biogeographical patterns and determinants of invasion by forest pathogens in Europe. *New Phytologist, 197*(1), 238–250.

Sheremet, O., Healey, J., Quine, C., & Hanley, N. (2017). Public preferences and willingness to pay for forest disease control in the UK. *Journal of Agricultural Economics, 68*, 781–800.

Shogren, J. F., & Crocker, T. D. (1991). Cooperative and noncooperative protection against transferable and filterable externalities. *Environmental and Resource Economics, 1*, 195–214.

Sims, C., & Finoff, D. (2013). When is a 'wait and see' approach to invasive species justified? *Resource and Energy Economics, 35*(3), 235–255.

Sims, C., Aadland, D., & Finnoff, D. (2010). A dynamic bioeconomic analysis of mountain pine beetle epidemics. *Journal of Economic Dynamics and Control, 34*(12), 2407–2419.

Sims, C., Finnoff, D., & Shogren, J. (2016). Bioeconomics of invasive species: Using real options theory to integrate ecology, economics, and risk management. *Food Security, 8*(1), 61–70.

Treasury, H. M. (2003). *The green book: Appraisal and evaluation in central government*. London: HM Treasury, The Stationery Office.

Waage, J. K., Fraser, R. W., Mumford, J. D., Cook, D. C., & Wilby, A. (2005). *A new agenda for biosecurity*. Horizon Scanning Programme, Department for Environment, Food and Rural Affairs, UK.

Ward, M. (2016). Action against pest spread—The case for retrospective analysis with a focus on timing. *Food Security, 8*, 77–81.

Wolf, K. L., Measells, M. K., Grado, S. C., & Robbins, A. S. T. (2015). Economic values of metro nature health benefits: A life course approach. *Urban Forestry and Urban Greening, 14*, 694–701.

10

Stated Willingness to Pay for Tree Health Protection: Perceptions and Realities

Colin Price

1 Introduction

As waves of insect attacks and tree diseases have surged across Europe and North America in recent years, the question has naturally arisen, of whether and to what extent it is worth expending resources to control or mitigate their effects. In a rational world, that would depend on the benefits of ecosystem services that are lost through tree diseases and pests and on costs incurred on control.

Whether trees are killed or only weakened, ecosystem services are compromised. Loss of the so-named provisioning service of timber production, central to traditional forest economics, has a market price. But disease also affects ecosystem services lying outside the market. It may reduce them in line with depressed growth of stem volume—which particularly correlates with the rate of carbon dioxide fixing (Price and Willis 2011); or according to loss of leaf surface area: "Leaf area and

---

C. Price (✉)
Colin Price Freelance Academic Services, Bangor, Gwynedd, UK

© The Author(s) 2018
J. Urquhart et al. (eds.), *The Human Dimensions of Forest and Tree Health*,
https://doi.org/10.1007/978-3-319-76956-1_10

tree canopy cover is [sic] the driving force behind tree benefits" (Rogers et al. 2015, p. 31). Assessing such values has become part of the general discourse of environmental economics (Watson and Albon 2011). Some regulating services, like carbon dioxide fixing, water regulation and microclimate amelioration, bring changes which do have market prices. However, for cultural services like aesthetic impact—which is particularly important for street and park trees—economists have in recent decades resorted to eliciting people's stated preferences. These are monetised through their willingness to pay (WTP) for environmental improvements or willingness to accept compensation for environmental deterioration. The approach has been termed "the contingent valuation method" (CVM). Specific applications have been made to the effects on trees of insect attacks (Crocker 1985; Moore et al. 2011) and fungal diseases (Areal and Macleod 2006; Mourato 2010; Notaro and De Salvo 2010; Meldrum et al. 2013). For brevity, both types of event will be referred to as "tree disease" throughout this chapter.

People seem willing to pay not only for outcomes—control or mitigation of the *effects* of tree diseases—but also for the *processes* whereby results are achieved. Jetter and Paine (2004) found an "overwhelmingly" higher WTP for control of urban tree pests by biological than by chemical means. A similar, though non-monetised, preference was found for control of *Dothistroma septosporum* on forest trees in the UK (Fuller et al. 2016).

Strong criticisms have been made of CVM and similar methods (Sagoff 1988; Kahneman and Knetsch 1992a; Diamond and Hausman 1994; Clark et al. 2000; Hausman 2012). This chapter does not attempt to review systematically the problems of stated preference and their potential solutions, on which many thousands of papers have been written. Instead, it focuses on some major issues which are particularly important in the context of tree disease mitigation. It does also refer to non-disease CVMs, where these throw light on the validity of valuations. Some studies referred to represent pretesting, aimed at exposing possible problems, rather than delivering applicable results.

2   Stated Preference for What?

Left to their own devices, what aesthetic effects of tree diseases might people imagine? The nightmare scenario delivers a landscape in which live trees no longer feature: this scenario fits consistently with valuations of trees that aim to determine the total value of their role (Rogers et al. 2015). Yet the size and distribution of the resource threatened by a particular disease, and the extent of effects on it, will not be known to most questionnaire respondents.

Some envisaged outcomes may be based on past experience. In the UK, the key tree disease of the twentieth century was Dutch elm disease (*Ophiostoma novo-ulmi*). My own recollection is not so much of an altered landscape, but of individual, individually known, dying trees, becoming leafless, twigless, branchless and eventually being removed by sanitation fellings; of a more abstract sense of the landscape's having become impoverished; yet of sharp images of "a resin-clogged unseasonably early autumn, followed by a late or never-coming spring" (Price 2011, p. 79); of an uneasy expectation that the prospective loss of hedgerow ash in the wake of ash dieback disease (*Hymenoscyphus fraxineus*), in a landscape already lacking elm, may be more than an additive change.

Obituary books were written contemporarily, recording not just the biological factors, but cultural ones too (Wilkinson 1978; Clouston and Stansfield 1979). Mourato et al. (2010) explored more systematically the role of tree diseases in people's consciousness, for example how memories of the Dutch elm outbreak conditioned WTP for control of current disease outbreaks.

In the attempt to envisage the effect of a further tree disease outbreak, unclarity, uncertainty and misconception may all play a part. Even pathologists and entomologists may be unsure of biological effects and more so of their landscape consequences. Ash dieback may render whole stands leafless, yet individual hedgerow trees may survive in aesthetically uncompromised condition. No-one is sure. When *Waldsterben* [forest death] was much in public discussion in the 1980s, foresters were sometimes asked "what have these larch trees died of?" and they sometimes replied succinctly "autumn": ironic, since larch is particularly valued for its seasonal colour changes. The eventual deathly grey of larch killed by *Phytophthora ramorum* is markedly different, but the most notable visual consequence of the disease has been premature felling and subsequent replanting, invariably with a visually distinct broadleaf species or a non-deciduous conifer. Is this actually what questionnaire respondents foresee?

Visual aids may be employed to create a more precise—not necessarily more accurate—indication of the effects. Areal and McCleod (2006) used digital manipulation to emulate the expected visual impact of *Phytophthora* disease on one scene with hedgerows. Notaro and De Salvo (2010) presented states of urban landscapes affected by cypress canker (*Seiridium cardinale*) through alternative photographic visualisations.

But some respondents may project wider damage: the depicted scenes are emblematic of a conceived, more general ecosystem malaise. It may even be that merely "trying to do something about the problem", rather than achieving an actual outcome, is what people are willing to pay for when they support environmental protection.

By contrast, other respondents may under-imagine the widespread and long duration of impact. They may fail to grasp that the arrival of a new disease may profoundly affect nearly all landscapes of the kind described or illustrated. How many, in the early 1970s, conceived the near-total elimination of English elm from the UK's lowland landscape, or the absence of an effective restorative measure?

> Black poplar's frivolous leaves and birch's light-twigged grace meant
> that they lacked required solemnity; nor yet
> were lithe-limbed lime or cloud-crowned ash a fit replacement
> for that heavy, high and hanging silhouette. (Price 2011, p. 79)

I surmise that emerald ash borer in urban settings in North America is presently having such a previously unconceived effect.

Follow-up surveys may help in resolving these issues (Johnston et al. 2017). For example, responses to disease in the light of what subsequently occurred to landscapes might reveal that the impact was more—or less—dramatic than had been supposed beforehand. And yet there is a danger that retrospective questionnaires will provoke constructed memories.

3 What Consequences Are Included?

The valuation required of respondents is for the cultural services threatened by disease. In addition to the aesthetic effects directly experienced in known landscapes, respondents may also include "non-use" or, more accurately, "passive use" values (Weisbrod 1964). These accrue to people who would not encounter changed landscapes physically, but who would be psychologically affected by *knowing* that a landscape was threatened by change, that it *had been* changed, or that it had been *preserved from* change. The usually recognised components of passive use value are: existence value (arising from knowing of the ecosystem's continuing presence); bequest value (to future generations, though this ought to be broadened to include vicarious value to contemporaries); option value (arising from maintaining the possibility of visiting the protected ecosystem in future). Of the studies of pests and diseases mentioned above, most were conducted off-site, so would have included, or even focused on, these values.

But when asked, do respondents confine their valuations only to these cultural services? Seemingly not, as shown by earlier studies outside the realm of tree disease. They make explicit judgements of what they cannot reasonably be expected to know: participants in Environmental Resources Management's (1996) survey seemed to be seeking a balance which included timber production. And "once the public take the commercial advantage of forestry into account, when delivering judgements on environmental effects, the balancing role of cost–benefit analysis has passed into other hands" (Price 1997, p. 183).

Moore et al. (2011, p. 36) record that "Mogas et al. (2006) use choice modelling and CVM to value recreational and ecological services in an afforestation program. Under both formats, respondents are presented with hypothetical scenarios in which different levels of ecological services, *such as carbon dioxide sequestration*, are combined with recreational opportunities" (my emphasis).

Any questionnaire which begins with a preamble such as "Trees are being threatened by X. They perform multi-purpose functions such as locking up carbon, reducing floods …" invites a response that includes such values. Areal and McCleod (2006, p. 121) reminded respondents "of the social benefits that trees provide". They refer to their results (p. 128) as "our estimate of £55 for an individual's *total economic value* (TEV) of trees susceptible to *P. ramorum* in the UK" (my emphasis).

Equally, offering a list of elements to which respondents might ascribe value leaves them free to include a broad range of ecosystem effects. In the 1980s, a major UK forestry controversy arose from afforestation of Scotland's Flow Country. As the landscape is rarely visited, the questionnaire on WTP to prevent afforestation was trialled in a classroom setting (Price 1999a). Respondents were asked to allocate their WTP across several categories of ecosystem services. Regulating services, as well as supporting services like nutrient cycling that underpin general ecosystem functioning, were together assigned 30% of total value. Cultural ones (largely passive use values) achieved an only slightly greater 34%. As it happens, the principal species in this afforestation, lodgepole pine (*Pinus contorta*), is the one now most seriously affected by *D. septosporum*. One possible result of the disease might therefore be to reverse the ecosystem effects of afforestation. Could this accidental restoration of a substantially treeless landscape be interpreted as a *benefit* of tree disease?

Dunn et al. (2017) recorded WTP an 18% premium for assurance of good tree nursery biosecurity. This would embrace not only potentially better aesthetic outcomes for purchasers, but also perhaps a sense of acting responsibly in the containment of plant diseases, with *all* their potential ecosystem impacts.

But the further question should be raised, as to how many respondents in such surveys are actually in a position to value any of these services. Public perceptions of the value of ecosystem services cannot rival those of *real* subject experts, even those who would, after a lifetime's work on these services, acknowledge that they knew very little about them. Despite the incompleteness of their knowledge, experts are the best people to judge the physical significance of regulating and supporting services. Questionnaires to the public can only reveal *perception* of such values, not the values *actually delivered* by a complex web of processes. Who, among respondents to questionnaires, would actually know the welfare significance of a tonne of carbon dioxide in the atmosphere? Who would know even how physical carbon transactions with the atmosphere would alter quantitatively, if disease were to kill trees or slow their growth?

By contrast, expert valuations predict, within the limit of current knowledge, the effects on carbon fluxes, then calculate economic consequences using government-mandated carbon prices (DECC 2013) and government-recommended investment appraisal protocols (HM Treasury, undated). Carbon effects so calculated have been dominant in the assessed costs of *Dothistroma* (Quine et al. 2015). They are among the benefits the public—in this case the global public—would actually get from mitigation of tree disease.

Contrariwise, according to other calculations based on UK government enactments, tree disease may in special cases have a positive value: this results from trees' decaying or being burned at a time when discounted carbon prices are relatively low, and their replacements' reaching maximum rate of carbon sequestration when carbon prices are relatively high (Price and Willis 2015). This result has surprised even subject experts and could hardly be suspected by members of the public.

As well as these anthropocentric values, there is strong evidence that respondents include—or believe themselves to include—what have been termed intrinsic or inherent values, those arising without human awareness or even human existence. These are the least-understood of values (Stenmark 2002), and cannot, of their nature, be properly assessed by humans: this fact has not stopped many respondents attempting to do so and thus making judgements of value that definitionally lie outside their competence.

- In the Flow Country study (Price 1999a), respondents explicitly ascribed 24% of their WTP to such values.
- Moore et al. (2011, p. 35) "find that there is substantial support for protection of hemlock stands providing ecological services with very little human-use value".
- Dyke et al. (2018) take a viewpoint more explicitly aligned with that of a tree, though they do not try to place a money value on it, instead attempting to broaden human perspectives.

Even the possibility that such valuations may be included needs a pre-emptive design to isolate the required cultural values. This may be attempted by:

(a) designing response options so as to separate the categories being valued (Moore et al. 2011);
(b) requesting explicit breakdown into categories (Price 1999a);
(c) cautioning against inclusion of non-cultural categories, on the grounds that there are more accurate ways of dealing with them (the interviewer should be able to provide respondents with examples of how this is done).

Focus groups may be engaged to consider whether respondents will understand the task that is actually required of them and whether elucidation is better done from the outset, or as an invitation to reconsider responses during interview. They may help in refining the proposed questionnaire into a more explicit and comprehensible format, especially for investigating the rather challenging concepts discussed above. Yet, if focus groups do think that all ecosystem services should be included in the stated valuation, does that sufficiently validate doing it this way? I personally think not. Discussion at the interview site may help to elucidate unclear issues (Philip and MacMillan 2005), but leaves it open for respondents to reject the proposal, that there are some elements of valuation better left to experts. Questions might then explore what basis respondents might actually have for making such valuations, yet these risk alienating respondents by implying their limited competence.

4 What Relevant Biases Might Exist?

It has often been proposed, notoriously by Scott (1965, quoted in Boyle 2017), that hypothetical questions get hypothetical answers. Respondents need to feel that their answers will be consequential in real-world decision-making (Vossler et al. 2012; Johnston et al. 2017). Otherwise, the cognitive effort of making a considered, truthful estimate of WTP may not seem justified, and a casual, "for instance" figure may replace it.

By and large, however, respondents to CVMs seem disposed to please interviewers, supporting any suggested valuations (Holmes and Kramer 1995). They feel good just by saying "the right thing" and enhancing their image in the eyes of the interviewer ("will a large stated WTP make her admire my magnanimity?"). There is some evidence of such benign motivation (Börger 2013). More sophisticated questionnaire design may reduce the tendency (Blamey et al. 1999).

Strategic bias has been a major preoccupation in CVM. Archetypally rational people may incorrectly state their WTP or to accept compensation, so as to seek greater value from the environment, while paying little for it, or to increase compensation for any loss. If, for example, tree disease was a particularly significant local problem, and it was thought that counter-measures would be funded by central government, it would in this narrow sense be rational to overstate WTP for such measures, to enhance the probability of their implementation, without incurring much individual cost. Refer to Price (2017, Chapter 11) for a fuller discussion of strategic incentives.

Countering such tendencies, many respondents might consider telling the truth as desirable in itself, imparting utility. Strategic bias may also be reduced by describing biases and the possible costs of untruthfulness (Murphy et al. 2005). Sometimes interview designs have included taking oaths of (intended) truthfulness (Jacquemet et al. 2013): but this seems to undermine a desirable base assumption of inherent trust.

WTP may be adjusted downwards for strategic bias by rules of thumb, a factor of 2 being suggested by Arrow et al. (1993). List and Gallet's meta-analysis (2001) found different degrees of overstatement: experimental procedures have derived factors ranging from 1 to 3.

The literature generally favours "incentive-compatible" forms of question to defeat strategic bias, so that respondents bear the consequences of not telling the truth. Truthful laboratory revelation seems to be encouraged in "contribution games" where something is collectively provided only if collective contributions reach a threshold. A similarly constructed incentive appears in Brookshire et al.'s (1976) format of question, in which the entry fee for a scenic park—if it is opened at all—will be set at the mean of all bids. However, if respondents believe that their WTP exceeds the general mean, there is still an incentive to overstate.

5   Elicitation Modes

A consistent finding in CVM has been that willingness to accept compensation for deterioration of environment is significantly greater than WTP to restore the original condition (Coursey et al. 1987). In justice, it seems that those whose interests are actively damaged should receive acceptable compensation. For control of diseases, however, the WTP measure has typically been used rather than willingness to accept: perhaps because disease incidence has been seen as "part of nature". Walker et al. (1999) found that the disparity between WTP and willingness to accept measures increased markedly when human decisions, rather than disease, caused loss of street trees. This may result in different attributed value loss, where spread of disease is perceived to result from human negligence—which it sometimes does.

Much discussion in the wider CVM literature has focused on format of elicitation, with biases variously attributed to open-ended (what do you reckon?) and discrete choice (take it or leave it) formats. Arrow et al.'s (1993) influential survey recommended discrete choice (it seems to parallel both market and public choice situations). Johnston et al. (2017), however, suggest that the most appropriate format depends on context.

Areal and McCleod (2006) used the discrete choice format, posed in the context of contingent referendum, as will be discussed in a later section. Moore et al. (2011) favoured payment cards that present a range of options to tick, which they consider efficient and reliable. However, a bias is introduced if respondents treat the offered options as defining the acceptable range of answers. A concentration of responses at the top end of the scale led Price (1999a) to suggest that many respondents may consider themselves *particularly* environmentally sensitive, so are drawn to locate themselves near the top of the range, creating an upward bias.

6 Embedding, Scoping and Symbolising

When asked about WTP to protect against a specific tree disease, people may envisage protection against a range of, or all, tree diseases. It is easy to show the existence of effects like these, where a particular instance is "embedded" by respondents in a package of wider issues (Kahneman and Knetsch 1992a; MacMillan 1999). It is not so easy to remove them—McDaniels et al. (2003) refer to "alleviating", not "eliminating" such embedding.

Kahneman and Knetsch (1992a) ascribed WTP which varies little with the offered size of protection package to "purchase of moral satisfaction": that is, people felt good about themselves, just through signing up to supporting some public cause. They further argued that the embedding effect casts doubt on the reliability of CVMs and suggested that tests are needed, before CVM results are accepted, to demonstrate that embedding is not important (Kahneman and Knetsch 1992b). Others (e.g. Harrison 1992; Smith 1992) have disputed Kahneman and Knetsch's interpretation and conclusions. Nonetheless, WTP is often found to increase much-less-than-in-proportion to the scope of benefits offered, even by such proponents of CVM as Mitchell and Carson (1989).

A similar phenomenon, involving different amounts of one type of product, is called "scoping". Meldrum et al. (2013) found that the area of forest offered protection from white pine blister rust (*Cronartium ribicola*) did not significantly affect WTP for the programme: actual aesthetic outcome seemed less important than symbolic support for taking some action against the disease.

There are other more subtle manifestations of scoping. Moore et al. (2011, p. 49) considered that "this [result] may indicate that people want to ensure some minimum level of ecological protection beyond which there is little marginal value [for extra sites with non-human value]". Nielsen et al. (2007) found positive WTP for dead trees in a woodland—but only a few. They attributed this to a general perception that some deadwood is "good for conservation". But it seemed a token preference, with larger amounts of deadwood—as would often result from tree disease—less preferred.

Further inkling of the potential importance of symbolic effects was given by exploring motivations for WTP, as expressed in another pilot questionnaire with follow-up questions (Price 2000). Respondents were asked to tick reasons for the value they gave for *Rafflesia arnoldii* (a plant producing the world's largest flower). Only two respondents claimed to know the importance of the species. Nine respondents believed that genetic resources should be maintained intact, although the questionnaire made no offer to maintain genetic resources intact (*Rafflesia* was used as a peg on which to hang general concern for genetic diversity). Similarly, no-one can offer the banishment of all tree diseases. Six respondents suspected *that R. arnoldii* did not really exist, distrusting the questioner's integrity, yet expressing WTP for something fictitious that acted as an emblem of conservation. By contrast, four thought that the interviewer would not have asked these questions if the species were not important, so implicitly mirroring that judgement of importance. Two respondents confessed that they wanted to be seen as concerned about nature conservation, turning benefit back on the individual's psyche, rather than on the importance of the issue in question. Thirteen admitted that they had no knowledge on which to base a response.

Sometimes, respondents seem to have no clear idea even of what outcome (success/failure) a protection project would have, or what was the do-nothing alternative (Price 2001). Nevertheless, they felt they should support it, as a symbol of commitment to the environmental cause.

At the extreme, Blamey (1998) suggests that stated WTP may reflect the value of all environmental concerns, as the particular questionnaire offers the only known opportunity to express these.

7   Disembedding

All these results pose problems, particularly in scaling up results from the sample to the wider population. Purchase of moral satisfaction and symbolic values are not available to, nor therefore relevant for, those who have not participated in the questionnaire. This element should hence be excluded in scaling up. As for values of a wider range of diseases in which those of a particular disease are embedded, they do not apply even to the sample, as an assessment of the value of particular disease protection.

It seems implausible that people would be willing to pay for "pure" moral satisfaction gained through some hypothetical programme that "protects zero area": thus this value cannot be established and excluded directly. Extrapolation of a regression of WTP on area protected might conceivably find a meaningful intercept at zero area.

Attempts may be made to isolate unwanted elements by scripts defining the task: "what is it worth, in terms of landscape protected, to prevent the effects of [this particular] disease?" Or else, follow-up questions may try to tease out elements of the valuation and remove those not germane to the task (e.g. Areal and McCleod 2006). Moral satisfaction is further treated below, under its alias "warm glow".

The scoping effect may, however, matter less than might be expected. A script appropriate to scoping might be: "imagine that the effects represented in these images may occur in all landscapes of this type which you may visit or visualise". A suggested alternative conception of the whole scoping problematic is this: respondents interpret a small area, when this is offered for protection, as being the area they themselves would normally experience. Increments of area for protection will be "elsewhere" and bring benefit "to other people", whose well-being, if considered at all, is encapsulated in the "moral satisfaction" element. Additionally, it may be considered that there is a small and decreasing probability that successive increments to the area offered protection *could* be experienced by the respondent, so some rapidly decreasing value is attributed to them. This would account for the small response that has been found to increments of protected area and accommodates some mainstream criticisms of Kahneman and Knetsch's view. If this perhaps surprising suggestion is valid—and it stands in need of testing—then the WTP for the maximum area of protection is a reasonable integrated value as perceived by the respondent, but it would not much matter if a smaller but similarly valued area was presented or envisaged in the questionnaire. The result just needs to be derived from a valid sample of affected population, without further adjustment. If, however, a protection programme is not expected to cover the whole national resource, the further difficulty arises of mapping the particular protected area onto the respondent's ambit.

8 Information Biases—Positive and Retrograde

A contention often made during contingent valuation's early days was that respondents needed information about species or habitats under threat, before their WTP would yield "informed judgements" and therefore be [considered] valid (Bishop and Welsh 1993). The belief remains: "Studies clearly indicate that specific information about the item being valued is required in order to elicit credible responses to contingent valuation questions" (Boyle 2017). Much of what has been said above appears to confirm the need for accurate and even detailed information (Johnston et al. 2017). Uncertainties and misconceptions surrounding tree disease seem to enhance the case for improved information.

But imparting knowledge, particularly of a scientific kind, also has a retrograde biasing effect. It may affirm in the respondent's mind an "expert" role. By presenting a nosegay of scientific facts, information seems to legitimise respondents' focus on "other-than-cultural" valuation, whereas cultural values are actually the ones on which the respondents have legitimate expertise, based on their own perception: these are in fact the only values which such investigations should be designed to reveal; they are the ones on which scientific experts have no other information.

Information may even create unhappiness in the minds of respondents, by notifying potential bad outcomes of which they might otherwise never have become aware. Seventy-three percent of respondents to the Flow Country questionnaire answered that, if the habitat was "lost", they would feel worse off because of the information they had been given, about an area which they had not previously known existed. In a further pilot survey of responses to a red squirrel (*Sciurus vulgaris*) conservation programme, the negative feelings projected in the event that the programme failed were: guilt for not supporting the programme [1 respondent]; sadness for the impoverished resource [12]; sadness for the squirrel [7]; anger at human apathy [4] (Price 2001).

Most seriously, giving information about one particular issue, species, habitat or disease "headlines" it as a priority. de Bruin et al. (2014)

obtained survey evidence that information about tree diseases really does modify people's perceptions and priorities (and presumably, had they been asked, their WTP). The process of "informing" thus creates a respondent body that has thereby been made precisely unrepresentative of the wider population. And yet this very small subset's values will, in the normal course of stated preference valuations, be scaled up to the uninformed population, whose own values may actually remain unchanged or little changed (Price 1999a). The "informed value" approximates what the wider population's values would be, were they to be informed in a headlining manner. But they will not be so informed. And, if in due course they experience actual effects of disease, that experience will be set in the context of other environmental changes, in which it is not headlined but takes its place among equally unheadlined concerns.

Moreover, there are the passive use values that informed people might hold in mind and that may be damaged if disease strikes at a particular location, without information reaching everyone who might hold such values. The wider population may never directly encounter actual effects.

Admittedly, there could have been no escaping the catastrophic landscape consequences of Dutch elm disease in the UK during the 1970s, whether questionnaires had been applied or not. Similarly, the lethal effects of *P. ramorum* on Japanese larch (*Larix kaempferi*) on Welsh hillsides are dramatic and unmissable. But it is quite possible that the effects of *D. septosporum*—which slows growth in some pine species and kills others that are usually well out of the public view—might pass almost unnoticed, except if attention were drawn to it—directly for respondents, but not for populations that they supposedly represent.

Even an "informed" sample is likely, over time, to lose the focus created by the method of informing. Sensationalist newspaper information along such lines as "Ash dieback will devastate England's landscape" creates this week's environmental cause. Time passes: perhaps next week it has become old news and any questionnaire has moved from near memory, together with the values transiently constructed. Even if it remains there, people become accustomed to the new state and lose less utility over it (Helson 1948). Other tree diseases arise and prompt reconstruction of respondents' headlines. A mental account of "concern" limits the number of diseases or other problems that they can care intensely about: if intense care simply transfers to a new headlined disease or issue, the overall sense of loss may be transient.

Similarly, anxious unhappiness created by questionnaire, about something that may or may not happen, will not be felt by the wider population.

A political point is sometimes made that the population has "a right to know" about issues of public concern. But, if a right to know exists, information should be given contemporaneously to all the population and on all environmental issues. Otherwise, a questionnaire will headline concern, as discussed. Alternatively, information should seek to put the sample of respondents at the same level of awareness as that which the wider population would actually obtain and maintain, in the event of an actual disease outbreak.

These points do not seem to have been much addressed in the literature of CVM.

9   On Being a Good Citizen

A substantial and influential body of thought rejects the "consumerist" perspective of CVM when applied in the context of public choice. Instead, it is claimed that, when public choices are made, people take decisions as citizens acting for the good of the community (Sagoff 1988). Meldrum et al. (2013) found "responsibility" and "future generations" characterising attitudes of respondents to questionnaires on white pine blister rust. Within this perspective, there may also flourish a regard for "apple-pie-and-parenthood values"—those values that "every right-thinking person subscribes to" and (in the frequent experience of those applying questionnaires) on which they may refuse to express any WTP (Price 2017, Chapter 7). The quintessential attitude—"How can you possibly put a money value on a child's life or health?"—evinces belief in lexicographical values (Sagoff 1988). The contention is that certain values—justice, beauty, health and perhaps even tree health—always take precedence over personal consumption values of the kind represented in WTP questionnaires. The self-seeking consumer is reconstructed as a public-spirited citizen. Refusal to respond to questions, or to negotiate, thereby becomes a public virtue rather than a private vice.

In practice, of course, society does make such trade-offs, through budgets for legal aid, national health services, art and landscape conservation. No unbiased person advocates that the entire national economic effort should be used to save one child's life, not least because other lives would thereby be forfeited. Nonetheless, interviewers may face responses such as "This place means the world to me! It's wrong to value it in monetary terms. Diseases threatening it should be prevented, whatever it costs". The problem then is to render trade-offs with money in a manner acceptable within the citizenly context, and disarming intransigent positions.

To embrace this understanding, and to create an incentive-compatible structure, CVM questions have been recast as though within a political realm, paralleling real public choices. "What would you be willing to pay for …?" is transmuted into "Would you vote for a programme to control this tree disease effectively, if that required an £X increase of taxation?" Areal and Macleod (2006) framed their evaluation of tree disease as though in this context. Scripts within the questionnaire may encourage such a broadened mindset, for example: "Consider the pros and cons of the alternatives as a citizen from the point of view of your own welfare *as well as the whole society*" (my emphasis) (Ovaskainen and Kniivilä 2005, p. 384).

Varying X in such a "contingent referendum" allows identification of the value at which the electorate would be split equally between willingly paying and not willingly paying the tax. At this point, the benefit of disease prevention is taken to equal that tax (alternatively, as the mean WTP the tax). Typically, such formats elicit higher values, with fewer protest bids from those having a lexicographic turn of mind (Ovaskainen and Kniivilä 2005).

But, while eliciting WTP where other formats fail, the meaning and validity of citizen responses may still be questioned (Price 2006a). How much tree disease mitigation *would* the good citizen desire and be willing to pay for? And does that differ from what the truly good citizen *should* desire? Logically, the good citizen should desire the greatest welfare of the whole body of citizens, summed across all goods (in the broadest sense of Broome 1995) and over all time periods. The US forester Gifford Pinchot (1910) considered that the maximand ought to be the greatest good of the greatest number in the long run. On the face of it, this "right response" accords with responses to a pilot questionnaire (Price 2006b) that sought motivations for buying certificated timber: 10 out of 18 respondents said that they had "a general commitment to doing what I think is right".

Suppose, however, that intendingly good citizens are misguided or insufficiently informed, through some of the following factors.

1. How do they know what is "the right response"? People don't even necessarily know what action is best for *themselves personally*, because of ecosystem and economic complexity. For example, if lodgepole pine is killed by *Dothistroma* and (as is the current policy) replaced by more productive Sitka spruce (*Picea sitchensis*), this may bring *greater* timber revenues and *better* long-term carbon dioxide sequestration: society and its constituents are thereby benefited.
2. If, moreover, citizens see their role as valuing aesthetic goods *on behalf of other citizens*, how do they know those citizens' own values for lost or altered landscapes? Have they had representative and quantitative discussions with fellow citizens? Or do they assume that their own values are themselves representative?
3. Even if they knew accurately those values, and to the extent that their motives were genuinely altruistic, they would merely double-count what other citizens *themselves* say of their own values. Perhaps through this reasoning Areal and McCleod (2006) eliminated the 19% of WTP for disease mitigation that was attributed to altruistic motives.

10   Warm Glows and the Cold Chill of Reason

By contrast, the warm glow element (similar to Kahneman & Knetsch's "moral satisfaction") felt for supporting communitarian benefits, rather than personal consumption, might seem a genuine addition to welfare.

Of the WTP recorded in the Flow Country study, 9% was explicitly attributed to this feel-good value. In the seditious words of Larcom (1931), a nineteenth-century hymn-writer:

> The grass is softer to my tread
> *because* it rests unnumbered feet;
> sweeter to me the wild rose red
> *because* she makes the whole world sweet.
> (my emphases)

And yet such values might equally accrue to all things that contribute to other citizens' welfare. I am willing to pay something *extra* for a public health service that gives everyone the same access to health care as I could provide for myself by private insurance. But I would also pay something for everyone to be able to provide food—purchased as private consumption—for themselves and family: hence, shoppers for private consumption goods will sometimes buy extra food and donate it to food banks; hence, there exists WTP a premium for fair trade goods (Price et al. 2008). If a communitarian premium exists, it applies to a wide range of economic goods, and to apply it only to the target of our particular evaluation tilts the playing field unjustifiably in its favour.

"Glow fatigue" might be expected, parallel to the compassion fatigue arising with charitable giving. Even if a warm glow value were to be reported in undiminished strength by one individual in a series of questionnaires (no apparent fatigue), it may be that it is the same benevolent value, transposed successively to new situations. This is equivalent to the successive transfer of (hypothetical) WTP through a succession of headlined environmental causes (Price 1999a). Thus, there would be no "incremental glow" from introducing more decisions about which a respondent could feel good.

A good citizen would surely not have a warm glow from supporting ethically neutral or negative things? But because of misperceptions arising in (1)–(3) above, and because of retrograde information bias, a supposed "right decision" based on such considerations may actually lead to a "wrong decision", so that the warm glow itself is misperceived. Ironically, it is even possible that the warm glow felt for supporting "the right decision" may load a pro-disease-prevention case sufficiently to cause a wrong decision to be made. The warm glow is either irrelevant (if it does not affect adoption of what is deemed to be the right decision), or improper and delusional (if its inclusion causes the wrong decision to be made).

Some commentators might argue from an ethical standpoint that benefit based on delusion should not be considered a benefit. Utilitarian economists would on the whole say that utility is utility, whatever its source, and might note the pleasures derived from role-playing and immersion in fantasy worlds. There is at the least a case to discuss.

Whatever the validity of delusional values, however, the warm glow felt—like purchased moral satisfaction and the markups for passive use and symbolic values arising from receipt of more information—is an artefact of the questionnaire, experienced only by the sample of the population who are asked to respond. It does not apply to those who are not asked and who thus are not involved in constructing the decision. Citizen-orientated questionnaires may induce a more reflective attitude to communitarian values. But, in doing so, they make respondents unrepresentative of the whole relevant population, the great majority of whom have not been so stimulated into this more reflective state. Thus, the reconstructed attitude ought not to be included when scaling up questionnaire responses, to whatever is deemed to be the wider population: the one that bears the outcome of the decision.

Nonetheless, the following question might be asked. In the reality of representative democracy, political referenda and (perhaps) contingent referenda, what does the [supposedly] good citizen actually vote for? Answer: maybe self-interest, tinged with consideration for public well-being, inasmuch as that may be induced by question format. Why, after all, would anyone really vote for anything except their own best interest—given that such interest may include the pleasure gained by providing benefit to others? Perhaps the "citizen" format accidentally does retrieve the desired individualistic "consumer" valuation? (Price 2006a)—provided that respondents do not actually vote in the (supposed) manner of altruistic "good" citizens, with their irrelevant or improper warm glows.

To avoid citizens' trying to guess other citizens' values, and to prevent expert valuations' being usurped, the role of the respondent should be clearly defined in the questionnaire script. "Please answer just from the perspective of how you *personally* might experience the aesthetic effects of tree disease". And "We shall be asking a cross-section of the population similar questions, so they will have the opportunity to represent their own interests" (Price 2006a, p. 293). "We shall also be making another study to value the effects of disease on timber production, carbon dioxide levels, water supply etc.: please do not consider them in your answer". If such guidance is accepted—and its efficacy does need pretesting—presenting the questionnaire in this way may uncover an individual's personal valuation of how tree disease might affect cultural values, less trammelled by apple-pie-and-parenthood thinking and less distorted by a falsely perceived need to value all ecosystem services on behalf of the community.

With such modification and interpretation, contingent referenda do offer a better way of determining WTP than do conventional CVMs. They can be constructed as "choice experiments", in which a variety of environmental "goods" and avoided "bads" and tax payments are offered in different permutations: so doing implies the value of changing the environmental mix by small degrees and should somewhat reduce headlining and symbolic responses.

11   The Outcomes of Disease in Reality

Responses unrepresentative of the wider population may nonetheless remain. The problem arises partly because questionnaires necessarily focus on *issues* rather than *well-being*; on *change*, not *states*; on *processes*, not *outcomes*. Of course, change and process have salience. The long-lasting response to loss of English elm from the UK landscape through Dutch elm disease still affects those who witnessed it, not just because of a no-longer-existing idiosyncratic hedgerow presence, but because of the process whereby loss occurred and perhaps because of its perceived genesis in human negligence.

> They *used* to stand alone, aloof, in sombre lustres,
> Englishly ungaudy in their lofty looks;
> parasol to languid sheep and cattle clusters,
> high-rise home for flocks of disputatious rooks;
> (Price 2011, p. 79) (my emphasis)

"*Used to* …". Not just the loss or the causes of loss, but the poignant comparison with a former state (Price 2011).

But process and change can take too-important a role. Process engages some usually small proportion of the population who both receive information and obtain warm glows, separating them from the population whom they supposedly represent. Respondents to questionnaires and other consultees are thereby encouraged to focus upon what is presently changing, rather than upon how others not consulted—including future others—will feel about the long-term outcome. The UK government's own dispensation on treating future costs and benefits (HM Treasury, undated) is but a pale facsimile of the oft-demonstrated tendency of individuals to give little importance to what happens beyond a short period of the future—to the outcome, contrasted with the process of decision. Shackle (1958, p. 13) characterises the overwhelming psychological importance of the moment in which decisions are taken thus: "There is for us a moment in being, which is the locus of every actual sense-experience, every thought, feeling, *decision* and action" (my emphasis). And so the values engaged and created through the process of consulting with decision constructors (who include questionnaire respondents) take undue importance, compared with values later experienced by the outcome bearers. It is this latter group with which economic cost–benefit analysis should largely be concerned (Price 1999b, 2000).

Once again, decision constructors have values which should not be transposed to the outcome bearers. Repeated evaluations have demonstrated the effect of minority knowledge on scores ascribed to a view in North Wales, in which a castle features. The castle is first perceived as medieval (those knowing Welsh history might respond adversely to that), which generally elicits a positive response for its grandeur and picturesqueness. Once it is known that it dates from a later period, partly financed by profits of the slave trade, valuations change markedly. Such shifts of value have been described as "aesthetic disillusionment" (Foster 1992). Final, post-information values of "what is" are conditioned by the perceived process of "coming to be" (which few of the population, however, know about) and by the process of evaluation (which few, however, are involved in).

It may be that, confronted with the immediate question, respondents will prefer "apple-pie" biological control of pests and diseases (Jetter and Paine 2004). But for the outcome bearers the effectiveness of the control, howsoever achieved, may count for more. Once again, the ecosystem side effects of alternative control measures are best assessed via the understandings of experts.

"Tree disease" as a general term in questionnaires has connotations of losing symbolic, apple-pie and citizen values. But the real importance of individual tree diseases is a function of their physical outcomes and the actual aesthetic effects as subsequently experienced by human populations. Predicting the physical outcomes is the province of scientists, and their welfare significance is for economists to value—for example by tracing the changes in utility resulting from diminished timber production and carbon transactions (Price and Willis 2011) and from hydrological impacts (Price 2010), as affected by tree disease.

Among the observable aesthetic outcomes of *D. septosporum* in the UK would be the following:

- In Cannock Chase, some early felling of diseased stands has opened up a previously enclosed landscape, revealing an interesting topography of ridges and valleys. Heavy thinning to promote air circulation has improved access and increased light at ground level.
- In Thetford Forest, early felling has been followed by diversification of the species palette. Some early underplanting has also diversified age-class structure.
- In mountain areas, salvage felling may improve the external shape of the forest and the harshness of internal species boundaries.

But no-one asks about WTP for such *benefits* consequent on disease.

How can the utilities of such outcomes be projected, without engaging the distorting results of *process*? Partly, presenting states of landscapes through alternative visualisations should help to focus on the actual aesthetic effects of a particular disease, rather than asking abstract and general questions about WTP to prevent "tree diseases". Arguably, there is no need even to mention disease as a cause of different visual conditions. Then, the questionnaire respondents will be more typical of the not-consulted population across which their values are scaled up.

12 On the Validity of Stated Preferences

Given all the biases and ambiguities discussed above, is "some number better than no number" (Diamond and Hausman 1994; Kling et al. 2012)? Not necessarily, unless the study is carefully designed. Johnston et al. (2017) make design recommendations, though these would not entirely disarm the critiques offered in this chapter. The recommendations are numerous and exacting, entailing major research effort.

Among desiderata found, in the literature and from personal experience and reflection, are these:

- The questionnaire should be extensively pretested, to identify areas of incomprehension and miscomprehension.
- The sampled population must be that which would suffer the predicted landscape change.
- The possible changes should be presented neutrally, with no hint on how the interviewer considers them, and as far as possible without value-loaded terms such as "disease".
- The investigated personal and cultural values should be specified: wider and symbolic values should be explicitly excluded.
- Phrasing should emphasise that expressing only self-interest furnishes the required answer and is not reprehensible.
- The impression should be conveyed that the questionnaire is purposeful and the responses significant in selecting options.
- The format should be one of incentive compatibility that avoids strategic bias.

To the enthusiast (e.g. Carson 2012), these may be part of a presumed norm: to the sceptic (e.g. Hausman 2012), they may be asking too much.

For the particular case of tree disease, there is a further exacting requirement. Any information given should emulate the truly expected effect on the wider population, of whatever incidence of disease is predicted to occur. This involves difficult speculation and is not readily achieved, especially when different sub-populations would experience change in different landscapes, according to their location and the expected spread of disease. Unlike in many CVMs, the effect is not confined to a single site with a particular population catchment.

Validity tests of stated preference valuations are widely promulgated, as follows.

- Criterion validity concerns whether questions are incentive compatible: does it further respondents' interests, to declare their valuations truly? It might be supposed that incentives in a contingent referendum would parallel those in an actual referendum. However, Schläpfer and Hanley (2006) found that the median value revealed in a contingent referendum was much greater than the median tax liability implied by actual voting behaviour in a political referendum.
- Construct validity depends on results' being as expected. For example, WTP questionnaires show that people prefer less tree disease: but that is hardly a remarkable conclusion or addition to useful knowledge. Nor is it surprising, if the higher the suggested WTP, the fewer are the respondents who agree to it: this does not authenticate the actual magnitude of recorded WTP. Cynically viewed, either a study accords with preconceptions one might have held anyway, and you accept it, or it does not and you reject it. There seems small space for surprising, contrarian conclusions.
- Content validity is judged according to perceived appropriateness of format of questions and contextualisation of the entire survey. It is assessed by expert peers, so may tend to conservatism.
- Convergent validity is evinced when stated preference results are similar to, or explicably deviant from, those of other means of evaluation.

Bishop and Heberlein (1979) and Christensen (1989) showed that willingness to accept compensation somewhat exceeded values derived from revealed preference, which in turn somewhat exceeded WTP. This is also an affirmation of expectations constituting construct validity. But in this case respondents have the guidance of payments already made, when formulating their stated WTP. Thus, any claim of convergence is not based on entirely independent valuations.

13 Actual Outcomes Envisaged

Otherwise, one may try to divorce landscape value from a particular issue, by investigating day-to-day decisions that focus on landscape states. Figure 1 shows cost of travel to five Welsh landscapes, manipulated by the once-popular travel cost method so as to give cash values (Bergin and Price 1994). Landscape quality was judged subjectively on a scale similar to that of Fines (1968), which had been in use, providing consistent results, over a period of 20 years (Price 2013).

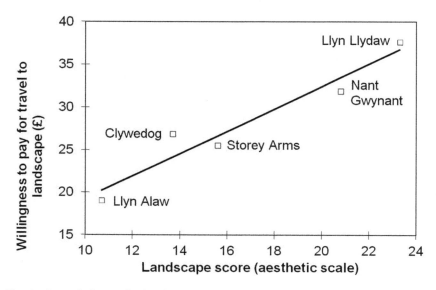

Fig. 1 Revealed WTP for landscape quality

Parallel research by Henry (1999) and by Ambrey and Fleming (2011) has related expert judgement of landscape quality, respectively, to house price and to differences in household income. Such monetary valuations for gradations of landscape quality could be transposed onto the projected effect of tree disease on landscape quality. Nowadays a similar process of transposition of stated preferences is advocated under the title "benefits transfer".

An illustrative evaluation based on that process indicated that the visual effects of tree disease might exceed, by an order of magnitude, the physical effects on timber production and carbon fluxes (Price 2010). This result arose without invoking the still-debatable value that probably arises from retrograde information bias and from headlining, symbolic, apple-pie and citizen values.

Stated preferences are, then, not the only way to assess the unmarketed values of cultural ecosystem services. On the contrary, stated preference approaches might gain in credibility, if they seem reasonably consonant with values derived in other ways, and thus evince convergent validity.

Acknowledgements The chapter has been extensively developed from a presentation made at the biennial conference of the Scandinavian Society of Forest Economics 2014 and published in *Scandinavian Forest Economics* (Price 2015). The author is grateful to the editor and members of the society for dispensation to make use of this material. He is also grateful to the two anonymous reviewers for suggestions on how to improve it.

References

Ambrey, C. L., & Fleming, C. M. (2011). Valuing scenic amenity using life satisfaction data. *Ecological Economics, 72,* 106–115.

Areal, F. J., & Macleod, A. (2006). Estimating the economic value of trees at risk from a quarantine disease. In A. G. J. M. Oude Lansink (Ed.), *New approaches to the economics of plant health* (pp. 119–130). New York: Springer.

Arrow, K., Solow, R., Portney, P. R., Learner, E. E., Radner, R., & Schuman, H. (1993). Report of the NOAA panel on contingent valuation. *Federal Register, 58,* 4601–4614.

Bergin, J., & Price, C. (1994). The travel cost method and landscape quality. *Landscape Research, 19*(1), 21–23.

Bishop, R. C., & Heberlein, T. A. (1979). Measuring values of extramarket goods: Are indirect measures biased? *American Journal of Agricultural Economics, 61,* 926–930.

Bishop, R. C., & Welsh, M. P. (1993). Existence values in benefit—Cost analysis and damage assessment. In W. L. Adamowicz, W. White, & W. E. Phillips (Eds.), *Forestry and the environment: Economic perspectives* (pp. 135–154). Wallingford: CAB International.

Blamey, R. K. (1998). Decisiveness, attitude expression and symbolic responses in contingent valuation surveys. *Journal of Economic Behavior & Organization, 34,* 577–601.

Blamey, R. K., Bennett, J., & Morrison, M. D. (1999). Yea-saying in contingent valuation surveys. *Land Economics, 75,* 126–141.

Börger, T. (2013). Keeping up appearances: Motivations for socially desirable responding in contingent valuation interviews. *Ecological Economics, 87,* 155–165.

Boyle, K. J. (2017). Contingent valuation in practice. In P. A. Champ, K. J. Boyle, & T. C. Brown (Eds.), *A primer on nonmarket valuation* (pp. 111–169). New York: Springer.

Brookshire, D. S., Ives, B. C., & Schulze, W. D. (1976). The valuation of aesthetic preferences. *Journal of Environmental Economics and Management, 3,* 325–346.

Broome, J. (1995). *Weighing goods: Equality, uncertainty and time.* Oxford: Blackwell.

Carson, R. T. (2012). Contingent valuation: A practical alternative when prices aren't available. *Journal of Economic Perspectives, 26*(4), 27–42.

Christensen, J. B. (1989). *An economic approach to assessing the value of recreation with special reference to forest areas.* Copenhagen: Skovbrugsinstituttet.

Clark, J., Burgess, J., & Harrison, C. M. (2000). 'I struggled with this money business': Respondents' perspectives on contingent valuation. *Ecological Economics, 33,* 45–62.

Clouston, B., & Stansfield, K. (Eds.). (1979). *After the elm.* London: Heinemann.

Coursey, D. L., Hovis, J. J., & Schulze, W. D. (1987). The disparity between willingness to accept and willingness to pay measures of value. *Quarterly Journal of Economics, 102,* 679–690.

Crocker, T. D. (1985). On the value of the condition of a forest stock. *Land Economics, 61,* 244–254.

de Bruin, A., Pateman, R., Dyke, A., Cinderby, S., & Jones, G. (2014). *Social and cultural values of trees in the context of the threat and management of tree disease.* York: Stockholm Environment Institute.

DECC (Department of Energy and Climate Change). (2013). *Updated short-term traded carbon values for policy appraisal.* London: The Stationery Office.

Diamond, P. A., & Hausman, J. A. (1994). Contingent valuation: Is some number better than no number? *Journal of Economic Perspectives, 8*(4), 45–64.

Dunn, H., Marzano, M., & Forster, J. (2017). *Consumer buying habits & willingness to support accreditation.* Paper presented at the IUFRO 125th Anniversary Conference, Freiburg.

Dyke, A., Geoghegan, H., & de Bruin, A. (2018). Towards a more-than-human approach to tree health. *The human dimensions of forest and tree health* (445–470). London: Palgrave (this volume).

Environmental Resources Management. (1996). *Valuing management for biodiversity in British forests.* Edinburgh: Environmental Resources Management.

Fines, K. D. (1968). Landscape evaluation: A research project in East Sussex. *Regional Studies, 2,* 41–55.

Foster, C. (1992). Aesthetic disillusionment: Environment, ethics, art. *Environmental Values, 1,* 205–215.

Fuller, L., Marzano, M., Peace, A., Quine, C. P., & Dandy, N. (2016). Public acceptance of tree health management: Results of a national survey in the UK. *Environmental Science & Policy, 59,* 18–25.

Harrison, G. W. (1992). Valuing public goods with the contingent valuation method: A critique of Kahneman and Knetsch. *Journal of Environmental Economics and Management, 23,* 248–257.

Hausman, J. (2012). Contingent valuation: From dubious to hopeless. *Journal of Economic Perspectives, 26*(4), 43–56.

Helson, H. (1948). Adaptation level as a basis for a quantitative theory of frames of reference. *Psychological Review, 55,* 297–313.

Henry, M. S. (1999). Landscape quality and the price of single family houses: Further evidence from home sales in Greenville, South Carolina. *Journal of Environmental Horticulture, 17*(1), 25–30.

HM Treasury. (undated). *The green book: Appraisal and evaluation in central government.* London: The Stationery Office.

Holmes, T. P., & Kramer, R. A. (1995). An independent sample test of yea-saying and starting point bias in dichotomous-choice contingent valuation. *Journal of Environmental Economics and Management, 29,* 121–132.

Jacquemet, N., Joule, R., Luchini, S., & Shogren, J. F. (2013). Preference elicitation under oath. *Journal of Environmental Economics and Management, 65,* 110–132.

Jetter, K., & Paine, T. D. (2004). Consumer preferences and willingness to pay for biological control in the urban landscape. *Biological Control, 30,* 312–322.

Johnston, R. J., Boyle, K. J., Adamowicz, W., Bennett, J., Brouwer, R., Cameron, T. A., ... Tourangeau, R. (2017). Contemporary guidance for stated preference studies. *Journal of the Association of Environmental and Resource Economists, 4,* 319–405.

Kahneman, D., & Knetsch, J. L. (1992a). Valuing public goods: The purchase of moral satisfaction. *Journal of Environmental Economics and Management, 22,* 57–70.

Kahneman, D., & Knetsch, J. L. (1992b). Contingent valuation and the value of public goods: Reply. *Journal of Environmental Economics and Management, 22,* 90–94.

Kling, C. L., Phaneuf, D. J., & Zhao, J. (2012). From Exxon to BP: Has some number become better than no number? *Journal of Economic Perspectives, 26*(4), 3–26.

Larcom, L. (1931). I learned it in the meadow path. In Anon (Ed.), *Songs of praise* (p. 199). Oxford: Oxford University Press.

List, J. A., & Gallet, C. A. (2001). What experimental protocol influence [sic] disparities between actual and hypothetical stated values? Evidence from a meta-analysis. *Environmental & Resource Economics, 20,* 241–254.

MacMillan, D. (1999). Non-market benefits of restoring native woodlands. In C. S. Roper & A. Park (Eds.), *The living forest: Non-market benefits of forestry* (pp. 189–195). London: HMSO.

McDaniels, T. L., Gregory, R., Arvai, J., & Chuenpagdee, R. (2003). Decision structuring to alleviate embedding in environmental valuation. *Ecological Economics, 46,* 33–46.

Meldrum, J. R., Champ, P. A., & Bond, C. A. (2013). Heterogeneous non-market benefits of managing white pine blister rust in high-elevation pine forests. *Journal of Forest Economics, 19,* 61–77.

Mitchell, R. C., & Carson, R. T. (1989). *Using surveys to value goods: The contingent valuation method.* Washington: Resources for the Future.

Mogas, J., Riera, P., & Bennett, J. (2006). A comparison of contingent valuation and choice modeling with second-order interactions. *Journal of Forest Economics, 12,* 5–30.

Moore, C. C., Holmes, T. P., & Bell, K. P. (2011). An attribute-based approach to contingent valuation of forest protection programs. *Journal of Forest Economics, 17,* 35–52.

Mourato, S. (2010). *Public knowledge, perceptions and who pays—Lessons from sudden oak death*. Paper presented at the Tree Diseases Conference, RASE Stoneleigh Park, 21 April 2010.

Mourato, S., Potter, C., Harwood, T., Knight, J., Leather, S., & Tomlinson, I. (2010). *Memory and prediction in plant disease management: A comparative analysis of Dutch elm disease and 'sudden oak death'*. Newcastle: Rural Economy and Land Use Programme.

Murphy, J. J., Stevens, T. H., & Weatherhead, D. (2005). Is cheap talk effective at eliminating hypothetical bias in a provision point mechanism? *Environmental & Resource Economics, 30,* 327–343.

Nielsen, A. B., Olsen, S. B., & Lundhede, T. (2007). An economic valuation of the recreational benefits associated with nature-based forest management practices. *Landscape and Urban Planning, 80,* 63–71.

Notaro, S., & De Salvo, M. (2010). Estimating the economic benefits of the landscape function of ornamental trees in a sub-Mediterranean area. *Urban Forestry & Urban Greening, 9,* 71–81.

Ovaskainen, V., & Kniivilä, M. (2005). Consumer versus citizen preferences: Evidence on the role of question framing. *Australian Journal of Agricultural and Resource Economics, 49,* 379–394.

Philip, L. J., & MacMillan, D. C. (2005). Exploring values, context and perceptions in contingent valuation studies: The CV market stall technique and WTP for wildlife conservation. *Journal of Environmental Planning and Management, 48,* 257–274.

Pinchot, G. (1910). *The fight for conservation*. New York: Doubleday, Page & Co.

Price, C. (1997). Twenty-five years of forestry cost–benefit analysis in Britain. *Forestry, 70,* 171–189.

Price, C. (1999a). Contingent valuation and retrograde information bias. In A. Park & C. Stewart Roper (Eds.), *The living forest: Non-market benefits of forestry* (pp. 37–44). London: The Stationery Office.

Price, C. (1999b). Stated and revealed preference analysis. In F. Helles, P. Holten-Andersen, & L. Wichmann (Eds.), *Multiple use of forests and other natural resources* (pp. 46–65). Dordrecht: Kluwer.

Price, C. (2000). Valuation of unpriced products: Contingent valuation, cost–benefit analysis and participatory democracy. *Land Use Policy, 17,* 187–196.

Price, C. (2001). Exact values and vague products? Contingent valuation and passive use value. In T. Sievanen, C. C. Konijnendijk, L. Langner, & K. Nilsson (Eds.), *Forest and social services—The role of research* (pp. 205–217, Research Paper: 815). Vantaa: Finnish Forest Research Institute.

Price, C. (2006a). Superficial citizens and sophisticated consumers: What questions do respondents to stated preference surveys really answer? *Scandinavian Forest Economics, 41,* 285–296.

Price, C. (2006b). Buying certification: Pigs in pokes, warm glows, and unexploded bombs. *Scandinavian Forest Economics, 41,* 265–272.

Price, C. (2010). Appraising the economic impact of tree diseases in Britain: Several shots in the dark, and possibly also in the wrong ball-park? *Scandinavian Forest Economics, 43,* 45–61.

Price, C. (2011). English elm. In *Tessellations*. Available electronically from the author at c.price@bangor.ac.uk.

Price, C. (2013). Subjectivity and objectivity in landscape evaluation: An old topic revisited. In M. van der Heide & W. Heijman (Eds.), *The economic value of landscapes* (pp. 53–76). London: Routledge.

Price, C. (2015). Perception of tree disease mitigation: What are people willing to pay for, and what do they actually get? *Scandinavian Forest Economics, 45,* 32–39.

Price, C. (2017). *Landscape economics* (2nd ed.). London: Palgrave Macmillan.

Price, C., & Willis, R. (2011). The multiple effects of carbon values on optimal rotation. *Journal of Forest Economics, 17,* 298–306.

Price, C., & Willis, R. (2015). Treating irregularities in carbon price and discount schedule: Resolving a nightmare for forest economics? *Scandinavian Forest Economics, 45,* 21–31.

Price, C., Cooper, R. J., & Taylor, R. C. (2008). Further thoughts on certification and markets. *Scandinavian Forest Economics, 42,* 66–74.

Quine, C., Marzano, M., Fuller, L., Dandy, N., Porth, E., Jones, G., … Brandon, G. (2015). *Social and economic analyses of Dothistroma blight management.* Edinburgh: Forest Research.

Rogers, K., Sacre, K., Goodenough, J., & Doick, K. (2015). *Valuing London's urban forest: Results of the London i-Tree Eco project.* London: Treeconomics.

Sagoff, M. (1988) *The economy of the earth.* Cambridge: Cambridge University Press.

Schläpfer, F., & Hanley, N. (2006). Contingent valuation and collective choice. *Kyklos, 59*(1), 115–135.

Shackle, G. L. S. (1958). *Time in economics*. Amsterdam: North Holland publisher.

Smith, V. K. (1992). Arbitrary values, good causes, and premature verdicts. *Journal of Environmental Economics and Management, 22,* 71–89.

Stenmark, M. (2002). *Environmental ethics and policy-making* (C. G. McKay, Trans.). Aldershot: Ashgate.

Vossler, C. A., Doyon, M., & Rondeau, D. (2012). Truth in consequentiality: Theory and field evidence on discrete choice experiments. *American Economic Journal: Microeconomics, 4*(4), 145–171.

Walker, M. E., Morera, O. F., Vining, J., & Orland, B. (1999). Disparate WTA–WTP disparities: The influence of human versus natural causes. *Journal of Behavioral Decision Making, 12,* 219–232.

Watson, R., & Albon, S. (2011). *UK National Ecosystem Assessment: Synthesis of the key findings*. Cambridge: UNEP-WCMC.

Weisbrod, B. A. (1964). Collective consumption services of individual consumption goods. *Quarterly Journal of Economics, 78,* 471–477.

Wilkinson, G. (1978). *Epitaph for the elm*. London: Hutchinson.

11

The Use of Rubrics to Improve Integration and Engagement Between Biosecurity Agencies and Their Key Partners and Stakeholders: A Surveillance Example

Will Allen, Andrea Grant, Lynsey Earl, Rory MacLellan, Nick Waipara, Melanie Mark-Shadbolt, Shaun Ogilvie, E. R. (Lisa) Langer and Mariella Marzano

1 Introduction

Tree health is an important factor for New Zealand's economic, social and cultural values. However, as a small island nation, New Zealand's forest conservation estate and primary production sectors are at risk from invading exotic plant pests (insects and pathogens). Moreover, the

---

N. Waipara—formerly Auckland Council.

W. Allen (✉)
Will Allen & Associates, Christchurch, New Zealand
URL: http://learningforsustainability.net

A. Grant · E. R. (Lisa) Langer
Scion, Christchurch, New Zealand

L. Earl · R. MacLellan
Ministry for Primary Industries, Wellington, New Zealand

scale of these biosecurity threats is escalating alongside the expansion of New Zealand's trade and tourism industries (Goldson et al. 2015). At the same time, there is a growing recognition that effective biosecurity in this challenged future calls for people to work together in a more coordinated, collective way, using partnership-based approaches rather than command and control approaches (Hellstrom et al. 2008). Successful biosecurity is inherently a collective endeavour. This is particularly true in terms of post-border operations where there are two main aims: (i) to reduce the likelihood of harmful pests and diseases from establishing; and (ii) to reduce or contain the harm from those that have established (MPI 2016). Activities in post-border operations include monitoring and surveillance, incursion response and sustained control. Policy makers and agencies cannot address New Zealand's biosecurity challenges in these areas without significant goodwill and collective action from Māori[1] and a range of key operational partners and associated stakeholders (including local communities).

A growing challenge for biosecurity management is to manage improved risk communication and engagement (RC&E) strategies that account for the range of different partnership and stakeholder perspectives (Enticott and Franklin 2009; Mills et al. 2011; Marzano et al. 2015). Recent research in this area highlights that agencies must step beyond a narrow technical operational focus that tends towards thinking only of RC&E as one-way delivery of information to engage more

---

N. Waipara
Plant & Food Research, Auckland, New Zealand

M. Mark-Shadbolt
Bio-Protection Research Centre, Lincoln University,
Lincoln, Canterbury, New Zealand

S. Ogilvie
Eco Research Associates Ltd, Burnside, Christchurch, New Zealand

M. Marzano
Forest Research, Edinburgh, UK

meaningfully with partners and key stakeholders and enter into dialogue based on participation, trust and understanding (Kruger 2011; Allen et al. 2014; Moser 2014). This recognises that managing an effective post-border biosecurity system—be it for surveillance, eradication or sustained control—relies on a range of activities that happen at a number of scales. Many activities are technical, but others are more about social processes (including management) and are difficult to observe or measure. Engagement and communication need to be viewed as an important part of the whole process; sharing and improving agencies' biosecurity intentions, actions and outcomes.

However, many managers do not have tools to involve the array of stakeholders in such a meaningful way. In particular, they do not have tools to easily set out, document and communicate complex pest and disease management programme activities and their intended outcomes (Allen et al. 2017). Against this background we explore the development of a rubric as a design and assessment framework for post-border biosecurity management. Rubrics are a device, originally used within education, to articulate key elements of a task or behaviour that can be evaluated against desired outcomes or demonstration of different levels of competence. Engaging practitioners in the development of rubrics, we propose, enables people working within a complex system (e.g. surveillance or eradication) to articulate and discuss the different social, technical and management dimensions (Allen and Knight 2009). In turn, this leads to a better appreciation of the different parts and how they interact. This contributes towards skills and pathways to help agencies to take an outcomes-based approach to assess and adapt their risk communication and engagement approaches to aid future response processes.

We begin this chapter by introducing the wider biosecurity setting, and the role of risk communication and engagement within that. We then outline our action research approach and introduce rubrics as an assessment tool. We indicate how action research and rubrics can be used in tandem to encourage a group to think more widely about the complex tasks and behaviours they may be engaged in. We then use the example of surveillance systems in biosecurity as a case study. We illustrate how the rubric can be used in practice by outlining how the authorship team tested its application against the potential introduction of myrtle rust[2] (*Austropuccinia psidii* (G. Winter) *Beenken comb. nov.*) in New Zealand. We end with a discussion of the benefits and challenges from using a rubric as a thinking technology, as both a process and a product.

2 Improving Risk Communication and Engagement in an Integrated Biosecurity System

New Zealand's biosecurity system has evolved to operate as a relatively integrated framework. As Jay and colleagues (2003) point out, the development of this system reflects New Zealand's history as a small island nation that has experienced significant biosecurity threats and problems. Biosecurity is implemented through a risk management system that involves many participants (MPI 2016). It involves different levels of government (national and regional), different biosecurity operations (surveillance, border control and pre- and post-border control) and different biosecurity objectives (control of economically significant pests and weeds, protection of native species and ecosystems, protection of health and the like) all working with some degree of interrelationship.

The Biosecurity 2025 direction statement for New Zealand's biosecurity system acknowledges a range of key players (MPI 2016). The Ministry for Primary Industries (MPI) is charged with overall leadership of the New Zealand biosecurity system and has a substantial operational role. At the same time, Biosecurity 2025 reminds us that an effective system will also require distributed leadership, in which other participants lead within their own parts of the system including active and general surveillance, incursion investigation and emergency response (MPI 2016). There is a wide range of other key stakeholders.

These include other government agencies and Regional Councils (local government). Māori or iwi (Māori tribal groupings) are partners with the Crown through Te Tiriti o Waitangi (1840), kaitiaki (guardians) of New Zealand's taonga (treasures) and increasingly have statutory roles in the management of natural resources. For any given pest or disease, there will also be a set of businesses and (conservation and production) land managers who have a responsibility and interest in managing risks directly related to their enterprises. Other key stakeholders include researchers (providing knowledge), and a wide set of community and other interest groups who come together to protect what they value.

The need for greater participation of stakeholders and communities in management of the environment and natural resources has become widely accepted in recent years (e.g. de Loë et al. 2009; Lockie and Aslin 2013). There are multiple rationales for this change in communication and engagement practice. It is in keeping with the democratic basis of local government internationally and in New Zealand that people should have an opportunity to take part in the decisions affecting them. Increasing stakeholder input can help ensure that the social and cultural impacts of decisions are considered (Hoppner et al. 2012), and better plans are generated (Burby 2003). There is also a realisation that scientific organisations and regulatory agencies are no longer regarded as the only source of what is to be considered in decision-making, and local and traditional knowledge needs to be recognised and considered as well (Weber et al. 2011).

Risk communication forms a key part of the biosecurity system in New Zealand and internationally, where a linear approach[3] to raising awareness of biosecurity risks is the most commonly utilised approach to increasing preparedness for newly introduced pests or diseases (Jay et al. 2003; Pegg et al. 2012; Perry 2014; Marzano et al. 2017). However, developing a closer interaction between agencies and other actors involved in these more collaborative biosecurity operations requires a different kind of understanding about risk communication and engagement. Typically, such differences from dissemination to interaction in communicating risk are described as one-way and two-way communication processes (Slovic 1986; Breakwell 2000; Frewer 2004). A growing challenge for biosecurity management is to also manage two-way risk communication and engagement strategies that account for multiple stakeholder perspectives (Mills et al. 2011).

Recent research in this area highlights that agencies must step beyond a narrow technical operational focus that tends towards thinking of communication as the one-way delivery of information to engage more meaningfully with stakeholders and take the opportunity to enter into dialogue based on participation, trust and understanding (Kruger 2011; Allen et al. 2014; Marzano et al. 2017). In this model, engagement and communication need to be viewed as an important part of the whole process; sharing and improving agencies' biosecurity intentions, actions and outcomes. As the continuum depicted in Table 1 points out, a primary difference between communication and engagement

Table 1 Seeing communication and engagement as a continuum (adapted from Morphy, n.d.)

Approach	What type of stakeholder engagement is required?
Partnership	• **Two-way engagement** as a priority.
	• Co-creation and co-development of activities as the goal/aspiration
Participation	• **Two-way engagement** within agreed limits of responsibility possible and appropriate in the particular task
	• The stakeholder can be viewed as one of the team. This can help to engage in delivering some tasks (e.g. co-design of operation)
Consultation	• **Limited two-way engagement**—Stakeholders are involved through discussion, but are not asked to be responsible for any element of delivery
"Push" communications	• **One-way engagement**—Used to tell stakeholders about agency or partnership activity
	• May involve broadcast information aimed at particular stakeholder groups—often using various Internet-based media channels
"Pull" communications	• **One-way engagement**
	• Information is made available, and stakeholders choose whether to engage with it, e.g. web pages

depends on whether the intent is to have largely one-way or two-way communication.

Our research was initiated via a New Zealand government contestable research funded programme—the Urban Biosecurity Toolkit—designed to deliver improved urban pest eradication biophysical and sociocultural technologies by looking at more targeted and socially acceptable approaches of dealing with biosecurity incursions (Scion n.d.). Two research objectives dealt with technological innovations in pesticide applications and in early detection while a third dealt with sociocultural innovation through agency-based learning. The starting point for this latter objective acknowledged that agency relations with stakeholders and communities relative to incursion response needed to be developed both during "wartime" (eradication and management) and "peacetime" (surveillance) operations. The development of the sociocultural research leading to this book chapter led to a joint MPI and research team project looking at improving risk communication and engagement in surveillance. This project enabled us to jointly reflect on the multiple elements that comprise an effective surveillance system, particularly one that involves partners and other key stakeholders reporting findings.

Our approach followed that of Mills and colleagues (2011), being a careful and considered engagement with agency professionals willing to reflect and learn about how they could create practical improvements in risk communication and stakeholder engagement. Such an engagement enabled views to be shared in a trusted environment that could critically reflect on current surveillance systems. We offered a process for engaging in a joint assessment that involved developing a rubric for identifying the elements of a surveillance system and measures of performance as a product of that engagement. We envisaged that such an integrated assessment could then be used as a device to facilitate a conversation about the performance of a surveillance system for a specific pest or disease concern, bringing in perspectives of other players or partners engaged in surveillance activities or operations.

3   Methods: Using Action Research to Co-Develop a Performance Rubric

Action research was used to guide the overall approach to learning from our case studies (Kemmis 2009; Allen et al. 2014). Action research is an approach that incorporates stakeholders as co-inquirers in processes designed to empower and change a set of circumstances in which a problem is identified. The researcher in these situations often plays a role of facilitator and collaborator rather than an expert observing and documenting phenomena (Kemmis 2012). Action research requires all those involved in the problem setting to improve their reflection and action. This approach links action, reflection, theory and practice to generate a practical solution or set of solutions (Reason and Bradbury 2008).

Our co-inquirers in the development of this performance rubric are agency staff involved in biosecurity operations. They comprise a multidisciplinary "team" of MPI scientific officers engaged in biosecurity surveillance and incursion investigations. Early discussions between the researchers and the team's manager led to an invitation to support the team in reflecting on and enhancing the agency's efforts in improving their surveillance systems. This recognised that an integrated system was required that linked both social and technical elements. A key idea behind this research is that one cannot be effective without the other (i.e. coming up with something that is technically very good won't necessarily be used if people do not like it, and vice versa). It also recognised that the research team brought complementary skills to the interactions in terms of communication and engagement expertise.

A rubric is an easily applicable form of assessment that can also be thought of as a guide or an evaluation tool that lists specific criteria for assessing performance. Rubrics are most commonly used in education and offer a process for defining and describing the important components of work being assessed (Allen and Tanner 2006). They are particularly useful in helping assess complex tasks or behaviours and are typically used by teachers or trainers to assess the competencies of learners. Rubrics offer an ideal approach to assessment that can lead to greater clarity of the area of competence being developed in a learner and therefore a basis for appreciating the desired elements of competence. Our approach was to engage our co-researchers in the design of rubrics that could capture key elements of a system of surveillance that could then be used as a basis for measuring the performance of that system. Co-developing rubrics was effectively a reflective approach to identifying elements of a system in which there was desired improvement. Although the format of a rubric can vary, they all have two key components (Andrade 2000):

- A list of criteria—or key elements that count in an activity or task; and
- Gradations of quality—to provide an evaluative range or scale.

Co-developing rubrics helps clarify the expectations that people have for different aspects of performance by providing detailed descriptions of collectively agreed upon expectations. They not only formulate standards for key areas of accomplishment, but they can be used to make these areas clear and explicit to all those with an interest in improving performance. It is important to involve programme participants, in our case MPI biosecurity surveillance and investigation team, in developing rubrics and helping define and agree on the criteria and assessment as something they feel is achievable and within the limits of normal operations. Different people within the system can offer different perspectives of what they do in the overall system to create a more complete picture of operations. This broad involvement increases the likelihood that different evaluation efforts can provide comparable ratings of performance. It is different from a simple checklist since it also describes the gradations of quality (levels) for each dimension of the performance to be evaluated.

Rubrics are often used to assess tasks and behaviours, but many authors argue that they can serve another, more important, role as well: When used by those undertaking the task or behaviour in question as part of a formative assessment of their works in progress, rubrics can instruct as well as evaluate (Reddy and Andrade 2010). Used as part of a practitioner-centred approach to assessment, rubrics have the potential to help learners understand the targets for their learning and the standards of quality for an assigned task, as well as make dependable judgments about their own work that can inform revision and improvement.

We have combined thinking about rubrics with science and technology studies concept of a boundary object. A boundary object is described as a visual representation that connects social worlds (Henderson 1991; Franco 2013). Typically, a free hand drawing or more openly conceptualised thinking platform is used to characterise a boundary object. Such an object enables a move away from rigidity of disciplinary modes of thinking to create a wider systems perspective of a problem situation (Checkland and Poulter 2006; Allen et al. 2017). In our case, the development of a rubric as a boundary object enabled people with different views of different elements of surveillance practice to come together to discuss, challenge and reconcile different appreciations of the same general concern.

We use the example of surveillance as a case study and demonstrate how a rubric can be used to develop an improved understanding around a general surveillance system. This understanding has linked broader social, technical and organisational functions that could then be appreciated as an integrated operational system.

4 Case Study Context

Surveillance is an essential component of New Zealand's biosecurity systems for the early detection of unwanted organisms and demonstration of freedom from pests and diseases. General surveillance is an important part of post-border pest and disease management. This type of surveillance (also known as passive surveillance and encompassing community surveillance) relies on members of the public, industry groups, plant or animal health professionals and their networks reporting suspected cases of plant or animal disease or the presence of a pest at their discretion (Hester and Garner 2012). General surveillance complements the targeted surveillance programmes managed by MPI as the lead agency for New Zealand's biosecurity system. As Cacho and colleagues (2012) point out, general surveillance cannot be controlled directly, rather it is activated by community communication and engagement programmes—with effectiveness dependent on a range of factors including pest attributes, the people involved and the wider sociocultural context of the area. While general surveillance has enabled the detection of many exotic organisms, MPI believes that there is room for improvement in how they engage New Zealanders to maximise the benefit of these surveillance systems and the value they offer (Earl et al. 2016).

5 Developing a Rubric

A draft rubric for improving a general surveillance system was developed during two workshops. The rubric was specifically developed from the perspective of how the MPI team could improve their surveillance system. Attendees consisted of two technical leads for the "animal" and "plant and environment" sectors and their managers, the project manager, the project executive and two independent engagement specialists. Prior to the workshop all participants were invited to write down and share two or three elements they considered essential to a well-functioning general surveillance system. These were subsequently discussed and collated into nine key elements during the workshop. It was noticed that different people emphasised different elements, depending on their area and experience. For example, some of the participants focused on the quality of inputs and how to get greater consistency of reporting records while others were concerned with the reporting experience of citizen observers and how to tailor reporting channels to suit their needs and enable feedback on reporting. This highlighted that both social and technical components are important to the functioning of surveillance. The rubric enables both to be recognised and evaluated.

The MPI attendees were then involved in defining an evaluative range or scale that could be used to assess performance in each element. Care was taken to formulate these in an appreciative way that encourages people to improve the outcomes of each performance dimension. The scale was defined using the labels: excellent, good and emerging. The workshop participants were then asked to describe how excellence would be defined for each of these elements. This provided an initial description of performance quality, and subsequent descriptions were also developed for "good" and "emerging" quality gradations. An abbreviated summary of the final rubric designed for a general surveillance system is shown in Table 2. This is adapted from the original rubric which looked at a general surveillance system specifically from an MPI perspective. This more generic rubric shown here has been slightly modified so the elements and descriptions can be used for consideration by a wider range of stakeholders.

The first three elements "Awareness and engagement", "Appropriate and well-functioning networks" and "Targets at-risk locations, industries and stakeholder groups" assess stakeholder awareness, engagement and to some extent motivation as well as efforts to enhance accuracy. It is assumed that early detection will occur if all relevant stakeholders are vigilant and willing to notify. However, the group identified that within each sector there naturally exists a network of stakeholders with varying levels of expertise who already exchange information about pests and diseases. The element "Appropriate and well-functioning networks" therefore aims to enhance this network to help enable accurate notifications. The element "Timely and accurate notifications" is a technical assessment of notifications made to MPI as the lead agency for biosecurity management. The communication channel between the notifier and MPI is assessed under "Notifying channels". To be effective, channels must be user-friendly, acceptable by the audience of potential observers and permit easy transfer of information, photos, videos and samples. The ability of MPI to respond effectively to notifications is captured specifically by "Notification data storage, retrieval and management". The "Resourcing" element looks at funding and other capacity issues such as training and skills. "Cross- and intra- organisational connections" focus on encouraging an awareness not only of direct actors in the system, but also of the importance of linking with a range of more indirect stakeholders. These include people without a direct role—but whose interests might be affected, and a range of related skill roles within key organisations such as policy makers, information technology (IT) teams and communication units. Finally, the performance element "Monitoring, evaluation and reflection" looks to indicate and assess the regular and meaningful evaluation of the surveillance

Table 2 An abbreviated summary of the final rubric developed by workshop participants (July–August 2016) to evaluate a general surveillance programme, modified to be applicable to a wider range of stakeholders

General surveillance system for biosecurity issues			
Elements	Excellent	Good	Emerging
Awareness and engagement	High audience awareness and motivation, consistent perception that MPI and partners handle biosecurity issues effectively. Builds on stakeholder engagement plan	Awareness and some established biosecurity activities, usually good perception of how MPI and partners handle biosecurity. Some stakeholder engagement planning	Low awareness and lack of biosecurity within the audience, poor perception of how MPI and partners handle biosecurity. No evidence of stakeholder engagement planning
Appropriate and well-functioning networks	Network clearly identified and each level engaged, consistently good trust and communication between the levels of the network	Network usually identified, some groups engaged, may be inconsistent communication	The network is not well identified, with few groups engaged and/or some distrust between the levels of the network
Targets at-risk locations, industries and stakeholder groups	Strategies consistently targeted to groups and locations likely to first incur new organisms. Wide participation by industry	Strategies are somewhat targeted, usually good level of participation from most areas, and key groups within the industry	Strategies are ad hoc and generalised. Participation is limited to certain individuals, groups or areas
Timely and accurate notifications	Notifications consistently timely and accurate, samples frequently available for diagnostics	Notifications are usually timely, and accurate, with samples usually available for diagnostics	Delayed/lack of reporting of incursions, low accuracy, samples often not available for diagnostics
Notifying channels	Users consistently report high satisfaction with the range of available notifying channels and all notifying channels provide good notifications	Users usually report satisfaction in channels, but prefer an alternative option and/or a notifying channel provides low quality notifications	Participants are reluctant to use the available reporting channel(s). ≥1 notifying channel does not provide useful notifications

(continued)

Table 2 (continued)

General surveillance system for biosecurity issues

Elements	Excellent	Good	Emerging
Notification data storage, retrieval and management	All core data is recorded and stored sufficiently and consistently, data is accessible and allows for meaningful interpretation	Data is usually recorded consistently with sufficient information yet may be difficulties in interpretation	Core data is stored inconsistently, not easily accessible, difficulties in interpreting the data meaningfully
Resourcing	Qualified, trained, motivated personnel, financial and other resources consistently available for surveillance activities. Ongoing training	Trained and motivated personnel. Usually sufficient resources available but stretched during peak times. Limited opportunities for ongoing training	Untrained/unmotivated personnel. Resources constantly stretched limiting ability to perform surveillance activities. No or little ongoing training
Cross- and intra- organisational connections	Consistently good relationships with other relevant teams in MPI and available resourcing for necessary activities. High awareness of the aims and functioning of general surveillance	Usually good relationships with other relevant teams yet can be a lack of resources available. General surveillance valued yet may be poorly understood	Low prioritising of MPI resources for general surveillance, relationships with other teams need developing. Low awareness and/or poor perception of general surveillance
Monitoring, evaluation and reflection	Performance of general surveillance is consistently monitored and assessed annually. Action taken to address areas of weakness	Performance is usually measured annually and action is taken to address the most important areas of weakness	General surveillance system is not regularly or incompletely evaluated, lack of action to address areas of weakness

system—involving stakeholders in assessing progress in both social process and technical elements.

All those involved recognise that this assessment (Table 2) represents a first version of a rubric that can be used to illustrate and discuss the key elements of a surveillance system, acknowledge the different actors involved and gain a better understanding of how their collective work contributes and performs to achieve the broader outcomes. In particular, the rubric enables those involved in its development an opportunity to consider a range of technical and social process elements in a system rather than to try and prioritise any one over the other. This increases the possibility that a rubric can be used to measure different areas of activity that contribute to the overall performance of a system even though they are doing different things.

6 Using a Rubric: Assessing the Surveillance System for Myrtle Rust

As a subsequent exercise, we (the authors) used the example of myrtle rust as a case study to examine how a surveillance rubric can contribute to assessment of a surveillance system. When we undertook this activity, myrtle rust had not been detected in New Zealand. Since this chapter was reviewed myrtle rust has been detected in a number of regions in New Zealand. Below we provide some background to the need to protect against the introduction of myrtle rust to New Zealand. This is followed by a brief illustration of how the rubric can be used for assessment, which we ran as a participatory exercise involving the co-authors as a multidisciplinary and cross-organisational team.

6.1 Myrtle Rust Context

The causal agent of myrtle rust (*Austropuccinia psidii* (G. Winter) Beenken *comb. nov.*) is an invasive pathogen of global significance that has rapidly expanded its international distribution and host range over the past decade. The pathogen was first described from common guava

(*Psidium guajava* (G. Winter)) in Brazil in 1884 and is believed to be native to South and Central America (Pegg et al. 2014). It was detected in Australia in 2010 and is now established along the east coast from southern New South Wales to far north Queensland (Carnegie et al. 2016). More recently that same invasive strain has been recorded in New Caledonia, Tasmania and Lord Howe Island (Pegg 2016). It was subsequently detected in New Zealand in May 2017 (although this chapter was submitted prior to this discovery).

Myrtle rust is known to have impact on young, developing tissue including infecting juvenile leaves and shoots, floral buds and/or fruit, with level of damage depending on the host (Tommerup et al. 2003; Zauza et al. 2010b). While infection can cause defoliation, twig mortality and abortion of flowers and fruits (Rayachhetry et al. 2001, citing Smith 1935), the rust affects different tissues on different species and some individual Myrtaceae plants have been found to have resistance to the damaging effects of the fungus (Zauza et al. 2010a). For some highly sensitive hosts such as rose apple *(Syzygium jambos)* plant mortality, including whole tree death, has been reported (Uchida and Loope 2009).

The long-term ecological implications of sustained rust outbreaks and damage are unclear for every host but some Australian experts have warned that severe damage to highly susceptible and vulnerable native species may even lead to extinction (Makinson 2016; Pegg 2016). Some of the more constructive representations of dealing with the disease include identifying and breeding plants with resistance to the disease and managing the disease through destroying infected plants before the disease spreads (Perry 2014). Measures for managing the risk of spread require very strict biosecurity practices (Pegg et al. 2012).

New Zealand Myrtaceae have been known to be potentially at threat from a biosecurity incursion of the rust for many years (Ridley et al. 2000). There is a growing acknowledgement that this will have negative economic, environmental and sociocultural impacts (Ramsfield et al. 2010; Clark 2011), including directly affecting Māori (Teulon et al. 2015). The rust is predicted to be able to survive in nearly all regions of New Zealand although warmer areas are more suitable. It poses a threat to our native myrtles such as rata (*Metrosideros robusta*), pohutukawa (*Metrosideros excelsa*), manuka (*Leptospermum scoparium*) and kanuka

(*Leptospermum ericoides*), as well as eucalypt growers and the honey industry.

As Bulman (2015) notes, the MPI has been active in putting several measures in place to reduce the risk of establishment. Shortly after its discovery in Australia import requirements of whole plants and cuttings from Australia were tightened. Cut flowers and foliage of the Myrtaceae family from New South Wales, Queensland and Victoria have been prohibited from importation into New Zealand due to the risk of transmission, and in February this ban was extended to Tasmania in immediate response to the discovery there (Bulman 2015).

6.2 Using a Rubric for Assessment

We (the chapter authors) brought an interdisciplinary and cross-organisational perspective to using the rubric—taking myrtle rust as our working example. We stress that our results are only intended to be indicative and were undertaken to provide a framework to help us think about the assessment process in practice. We used an iterative and facilitated approach. We began with those of us most knowledgeable about myrtle rust beginning the process and then involved the remaining co-authors in subsequent sessions that created further discussion and filled the table out more completely (see Table 3). We also shared successive drafts of this paper which enabled everybody to see where the discussion and table had got to in each iteration, and also provided opportunities for discussions on contested areas. The only guidance we used for our contributions into our example assessment was to: (i) look at the guide provided in the general surveillance rubric; and (ii) think of an example and indicator that could be used to demonstrate performance in that general area.

This initial exercise provided us with an appreciation of the utility of using a rubric to develop a discussion around the wider surveillance system. The framework proved useful in enabling different people (from our different stakeholder groupings) to add in a range of activities that they knew about, and collectively this helped everyone gain a better appreciation of the bigger picture. The approach supports an appreciative inquiry approach by asking people to think about an activity

Table 3 Assessment of the myrtle rust surveillance system in New Zealand (undertaken before its 2017 identification in the country)

Element	Assessment	Evidence
Awareness and engagement	Good (Better than good with pamphlets and training)	The industry is aware of the need for surveillance and its importance. There have been several workshops over the past 5 years. Ministry of Primary Industries (MPI) has worked with the forest industry and Department of Conservation (DOC) to promote industry and public awareness. Auckland Council (AC) has developed a pamphlet and also runs awareness days for community and environmental groups
Appropriate and well-functioning networks (linking stakeholders)	Emerging to good	A number of network members can be identified, in addition to MPI, DOC and AC. These include primary foresters, forest health specialists, researchers, forestry product end users, iwi, community groups. A Māori Biosecurity network is being formed. Some sectors of the network are more informed than others. Forest health specialists have annual training about emerging risks such as myrtle rust
Target at-risk locations, industries and stakeholder groups	Emerging to good	Myrtle rust signs and symptoms are included within the MPI forestry High Risk Site Surveillance (HRSS) programme. Auckland Botanic Gardens also maintain a programme. The awareness of other groups is being actively developed by biosecurity specialists in MPI, DOC and AC as resources and time permit
Timely and accurate notifications	N/A	Unknown. There have been no positive detections of myrtle rust. There are a low number of negative notifications annually
Notifying channels	Good	The primary notifying channel remains MPI's 0800 phone number. This is well known and regularly used for biosecurity threat reports. There are also a growing number of credible intermediaries who may receive initial reports and then pass them on via MPI's 0800 number. These include appropriate units in DOC and AC, and the forestry industry

(continued)

Table 3 (continued)

Element	Assessment	Evidence
Notification data storage, retrieval and management	Good	MPI has infrastructure set-up generically for surveillance notifications. This is currently being redeveloped to make it more accessible, interoperable and improve ability to monitor, interrogate and analyse data
Resourcing	Emerging to good	Myrtle rust has not yet been identified, but remains under monitoring surveillance. Funding is provided through the MPI High Risk Site Surveillance programme, and by other parties as outlined above
Cross- and intra- organisational connections	Good	Within MPI, functional relationships are actively maintained between relevant operational units. A Māori biosecurity network is being formed, and this will support both agency-Māori and cross-iwi communication
Monitoring, evaluation and reflection	Emerging	No formal reflection on myrtle rust is occurring. This is addressed through occasional workshops (e.g. a 2016 national workshop involved a range of stakeholders)

element and then to identify specific actions that they were aware of. They are also asked to provide evidence of those actions. Discussions around the validity of what constitutes evidence provide the opportunity for those involved to assess how well (or poorly) an action is being implemented. In turn, this enabled people to start their discussions about the bigger system with a more grounded understanding of what different groups were doing. As those involved repeat these assessments (and compare them), they gain an opportunity to identify and track where key activities may be reducing over time.

Our exercise served to highlight that rubrics will always need to be tailored to the context and the people involved, and be part of an ongoing process. For example, from our initial workshop we had written the first element as "awareness and motivation". However, we found that the term motivation meant very different things to different people—and so was difficult for people to agree on the level of performance. For this exercise, we changed motivation to engagement, which seemed to work in a more complementary way with awareness. The term motivation, in hindsight, seems to be better thought of as an outcome of awareness and engagement. If new stakeholders are to be involved, they will need to have the opportunity to redefine the rubric through these types of dialogic discussions. In this way, they will often be able to add to the performance descriptors and create a richer picture of how the system is operating—bringing in the perspectives of different cultural and knowledge systems. It helped us to collectively raise our awareness of these challenges to collaboration early in the programme, and in so doing we have begun working on ways to provide for better communication across different stakeholder groups, and foster a more coordinated approach to collective action.

7 Discussion

Through this process, a number of benefits of using rubrics to help design, evaluate and improve surveillance systems began to emerge. Although rubrics are a comprehensive performance measure for use with complex systems and behaviours, they are easy to use and explain.

They help multiple stakeholders make sense of how a range of different elements fit together in one system from different perspectives. This helps experts in different areas appreciate the importance of technical, social and organisational aspects—and how they link together. In turn, stronger collaborations support the range of research disciplines and end users to engage more effectively in discussions around the different areas involved. The early indicators of progress in these endeavours are supported by our reflections as a multi-author writing team who collectively cover agency, Māori and different disciplinary perspectives.

In this regard, rubrics should be seen as both a process and a product (Vogel 2012; Taplin et al. 2013). Their development involves practitioners and stakeholders in a facilitated dialogic process of analysis and reflection about the system in question. At the same time, the inquiry results in a table (or rubric) that articulates the key elements and their assessments for the project team and stakeholders. Developing a rubric should not be a one-off exercise to be used in the design (or evaluation) phase of a biosecurity initiative, but implies that those involved are entering into an ongoing process of learning and adaptive management that continues throughout the life of the initiative (Ison and Russell 2011; Cook et al. 2010).

Currently existing biosecurity programmes often fail to effectively engage their key stakeholder groups and emphasise one-way and top-down communication approaches that tend to see engagement as additional to other programme areas (Kruger et al. 2009) rather than embedded within them. The use of rubrics provides a tool that can help address this and provides a framework to guide more two-way or dialogic communication that is required to support more participatory and partnership modes. Developing the rubric helps people understand the bigger picture, and the way in which assessments are conducted invites people to explain in objective terms what is happening from their perspective, and supports an outcomes orientation.

Similarly, biosecurity programmes often lack participatory monitoring and evaluation components that could show the way to more effective engagement (Ison and Russell 2007; Kruger et al. 2009). Few biosecurity system surveillance evaluations provide any guidance or tools that help understand key stakeholder perceptions and expectations, or how to acknowledge the efforts of members of the public (Calba et al. 2015; Hester and Cacho 2017). As performance frameworks, rubrics such as that illustrated in this paper provide more informative feedback about strengths and areas in need of improvement than traditional forms of assessment do. A well-formulated rubric supports a partnership approach by helping stakeholders articulate system shortcomings in a concrete way—and provides guides to look for improvement, as well as ways in which elements are well managed. System practitioners and their partners can learn from developing and using a performance framework in a way they cannot learn from just measuring outputs or other narrow performance measures. Some newer evaluation frameworks take a more comprehensive approach which includes the need for more participatory approaches (e.g. Muellner et al. 2016), and in these cases rubrics can provide a useful tool to engage stakeholders in some of the needed conversations.

Our project took a broad view of evaluation as a starting point for helping the MPI team think about how to assess the wider surveillance system they operated within. The literature on evaluating surveillance systems is, in the main, limited to an assessment of one or two key elements in the wider system (Drewe et al. 2012). There is also a lack of consideration of the sociological aspects that may be involved for any particular setting (Calba et al. 2015). While an effective surveillance system is one that enables early detection, this effectiveness is most commonly only assessed after an incursion has occurred. There is, for example, little attention in the literature as to how we might demonstrate the presence (or lack of) appropriate surveillance capability. Measuring general surveillance during "peace time" is more difficult and is often done by measuring the quantity of notifications. However, number of notifications does not by itself provide a useful indication of vigilance across key stakeholder groups. The rubric element of intra-organisational connections (Table 2) provides an alternative point of evaluation which encourages us to look at the capacity of the networks to actively contribute to a surveillance system, and the quality of those networks to effectively detect an incursion. An evaluation framework which encompasses the multiple aspects of general surveillance was therefore helpful for those looking for appropriate performance measures that could be used and reflected on as achieving desired outcomes.

Moreover, for dynamically evolving contexts in which the effectiveness of a performance cannot be known in advance it is important to develop draft rubrics and then to periodically revise them. The context of our rubric development has been one involving different disciplinary perspectives and different organisational capacities coming together to articulate the many elements that make up a surveillance system. This gets away from a tendency to prioritise one element over another and recognises that the system works because so many elements contribute to its effective performance. We are not only involved in defining the elements of such a complex system with each other but are then able to use the development process to engage others in a broader assessment of the performance of that system. In our case, we have used the development of the rubric as a "thinking technology" where we have reflected on the process and product of rubric development. Here, we have found that the discussion (process) that goes into the development of the rubric is as important as the rubric itself once developed (product). In fact, we have found that the rubric acts as a boundary object or technology that can be used to mediate an ongoing conversation about performance (Ison and Russell 2007; Franco 2013), including discussion of what is desired—as well as discussion of different ways of achieving desired outcomes.

In these ways, rubric development can open up robust conversations about the way we see our biosecurity systems in the world and provide a space for people to offer evidence about the way these systems work. When people in a multi-stakeholder group demonstrate that they can hear the different perspectives in the group, then they are building capacity for trusting those they interact with. In this way, we create the likelihood that our diverse partners can see that they are being heard and included in the framework for system design and performance measurement. Effective risk communication ideally results from engagement with the key communities that you want to involve before, during and after emergency responses and involving them in the discussion on choices about a range of safety and wider surveillance options.

8   Final Comments

Rubrics help provide a means for reaching a shared understanding of what matters, and how to assess that in terms of what can be confidently regarded as good practice—and equally what can be agreed on as emerging practice. As Allen and Knight (2009) state, the process is neither complicated nor unduly time consuming, and benefits of collaborating are available to all participants. We have engaged with current literature and approaches to biosecurity risk communication and engagement. Through this we have recognised the need for tools that can support a range of engagement practices that can communicate complex pest and disease management programme activities and their intended outcomes. We have used a participatory action research approach to the development of rubrics as a design and assessment approach. As a tool, the development and application of a rubric can help agencies move beyond a narrow operational focus that deals with technical aspects to engage more meaningfully with partners and stakeholders and enter dialogue based on participation, trust and understanding. This can be seen to have contributed towards skills and pathways to help agencies use rubrics to assess and adapt their risk communication and engagement approaches.

Our approach sees the product of interactive processes as worthy of reflection, highlighting that processes are generative and open to review. A useful product can be operationalised but it also needs to be open to scrutiny at appropriate times (e.g. when engaging new stakeholders). A remaining challenge is to get agencies and other key stakeholder groups to see rubrics as both process and product and to move beyond a metric of evaluation to increase capacity to work more collectively. In turn, this will require operational biosecurity teams to move beyond their current focus on technical expertise to also include people with skills in surfacing other perspectives, listening and actively engaging with a range of partners.

Acknowledgements This chapter is made available under the Creative Commons Attribution 4.0 (CC BY) license with funding provided by Ministry of Business, Innovation and Employment for the Scion-led research programme: "Protecting New Zealand's primary sector from plant pests; a toolkit for the urban battlefield" [2016–18].

Notes

1. The indigenous people of New Zealand.
2. Myrtle rust has been detected in New Zealand subsequent to the completion of the workshops described here and the development of the accompanying tables included in this paper.
3. A linear approach refers to the one-way dissemination of information or knowledge that fails to appreciate that audiences are not a "blank slate" to have ideas written on but bring their own experiences, values and judgements to understanding risk, through which new information is interpreted.

References

Allen, D., & Tanner, K. (2006). Rubrics: Tools for making learning goals and evaluation criteria explicit for both teachers and learners. *CBE-Life Sciences Education, 5*(3), 197–203.

Allen, S., & Knight, J. (2009). A method for collaboratively developing and validating a rubric. *International Journal for the Scholarship of Teaching and Learning, 3*(2), 10.

Allen, W., Ogilvie, S., Blackie, H., Smith, D., Sam, S., Doherty, J., et al. (2014). Bridging disciplines, knowledge systems and cultures in pest management. *Environmental Management, 53,* 429–440.

Allen, W., Cruz, J., & Warburton, B. (2017). How decision support systems can benifit from a theory of change approach. *Environmental Management, 59*(6), 956–965.

Andrade, H. G. (2000). Using rubrics to promote thinking and learning. *Educational Leadership, 57*(5), 13–19.

Breakwell, G. M. (2000). Risk communication: Factors affecting impact. *British Medical Bulletin, 56*(1), 110–120.

Bulman, L. (2015). *Latest on myrtle rust*. Forest Health News, No. 255:1. Scion, Rotorua. Available at https://www.scionresearch.com/__data/assets/pdf_file/0006/45690/FHNewsApril2015.pdf.

Burby, R. J. (2003). Making plans that matter: Citizen involvement and government action. *Journal of the American Planning Association, 69*(1), 33–49.

Cacho, O., Reeve, I., Tramell, J., & Hester, S. (2012). *Post-border surveillance techniques: Review synthesis and deployment—Subproject 2d. Valuing community engagement in biosecurity surveillance* (Final report, ACERA Project No. 1004 B 2d). Melbourne: University of Melbourne.

Calba, C., Goutard, F. L., Hoinville, L., Hendrikx, P., Lindberg, A., Saegerman, C., et al. (2015). Surveillance systems evaluation: A systematic review of the existing approaches. *BMC Public Health, 15*(1), 448.

Carnegie, A., Kathuria, A., Pegg, G., Entwistle, P., Nagel, M., & Giblin, F. (2016). Impact of the invasive rust *Puccinia psidii* (myrtle rust) on native Myrtaceae in natural ecosystems in Australia. *Biological Invasions, 18,* 127–144. https://doi.org/10.1007/s10530-015-0996-y.

Checkland, P., & Poulter, J. (2006). *Learning for action: A short definitive account of soft systems methodology and its use for practitioner, teachers, and students.* Chichester: Wiley.

Clark, S. (2011). *Risk analysis of the Puccinia psidii/Guava Rust fungal complex (including Uredo rangelii/Myrtle Rust) on nursery stock.* Biosecurity Risk Analysis Group. Wellington, Ministry of Agriculture and Forestry.

Cook, D. C., Liu, S., Murphy, B., & Lonsdale, M. W. (2010). Adaptive approaches to biosecurity governance. *Risk Analysis, 30*(9), 1303–1314.

de Loë, R. C., Armitage, D., Plummer, R., Davidson, S., & Moraru, L. (2009). *From government to governance: A state-of-the-art review of environmental governance* (Final Report prepared for Alberta Environment, Environmental Stewardship, Environmental Relations). Guelph, ON: Rob de Loë Consulting Services.

Drewe, J., Hoinville, L., Cook, A., Floyd, T., & Stärk, K. (2012). Evaluation of animal and public health surveillance systems: A systematic review. *Epidemiology and Infection, 140,* 575–590.

Earl. L., Gould, B., Bullians, M., Vink, D., Acosta, H., Stevens, P., & Bingham, P. (2016). Strengthening New Zealand's passive surveillance system. In *Proceedings of the Food Safety, Animal Welfare & Biosecurity, Epidemiology & Animal Health Management, and Industry Branches of the NZVA, 2016 FAB Proceedings* (pp. 81–85).

Enticott, G., & Franklin, A. (2009). Biosecurity, expertise and the institutional void: The case of bovine tuberculosis. *Sociologia Ruralis, 49*(4), 375–393.

Franco, L. A. (2013). Rethinking soft OR interventions: Models as boundary objects. *European Journal of Operational Research, 231*(3), 720–733.

Frewer, L. (2004). The public and effective risk communication. *Toxicology Letters, 149*(1–3), 391–397.

Goldson, S., Bourdot, G., Brockerhoff, E., Byrom, A., Clout, M., McGlone, M., et al. (2015). New Zealand pest management: Current and future challenges. *Journal of the Royal Society of New Zealand, 45,* 31–58.

Hellstrom, J., Moore, D., & Black, M. (2008). *Think piece on the future of pest management in New Zealand* (76 pp). LECG, Wellington.

Henderson, K. (1991). Flexible sketches and inflexible data bases: Visual communication, conscription devices, and boundary objects in design engineering. *Science, Technology and Human Values, 16*(4), 448–473.

Hester, S., & Garner, G. (2012). *Post-border surveillance techniques: Review synthesis and deployment* (ACERA Project No. 1004 B). Australian Centre of Excellence for Risk Analysis, University of New England.

Hester, S. M., & Cacho, O. J. (2017). The contribution of passive surveillance to invasive species management. *Biological Invasions, 19,* 737. https://doi.org/10.1007/s10530-016-1362-4.

Hoppner, C., Whittle, R., Brundl, M., & Buchecker, M. (2012). Linking social capacities and risk communication in Europe: A gap between theory and practice? *Natural Hazards, 64,* 1753–1778. https://doi.org/10.1007/s11069-012-0356-5.

Ison, R., & Russell, D. (2007). Part1: Breaking out of traditions. In R. L. Ison & D. B. Russell (Eds.), *Agricultural extension and rural development: Breaking out of knowledge transfer traditions.* Cambridge: Cambridge University Press.

Ison, R., & Russell, D. (2011). The worlds we create: Designing learning systems for the underworld of extension practice. In J. Jennings, R. P. Packham, & D. Woodside (Eds.), *Shaping change: Natural resource management, agriculture and the role of extension* (pp. 64–76). Wodonga: Australasian-Pacific Extension Network (APEN).

Jay, M., Morad, M., & Bell, A. (2003). Biosecurity, a policy dilemma for New Zealand. *Land Use Policy, 20*(2), 121–129.

Kemmis, S. (2009). Action research as practice-based practice. *Educational Action Research, 17*(3), 463–474.

Kemmis, S. (2012). Researching educational praxis: Spectator and participant perspectives. *British Educational Research Journal, 38*(6), 885–905.

Kruger, H. (2011). Engaging the community in biosecurity issues. *Extension Farming Systems Journal, 7*(2), 17–21.

Kruger, H., Thompson, L., Clarke, R., Stenekes, N., & Carr, A. (2009). *Engaging in biosecurity: Gap analysis*. Canberra: Australian Government, Bureau of Rural Sciences, No. 39.

Lockie, S., & Aslin, H. J. (2013). Citizenship, engagement and the environment. In H. J. Aslin & S. Lockie (Eds.), *Engaged environmental citizenship* (pp. 1–18). Darwin, NT: Charles Darwin University Press.

Makinson, R. O. (2016). *Myrtle Rust epitomises a critical challenge for biodiversity conservation*. Paper presented at National Myrtle Rust workshop "The Threats Posed to New Zealand from Myrtle Rust—International Perspectives, Potential Impacts and Actions Required", 6–7th December 2016, Brentwood Hotel Conference Centre, Wellington, New Zealand.

Marzano, M., Dandy, N., Bayliss, H. R., Porth, E., & Potter, C. (2015). Part of the solution? Stakeholder awareness, information and engagement in tree health issues. *Biological Invasions, 17*(7), 1961–1977.

Marzano, M., Fuller, L., & Quine, C. P. (2017). Barriers to management of tree diseases: Framing perspectives of pinewood managers around Dothistroma Needle Blight. *Journal of Environmental Management, 188,* 238–245.

Mills, P., Dehnen-Schmutz, K., Ilbery, B., Jeger, M., Jones, G., Little, R., et al. (2011). Integrating natural and social science perspectives on plant disease risk, management and policy formulation. *Philosophical Transactions of the Royal Society B: Biological Sciences, 366*(1573), 2035–2044.

Morphy, T. (n.d.). Stakeholder analysis, project management, templates and advice. Source: Engaging Stakeholders—A strategy for Stakeholder Engagement. Available from https://stakeholdermap.com/stakeholder-engagement.html (Accessed February 23, 2017).

Moser, S. C. (2014). Communicating adaptation to climate change: The art and science of public engagement when climate change comes home. *Wiley Interdisciplinary Reviews: Climate Change, 5*(3), 337–358.

MPI. (2016). *Biosecurity 2025 direction statement*. Wellington: Ministry of Primary Industries.

Muellner, P., Stärk, K., & Watts, J. (2016). *Surveillance Evaluation Framework (SurF). Report prepared for Investigation and Diagnostic Centres and Response Directorate*. Wellington: Ministry for Primary Industries.

Pegg, G., Perry, S., Ireland, K., Giblin, F., & Carnegie, A. (2012). *Living with Myrtle Rust—Research in Queensland*. Department of Agriculture, Fisheries and Forestry, Queensland Government, Myrtle Rust Program. Available from http://www.wettropics.gov.au/site/user-assets/docs/6-myrtle-rust-r&d-workshop-pegg.pdf (Accessed February 23, 2017).

Pegg, G. S. (2016). *Myrtle Rust in Australia*. Paper presented at National Myrtle Rust workshop "The Threats Posed to New Zealand from Myrtle Rust—International Perspectives, Potential Impacts and Actions Required", 6–7th December 2016, Brentwood Hotel Conference Centre, Wellington, New Zealand.

Pegg, G. S., Giblin, F. R., McTaggart, A. R., Guymer, G. P., Taylor, H., Ireland, K. B., et al. (2014). *Puccinia psidii* in Queensland, Australia: Disease symptoms, distribution and impact. *Plant Pathology, 63,* 1005–1021.

Perry, S. (2014). *Myrtle Rust—Lessons from Australia*. Presentation to the Better Border Biosecurity Conference, New Zealand, Biosecurity Queensland, Department of Agriculture, Fisheries and Forestry. Queensland Government. Queensland, Australia. Available from http://www.b3nz.org/sites/b3nz.org/files/conferencefiles/day2key/140529%20Suzy%20Perry%20Lesson%20from%20Australia-%20Myrtle%20Rust.pdf (Accessed February 23, 2017).

Ramsfield, T., Dick, M., Bulman, L., & Ganley, R. (2010). *Briefing document on myrtle rust, a member of the guava rust complex, and the risk to New Zealand*. Scion Report, Rotorua.

Rayachhetry, M. B., Van, T. K., Center, T. D., & Elliott, M. L. (2001). Host range of *Puccinia psidii*, a potential biological control agent of *Melaleuca quinquenervia* in Florida. *Biological Control, 22,* 38–45.

Reason, P., & Bradbury, H. (2008). Introduction. In P. Reason & H. Bradbury (Eds.), *Sage handbook of action research: Participative inquiry and practice* (2nd ed.). London: Sage.

Reddy, Y. M., & Andrade, H. (2010). A review of rubric use in higher education. *Assessment & Evaluation in Higher Education, 35*(4), 435–448.

Ridley, G., Bain, J., Bulman, L., Dick, M., & Kay, M. (2000). *Threats to New Zealand's indigenous forests from exotic pathogens and pests*. Science for Conservation 142 (68 pp). Wellington, New Zealand: Department of Conservation.

Scion. (n.d.). *About the programme*. Webpage for the Biosecurity Toolkit Programme. Available from https://www.scionresearch.com/research/forest-science/biosecurity/urban-biosecurity-toolkit/about-the-programme (Accessed March 17, 2017).

Slovic, P. (1986). Informing and educating the public about risk. *Risk Analysis, 6*(4), 403–415.

Taplin, D. H., Clark, H., Collins, I., & Colby, D. C. (2013). *Theory of change technical papers: A series of papers to support development of theories of change based on practice in the field*. New York: ActKnowledge.

Teulon, D., Alipia, T., Ropata, H., Green, J., Viljanen-Rollinson, S., Cromey, M., & Marsh, A. T. (2015). The threat of Myrtle Rust to Māori taonga plant species in New Zealand. *New Zealand Plant Protection, 68*, 66–75.

Tommerup, I., Alfenas, A., & Old, K. (2003). Guava rust in Brazil—A threat to Eucalyptus and other Myrtaceae. *New Zealand Journal of Forestry Science, 33*, 420–428.

Uchida, J. Y., & Loope, L. L. (2009). A recurrent epiphytotic of guava rust on rose apple, *Syzygium jambos*, in Hawaii. *Plant Disease, 93*, 429.

Vogel, I. (2012). *ESPA guide to working with theory of change for research projects*. ESPA programme. Available from http://www.espa.ac.uk/files/espa/ESPA-Theory-of-Change-Manual-FINAL.pdf (Accessed February 23, 2017).

Weber, E. P., Memon, A., & Painter, B. (2011). Science, society, and water resources in New Zealand: Recognizing and overcoming a societal impasse. *Journal of Environmental Policy & Planning, 13*(1), 46–69.

Zauza, E., Couto, M., Lana, V., & Maffia, L. (2010a). Myrtaceae species resistance to rust caused by *Puccinia psidii*. *Australasian Plant Pathology, 39*, 406–411.

Zauza, E., Couto, M., Lana, V., & Maffia, L. (2010b). Vertical spread of *Puccinia psidii* urediniospores and development of eucalyptus rust at different heights. *Australasian Plant Pathology, 39*, 141–145.

Open Access This chapter is licensed under the terms of the Creative Commons Attribution 4.0 International License (http://creativecommons.org/licenses/by/4.0/), which permits use, sharing, adaptation, distribution and reproduction in any medium or format, as long as you give appropriate credit to the original author(s) and the source, provide a link to the Creative Commons license and indicate if changes were made.

The images or other third party material in this chapter are included in the chapter's Creative Commons license, unless indicated otherwise in a credit line to the material. If material is not included in the chapter's Creative Commons license and your intended use is not permitted by statutory regulation or exceeds the permitted use, you will need to obtain permission directly from the copyright holder.

12

Enhancing Socio-technological Innovation for Tree Health Through Stakeholder Participation in Biosecurity Science

Mariella Marzano, Rehema M. White and Glyn Jones

1   Introduction

Tree health has become an urgent issue across international, national and local borders. Increasing incidences of tree pest and pathogens have caused losses in commercial forestry and present a significant threat to the ecological integrity of many forests, impeding provision of ecosystem services (Boyd et al. 2013). Underlying causes include globalisation, widespread trade and increased movement of goods, along with climate change and socio-cultural expectations that there should be the

---

M. Marzano (✉)
Forest Research, Edinburgh, UK

R. M. White
School of Geography and Sustainable Development,
University of St. Andrews, Fife, UK

G. Jones
Fera Science Ltd., National Agri-Food Innovation Campus,
York, UK

availability of exotic tree species for horticulture (Brasier 2008; Everett 2000; Webber 2010; Perrings et al. 2010; Marzano et al. 2015). In response to biosecurity challenges, there has been increasing focus on statutory frameworks and monitoring and awareness raising schemes across a broad range of stakeholders (Dandy et al. 2017). Within the UK, this has manifested as a shift from government control, primarily through inspectors, to wider governance of tree health through the inclusion of additional actors such as the nursery or landscaping sector (Dandy et al. 2017). Tree health is thus recognised as being an issue of public interest, but also one in which public, private and third sectors are involved in detection and management.

It is recognised that the early detection of harmful organisms will be essential if new outbreaks are to be identified and contained or managed. Until now, biosecurity systems have largely relied on trained inspectors to find pests and pathogens through visual inspections of imported plants and wood-based product. Inspectors face huge challenges given the volume of traded plants and the finite amount of time and resources available to them. Robust science is required to develop new and innovative technologies, including those required for surveillance and eradication of tree pests and diseases. However, evidence suggests that technological innovations also demand interactions across academics, end-users and those involved in commercial development and marketing if they are to be 'fit for purpose' and have a public value (see Fagerberg and Verspagen 2009). Detection and management of both known and unknown agents of bacterial, viral, fungal and insect forms will require a complex suite of technologies to be used by a wide range of actors within novel detection regimes. The wider socio-cultural and ecological context of tree health demands that technology deployment and environmental complexities also be considered. Thus, whilst early innovation literature focuses on enhancement of single products and innovation in their application and use, we focus on socio-technological innovations of the broader system of early detection of tree pests and pathogens.

In this chapter, we discuss the process and review outcomes from our recent work designed to stimulate socio-technological innovation in tree health monitoring and detection through stakeholder engagement. The research project[1] aimed to facilitate the transition of five novel technologies for the early detection of tree pests and pathogens from concept to implementation. The technologies involved: (1) the use of volatile organic compounds in 'sniffer technologies' to identify the differences in volatile chemicals given off by infected plants; (2) multispectral imaging for the detection of biotic and abiotic stress in plants beyond the range of human vision; (3) an air and (4) a waterborne spore trapping network coupled with high throughput sequencing to identify new as well as known pathogens; and (5) improved trapping mechanisms for wood-boring beetle tree pests. Our aims as social scientists were both pragmatic and intellectual: to support the technological development, facilitate stakeholder engagement to encourage co-development of the technologies but also uptake of the final product and at the same time undertake novel research to investigate the role of stakeholder engagement in enabling socio-technological innovation.

1.1 Socio-technological Innovation

Socio-technical systems recognise the parameters and interactions between those actors and processes involved in the production of artefacts (technologies), their distribution (markets, networks and infrastructure) and their use (application domain) (Geels 2004). However, such systems are not clearly bounded and include networks of state, market-based and civil society actors (Smith et al. 2005). Socio-technical system changes not only require new technologies but also changes in markets, user practices, policy and cultural understandings (Geels 2004). System changes can occur incrementally or can flip into new configurations. Three types of socio-technological system dynamics have been identified: 'reproduction' is incremental change along existing trajectories of development; 'transformation' a shift in the direction of trajectory and 'transition' a discontinuous change to a new trajectory or system (Geels and Kemp 2007). These types of system innovation can form part of wider socio-technical transitions to sustainability, which imply large-scale social change in, for example, the provision of systems for production of food and energy (Geels 2010).

In addition to the rate and extent of system change described above, we can also differentiate the part or level within the system that might change. One model for socio-technological system innovation change is the multi-level perspective (MLP), which focuses on three levels within any societal system (e.g. tree biosecurity) and the interactions that occur between them to produce (radical) change. For example, one level involves *niches* within the socio-technical system required for radical innovation; another level the socio-technical *regime* that include institutions and rules (such as tree health policy); third, the external socio-technical *landscape*, including actors and groups embedded in networks and barriers to change (Geels 2010; Geels and Kemp 2007; Whitmarsh 2012). Whilst this MLP aligns well with some social theories, Geels (2010) recognises that other modes of recognising innovation transitions have validity, particularly in relation to agency and structure (see Smith et al. 2005). Actors do not act entirely autonomously but in the context of social norms and regulatory environments. Other related models include the creative factory model that focuses on aspects of a commercial firm, core innovation practices and national innovation support to define characteristics and levels of the system (Galanakis 2006). Nevertheless, this MLP provides an appropriate framework for us to examine how changing interactions amongst actors and groups might influence socio-technological innovation in the context of tree health. We explore how creating a technological niche (the THAPBI research programme—see Footnote 1) and disrupting the socio-technical regime (by changing the rules and interactions between scientists, policy makers and users) within the context of a changing socio-technical landscape (the imperative for early detection of tree pests and pathogens) could enable *transformative innovation*, if not a *transition* to a new system. This transformation is demanded because of the current paradigm of rapid social and ecological change, including environmental challenges such as climate change and biodiversity threats.

1.2 Stakeholder Participation, Engagement and Co-design

Participation can lead to more effective solutions, increase buy-in of outputs and create better-shared understanding and social learning amongst participants (Blackstock et al. 2007). Building relationships to increase 'preparedness' can thus influence the social acceptability of technologies needed to improve biosecurity responses. There are different forms and intensities of participation, ranging from the provision of information, through to consultation with, and then involvement of, participants and concluding with the empowerment of participants, often with the understanding that they may thus be enabled to take on responsibility themselves in some way (cf. Arnstein's 1969) 'ladder of participation'; (Davies and White 2012). 'Stakeholders' constitute individuals or groups with an interest in and usually some influence over an aspect of management (Prell et al. 2009; Dandy et al. 2017). Stakeholders differ depending on their degrees of interest, roles and responsibilities. The intention to engage stakeholders in tree health, and specifically in early detection of tree pests and pathogens, thus involves decisions around which stakeholders are invited to participate, at what intensity and for what purpose. In this project, we intended to change the way, in which people interacted; to move beyond the provision of information or consultation towards involvement and empowerment, in order to shift the current system of detection and monitoring; and to create a Learning Platform for future knowledge exchange. We preferred to use the term 'Learning Platform' to signify a large communication network but we drew on the concept of a Learning Alliance, an innovative multi-stakeholder process to encourage engagement and innovation (Sutherland et al. 2012). Learning Alliances are based on the concept that participation and collaboration are likely to engender better and more creative solutions for complex environmental contexts (Sutherland et al. 2012). Learning Alliances are centred on social learning and knowledge exchange and enable this change from an information mode of stakeholder engagement and participation to greater involvement, encouraging dialogue and interaction. Stakeholders build their capacity through active learning. Other important facets of this approach are to ensure that research generates real impacts and that Learning Alliance projects promote greater accountability, transparency and equity (Pahl-Wostl 2006; Sutherland et al. 2012).

Enhancing stakeholder engagement could have several influences on socio-technological innovations. The social structure, including actors and organisations, is generally studied more than the dynamics and interactions between social actors (Hekkert et al. 2007). Different actors might change the cognitive rules limiting scientist focus to particular directions; cross-actor discussion might shift normative rules and policy makers might be influenced to redirect national strategies and change formal and regulative rules (Geels 2004).

2 Methods: Creating the Learning Platform

We mapped stakeholders involved in the detection of tree pests and pathogens, drawing on previous research (Dandy et al. 2017) but also ensuring applicability by brainstorming across our project team. Further identification of stakeholders occurred through snowball sampling (Bryman 2001). Two of the authors were embedded in past and multiple ongoing projects on tree health and so brought an existing awareness of activities and groups whilst being open to new constellations, and the project's Expert Advisory Group offered additional suggestions. We developed a database of stakeholders, acknowledging that some roles were multifunctional and crossed group boundaries (e.g. policy, non-government organisations, industry and woodland managers), ensuring that all identified roles were represented by at least one individual, but also recognising that not all individuals could be regarded as 'representatives' of their groups. Our stakeholders were interested participants, and we accepted the messy context of stakeholder engagement across space and time. We appeared to reach saturation in the second year of the project, with some new individuals emerging but few additional groups or roles.

These stakeholders were engaged in different ways throughout the project. In our (social scientist) dual roles as facilitators and researchers, we used a range of innovative engagement tools to attract interest from a wide range of stakeholders, to encourage scientists to present their technological ideas and developments in appropriate ways for a diverse audience and to elicit responses and active engagement from stakeholder participants. We aimed to create a context for co-design, in which scientists and stakeholders could freely discuss and develop ideas in theory and practice. The Learning Platform helped promote our participatory and interdisciplinary approach to technology development and implementation. The Learning Platform was supported firstly by annual Learning Platform workshops, designed to engage various groups across the country and to facilitate different types of interaction between scientists and other stakeholders. Activities included a 'Dragon's Den' approach in which scientists had to pitch their technology in three minutes and receive audience feedback; Pecha Kucha[2] presentations of technologies development; short videos illustrating technology trialling or use in the field; a lunchtime technology fair showcasing relevant technology instruments; keynote listeners who provided a summary and their own professional perspective on discussions held throughout the day; and World Café discussions.[3] We produced detailed reports documenting discussion points for each workshop to create a shared record of interaction.

In addition to the backbone of Learning Platform workshops, we encouraged Socio-technological Learning Laboratories (SLLs). SLLs were intended to be a participatory process to facilitate knowledge exchange and collaboration between the teams developing specific technologies and stakeholders interested in those technologies. These small networks would then feed in their experiences to the larger Learning Platform. It was envisaged that each technology team would organise engagement with relevant groups of stakeholders, but scientists reported a lack of time, confidence and capacity to do so. We therefore focused mainly on creating opportunities for the technology developers within the project team to meet key stakeholder groups such as nurseries and inspectors in their places of work to promote a better understanding of technology needs and constraints in practice. For example, we undertook a visit to Heathrow, Southampton docks and at several horticultural nurseries where the project team learnt more about the existing operational structures and identified how their technologies might enhance or improve current detection systems (or whether they needed to review assumptions and plans for their use). This process developed experiential learning exchanges between the team and potential end-users, and a report was produced from each event.

Scientists were encouraged to contact stakeholders throughout the process of development of technologies. In order to clarify why and how this could facilitate technology development, we held focus group sessions with the project team during team meetings, we offered support and we developed a detailed Stakeholder Analysis and Engagement Plan Template (SAEPT). This model document supported focus group and meeting discussions; it explained the rationale for engagement, offered advice on how to engage people and presented templates to record the incidence, format and impact of interactions.

We used the concept of Technology Readiness Levels[4] (TRLs: from 1—concept to 9—application) to help us think about and describe innovation and technology development (see, for example, EARTO 2014). Whilst recognising that progression along this scale is iterative, it offered a framework within which to discuss with project scientists how key stakeholder interactions influenced development at any point, and it demanded that scientists consider application as well as theory.

Finally, we undertook a series of semi-structured interviews (see Table 1) with the science leaders for each of the five technologies (two interviews per science leader) and with a broad range of stakeholders that were randomly selected from the stakeholder database ($N = 16$). All interviews were recorded, professionally transcribed and thematically coded (Bryman 2001). Themes included concerns around tree pests and diseases, technologies currently used to detect/manage pests, challenges to using technology and level of engagement with the early detection technologies project. Not every stakeholder interviewed had participated in the Learning Platform workshops, but their roles in tree health biosecurity covered public and private sector forestry, conservation, industry (specifically nurseries), policy and community woodlands. Questions revolved around the extent of engagement in our early detection technologies project and in detection of tree pests and pathogens in general as well as barriers and opportunities for use and development of

Table 1 Data sources

Type	Format
Learning Platform 1 (November 2014)	Report http://protectingtreehealth.org.uk/
Learning Platform 2 (October 2015)	Report http://protectingtreehealth.org.uk/
Learning Platform 3 (November 2016)	Report http://protectingtreehealth.org.uk/
SLL—Heathrow airport and Southampton docks (November 2015)	Report http://protectingtreehealth.org.uk/
SLL—three nurseries in York (June 2016)	Report
5 Semi-structured interviews with project work package leaders and project leader (PI) (June–September 2015)	Interviews recorded and transcribed
5 Semi-structured interviews with project work package leaders and project leader (PI) follow-up (June–September 2016)	Interviews recorded and transcribed
16 Stakeholder interviews (May 2016–July 2017)	Interviews recorded and transcribed

technologies. Our analysis included both deductive and inductive elements, drawing on themes derived from our predetermined questions as well as the development of themes shaped by participants. We present below indicative quotes and other evidence from the project to address our aims and illustrate these themes. Our extensive use of quotes permits the voices of the stakeholders themselves to be heard, providing an empirical insight on a socio-technological innovation system.

3 Results

3.1 Scope of Results

In this chapter, we present the stakeholder map underpinning system groups and relationships, reinforce perceived needs for new technologies, outline current technology deployment practices and illustrate aspects of innovation offered by scientists and other stakeholders. We refer to 'technology' rather than to specific technologies within this or other projects unless it was relevant to highlight a particular example. However, we also demonstrate wider impacts of stakeholder interactions. Whilst reflecting on our role as facilitators, we also develop new insights and argue for a transformational socio-technological change seeing technology innovation as a cumulative process rather than a series of iterative outcomes.

3.2 Stakeholder Mapping

Our stakeholder mapping represented the diversity of stakeholders involved in the early detection system, from specialist inspectors with statutory duties, through a range of forester roles to organisations with specific interests around tree health or non-specific interests in woodlands (see Fig. 1). In this project, we mainly engaged those stakeholders with a more clearly defined role in technology use, such as inspectors, foresters and horticulturalists.

4 Perceived Needs for Technology Development

Stakeholder interest in the project was related to recent personal experiences of tree pests and diseases but also a concern about tree health more generally. Several respondents indicated that they believed the scale and extent of the global plant trade was the biggest threat to domestic tree health. A wider need for early detection of tree pests and pathogens was identified by several stakeholders with concerns over future potential incursions, both the 'knowns' and the 'unknowns':

> …because of these diseases and pests you know, we have been losing quite a lot of species and business for that matter. (I3: Nursery)

> I think one of the things that limits the ability for us to detect new diseases is lack of awareness of what those new diseases are likely to be. (I4: Policy)

12 Enhancing Socio-technological Innovation for Tree Health ...

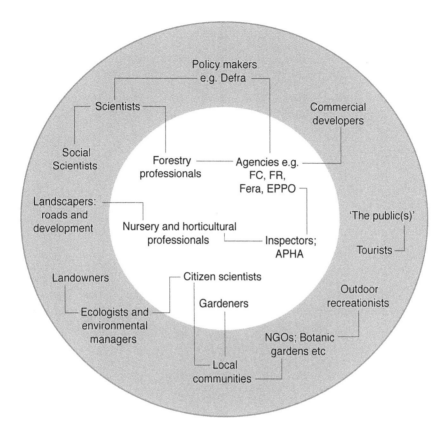

Fig. 1 A schematic representation of the stakeholder map for the early detection system for tree health. Stakeholder groups were initially identified from Dandy et al. (2017), but further groups were categorised during this project. The inner circle denotes groups with more specific interests and responsibilities in tree pest and pathogen detection. Lines illustrate strong relationships between groups, but additional connections between individuals and groups were identified

There were interest and support for the development of new or improved technologies with two stakeholders emphasising the need for a better system to find and identify diseases.

> There's 40,000 hectares of that which…you can't walk every square inch… Particularly given that I'm relatively remotely based away from most of them, so I think [spectral imaging technology] would be interesting. (I5: Forestry)

> I'm not sure we can stop [pests and diseases] coming in but what we need to be able to do is identify them. And we need to know more about trees that we're growing to find more resistance, or solutions to the pathogen's coming in, more quickly… (I6: Forestry)

However, there was some scepticism around technology development with concerns being raised about a potential reliance on detecting incursions instead of focussing on preventing new incursions:

> …I wouldn't want detection to distract from the need to actually reform trade, to stop things coming in. (I8: NGO)

> Whilst it's important that we have early detection of things arriving here, it's probably equally important that we are talking to colleagues abroad about the things they've got that may pose a threat to other countries. (I12: Forestry)

Some doubted whether the priority should be on technologies rather than spending time in the forest to check on tree health and one stakeholder worried about the consequences of finding pests:

> …the reason that people are investing in remote sensing is because they don't want to leave their desk (laughs)… or it seems to be that they can't leave the desk and so they have to react with their forestry community through a computer screen. (I10: NGO)

> Well I'd only want [technologies] if maybe I could do something about it [the pest or disease]. (I7: Nursery)

4.1 Current Technology Deployment

When discussing which technologies are currently used in detecting tree pests and diseases, most forestry personnel indicated that they rely primarily on visual assessments of plants, either because there is limited access to suitable technologies, reliance on official inspectors or issues with access to and the size of woodland areas covered. Some stated a need for additional resources or for more integrated approaches to pest and pathogen detection.

> Unfortunately our biggest problem now is…a lot of our work is done by contractors, and that's a huge difference, because field staff… were in there doing work, driving through a forest to get to their working site, if they see something that's wrong in the forest they'll take note and they'll come back to the office and they'll say 'oh by the way I noticed this over here, I think something's up'…contractors don't really have that interest… (I9: Forestry)

> …it's just by keeping our eyes open for anything, anything unusual. And…that isn't an easy thing to do, because if you're out in the field, working, you're thinking about 1000 different things at any one time and tree diseases is only likely to be a small part of that. (I12: Forestry)

One respondent identified the utility of helicopters to assist with monitoring of large forests, highlighting the demand for technologies to facilitate surveillance over sizeable areas, but noted that cost limited its use.

> … it is that early identification before impact…it was just two trees that were noticed…that helicopter trip managed to pick that up in of thousands of hectares…they were in inside a crop, which you would never be able to do…but helicopters' costs are too expensive… (I9: Forestry)

During an SLL visit, inspectors noted that *'detection is labour-intensive'* (SLL 091115). A senior inspector highlighted that inspection, especially of imported materials, requires time to *'build up the eye'* and experience. Rather than relying only on technology, in most cases *'the capability is in the inspector'* to identify possible problems. They then need technologies that are sensitive enough to detect if there is a problem with samples being sent to the laboratory to determine exactly which harmful organism is present. In parallel with the foresters, inspectors indicated that detection effort was constrained by resource availability.

4.2 Process of Technology Development

A range of responses revealed frustration with the timescales and applicability in producing technologies:

> It's the time scale. Whether you can shorten that time-scale in some way…I think it's [also] about joining up as well, to some extent. The actual testing out the development bit with the scientific bit? So whether we can short cut some of that. (I15: Inspector)

> …in some respects being cynical, it's perhaps not in the researcher's interest to finish, to provide the solution…sometimes you do get that feeling that they go from one project to the next project to the next project… So what I would like to think that they are doing is breaking the problem down, into a number of bite-sized chunks, solving that and then going on to that… but of course what's happening is you never get from here to there, because something is changed in the middle. (I6: Forestry)

Others were concerned about the specificity of detection technologies and what they can be used for:

> …you would need to have a very wide range of different early detection technologies to cover all the potential, or the highest priority pests and diseases…that we might expect for our native plants. (I4: Policy)

Questions were also raised around whether current detection technologies are reliable or quick enough (e.g. provide results in the field) to warrant a change in practices.

> …how effective are the early detection technologies and what's the risk of false negatives and how do they compare with other methods of detection? (I4: Policy)

> …there was a long time ago a discussion about volunteers using the [name] machine that had been developed by [name], and we did do some pilot work to see if the volunteers could effectively use it, and they could. The sticking point was actually that it was showing that the machine itself wasn't massively reliable… (I14: NGO)

4.3 Technology Distribution Infrastructure and Networks

It proved difficult to engage many commercial development companies in the Learning Platform, although representatives from two companies did attend one event and interacted with scientists. However, stakeholders noted different aspects of distribution processes, notably that commercial uptake would be required and costs would have to be reduced for some technologies to be used in practice.

> You know, these things will only succeed if there is a commercial angle… (I3: Nursery)
>
> At the moment you're using contractors, but I can see the stage when the technology and the upfront costs are reduced and every forester will have one of them in his car … to look at crop health. (I6: Forestry)

At one of the Learning Platform workshops, a speaker noted that it will be important to understand '*who pays for technology use*' (LP2). There were differences of opinion on investment for tree health technology and whether there should be sufficient freedom allowed for innovation:

> …how can these technologies help us in our day job…I also recognise that sometimes there [is] wonderful technology and you have no idea what you're going to do with it, but ultimately you'll find something. So sometimes it's a bit of serendipity, you can't always specify what you're going to do with the technology. (I15: Inspector)
>
> …is it money being spent wisely, you know what I mean? …A number of projects and things I've seen go through go through the system and you think, 'what's it going to be used for - is this going to end up in the drawer'… Yeah, seven million [funding for THAPBI projects]. I think 'ooh, that's a lot of Rhododendron clearance for that. (I2: Policy)

Stakeholders had different views on their role in technology development. Some indicated that seeing was believing and, although development was for scientists, early testing by practitioners was important:

> …it's actually seeing it working and to…have the confidence that it does work… (I2: Policy)

> The actual technology, I think that's for other experts. I would be at the front end of testing … so the likes of…[technology], you know I felt when it came out, it would have been better to have it tested…and given to more practitioners because it lost a bit of credibility early on. (I11: Forestry)

The gap between concept and application was noted by one stakeholder:

> There was one possible practical action that came out of that first platform that I then didn't follow up…I think I probably felt that there was still a big gap between all these brilliant technologies and them actually getting in the hands of the people on the ground and actually working. (I14: NGO)

Stakeholders suggested that not only should they engage with scientists, but also that scientists should understand practice contexts:

> I will definitely say, the scientists…they should come and visit the real world. (I3: Nursery)

Scientists in the project indicated that it is unlikely that a technology will be conceived, designed, tested and ready for market in one project life cycle of 3 or 4 years, despite the promises within funding proposals or expectations from funders and potential end-users.

> I mean I've done just a thumb-nail study looking at the stuff that we've done…from the first time you see something published to the first time you get a commercial service or a product out at the other end and you are talking a 10 to 15 year process. (S1)

Two of the project scientists highlighted the complex, serendipitous and cumulative nature of the development process:

…Quite often it's not necessarily a nice discrete project, things overlap and you end up combining the experiences from several projects. (S3)

…I suppose we see this as one project and how we develop these things in three years but of course, all of those technologies have actually been developed as they're being supported by ongoing other projects, previous projects and potentially now future projects as well. I suppose the one thing we've also seen from doing some of this work and looking at how you progress a technology through, is that it never seems to be related to a single project that takes something and pushes it through to the end. (S1)

Another team member highlighted how the technology developed through the early detection technologies project programme had proved to be useful in horticulture. This scientist highlighted that innovations are more likely to be achieved if the process of technology development is not strictly defined and controlled but it may only be achieved through multiple funding sources:

Money's tighter and tighter, and it's what are you going to invest in? I have also recognised that it's difficult to back winning horses doing this, so you know, the fact that we invested in that work for tree health, but the benefits might be to horticulture. You know, it's difficult to predict and so down that route you may end up only doing only very safe innovation. Or is that innovation? If it's very safe…you won't really innovate, and you won't do anything very different. (S5)

It was clear that technology development was not linear, but rather was messy and iterative, and that the final stages of development took proportionately longer:

…getting over the last TRLs is quite surprisingly hard. When you get that close you think 'we've just got to do those experiments and then we are done'. And that's where you then start to get into training and users for example. And suddenly they're like 'ooh, this isn't quite what we wanted to do', or 'we didn't want that target, we wanted you to look at this target'…that was best furnished with lots of little projects trying to tick off all of these little problems, be they technical, or more about what you are using it for… (S5)

5   Additional Outcomes from the Learning Platform Approach

One of our aims was to investigate the role of stakeholder engagement in enabling socio-technological innovation and changing the wider tree health system. Hence, we also identified various mechanisms by which this had occurred, largely through the Learning Platform approach. Two stakeholders highlighted that attendance at a host of tree health-related meetings and workshops were influential in helping them prioritise biosecurity:

> Well discussing [and] going along to all these things certainly has an influence on us. I've now set a budget to increase our biosecurity. (I7: Nursery)

> …I think this project certainly kick started a sort of discussion…think it's helped to bring it more down to more the operational level. (I15: Inspector)

A speaker at a Learning Platform commented that there was a need for balance between stakeholders indicating what they needed and scientists indicating what is possible suggesting that '*Technology push is… a valuable tool to inform users of new opportunities*' (LP1). However, the project concept that placed stakeholder engagement at the heart of the technology development also presented challenges. Early involvement at the conceptual stage was not always considered feasible or desirable; making sure the technology worked was a higher priority:

> …We've seen it lots of times before where people…will not engage until they've got something in their hands or a prototype or something like that. (S1)

> I feel that we're at quite an early stage… I'm not desperate to go out and find lots of stakeholders because I think I've got other issues within the project which…I need to address. (S7)

Discussions through the Learning Platform also highlighted potential conflicts over technological innovations in terms of the perceived threats to job security:

> One of the things that I have felt a bit with some of the stakeholders, is they feel like, with this early detection stuff [and] I don't know whether it's limited to [technology]…but they feel that we are trying to replace them, as inspectors…I more feel that we're trying to help them, and make their job easier rather… than trying to make them obsolete. (S7)

However, involvement in the Learning Platform allowed research scientists the opportunity to talk to relevant end-users and to think through practicalities:

> They've definitely helped in terms of making contacts and talking to the right people and getting an idea of what actually needs to be done and how it can be done, the practicalities, the logistics. (S3)

The Learning Platform workshops enabled general and non-specific comment on detection technologies. Foresters in particular wanted *'portable technologies and … things that will last and be simple to use and robust and rugged for different conditions'* (LP1). In contrast to the inspectors, who cited sensitivity as being critical, foresters suggested that whilst sensitivity and efficacy should be known, 'pragmatism is also necessary', and both groups agreed that reliability was the key (LP1). With regard to particular technologies, participants pointed out that placing an instrument to detect volatile compounds in every container and gaining access to and checking each one would not be logistically possible, prompting discussion of other use contexts. They asked if a spectral imaging instrument could be airborne instead of hand-held. There was curiosity about the utility of trapped spore assays for specific pathogens (e.g. ash dieback *Hymenoscyphus fraxineus*) and for broad spectrum unknowns. There were questions about the ability of water surveillance technologies to handle 'murky [muddy] samples' that prompted intention to try different filter sizes and possibly cyclone processing of samples. Participants raised additional potential uses for technologies beyond those originally conceived by scientists and also suggested how they might be deployed and even who might manage them (e.g. '*I… see this as being part of a network managed by a host agency … to track progress of pathogen/s across and between control zones to better target … detailed surveillance*' [LP1]).

Discussion at Learning Platform workshops also highlighted additional potential users of technologies and the challenges of keeping them informed and engaged and able to contribute meaningful results to national surveys and receive local feedback. For example, it was commented that whilst community woodland groups like to do surveys, *'they often don't know what to survey or what to do with the results'* (LP2); and some surveys rely on observation rather than technology use. Citizen science was reported to be useful, especially in monitoring technologies such as beetle traps (LP1 and LP2). For both groups, questions arose about the interpretation of technology and survey results such as whether a zero result meant no effort or no detection.

One project scientist highlighted a step change in the process of developing technologies for early detection, which has come about through involvement in the project and Learning Platform. It marks a shift from a singular focus on the product to wider thinking about the requirements of early detection for better tree biosecurity systems.

> So I suppose the co-design has come from… two things, so from [government]… re-evaluating the program, and thinking 'are we doing the right thing, are we commissioning the right kind of work, are they proposing the right kind of work?' And then from our side, questioning 'what are your priorities'…and I don't mean that in terms of 'come to us and tell us what technology you want developed, or what work you want doing'. Can we collectively take a step back and say… 'what are the problems that we are trying to solve'? (S5)

There was also evidence of stakeholders seeing the detection of tree pests and pathogens as a system within which integrated collaborative working was required:

> What's really useful is to have shared priorities for monitoring…I think what is also useful is to have a lead body, or member of staff to coordinate all of this. You need somebody that has sort of ownership and that can facilitate things and updates and steer the effort. (I4: Policy)

Within this context, there was recognition of a responsibility for stakeholders to contribute to technology development:

…The industry practitioners have to take responsibility for helping to find solutions to things and… not to just expect someone else to do that for us, and I think it is really important that we make time to support and to provide input into initiatives when they happen, instead of just saying, 'I'm too busy'. Actually I don't think that's good enough. (I5: Forestry)

6   Discussion

Throughout the project, the authors explored opportunities for transformational socio-technological change in the development of early detection technologies and their use. This chapter has demonstrated the complexity of developing socio-technological innovation for tree health, drawing on an interdisciplinary project in which we pursued the co-design or co-development of five pre-identified technologies through stakeholder engagement. We reflect below on the extent and forms of engagement attempted, discuss our product (technology) and wider system (detection of tree pests and pathogens) innovations, consider our contributions to the theory of socio-technological system transitions, outline some insights and recommendations for tree health and finally offer some conclusions.

6.1   Forms, Extent and Timing of Stakeholder Engagement

Stakeholder mapping illustrated the complexity of stakeholder engagement in the early detection of tree pest and pathogens, with additional challenges regarding engagement around technology innovation. As social scientists, we had anticipated that scientists would demonstrate greater independence in pursuing technology-specific SLLs but project team members were more comfortable with sector-based (to airport, seaport and nurseries) visits. The need for interdisciplinary projects including social scientists was thus reinforced (see O'Brien et al. 2013), but we also note our sometimes conflicted roles as facilitators (striving for good practice in engagement methods) and also researchers (attempting to address theoretically rigorous research questions around socio-technological innovation). The multiple routes by which social scientists can contribute to conservation and environmental management have been analysed by a number of scholars (Bennett et al. 2017). For example, O'Brien et al. (2013) suggest that engagement with stakeholders through an interdisciplinary team of academics ('participatory interdisciplinarity') creates better trust and understanding as well as some instrumental benefits. Indeed, the Learning Platform workshops proved to be good venues to engage in debate across the field of tree health and incorporated diverse successful tools for engagement. Towards the end of the project, a senior member of the Expert Advisory Group publicly stated that the diverse modes of engagement employed in Learning Platform workshops helped maintain interest, kept the discussion fresh and caused different ways of seeing the technologies (LP3). The use of novel and different approaches proved entertaining and effective for stakeholders, although challenging for some of the team scientists participating. Whilst we demonstrated evidence of mutual learning and relational benefits from this project, we also found that these outputs required resource investment in terms of time, energy and social science expertise.

Many of the benefits of engagement were due to engagement processes, rather than planned engagement outcomes. Whilst there may be a tendency to prefer instrumental impacts from engagement (such as specific changes in technologies), our results indicate additional types of impact as well such as human capacity, connectivity between people and new conceptual framings.

6.2 Product and System Innovation

The effects of stakeholder engagement on individual technologies differed depending on the specific technology, its stage and the perspective of the scientists involved. For example, stakeholder input to one technology consistently highlighted practical problems with the intended use application, but these were never resolved. For others, modifications were made.

There was a discussion about the stage at which engagement was most beneficial. Some scientists felt that they needed to progress technologies to a certain stage before the utility of that technology became apparent and co-design of the use format could occur. There were some specific points made through engagement that caused modification of technology adaptation. Several scientists and stakeholders noted the importance of co-design across the system of tree health as a whole, identifying needs and priorities and monitoring requirements (prior to the emergence of specific technologies). It seems that stakeholder engagement can be useful in co-design particularly at certain stages (TRLs) of development: at initial broad prioritisation of needs, at sporadic phases throughout development and mainly in the final stages in which application is refined. It sometimes proved difficult to separate product (technology) and system innovation; finding sustainable innovation solutions depended on more than the technology but also the context within which the technology was to be used.

Although we broadly outline roles of different 'stakeholders' and scientists, we identified overlapping and hybrid roles in practice in technology innovation through this engagement process. For example, scientists working in government agencies had a good realisation of the sector challenges in practice, existing relationships with many stakeholders and a responsibility for application and most entered enthusiastically into the engagement processes. Many of the scientists working in university or research centres not formally linked to government had less contact with stakeholders prior to the project. Some of them welcomed engagement, but others were more reluctant contributors. In order to benefit from the strengths of project team members, we thus needed to respect the diversity of routes by which each could contribute, creating an 'ecosystem of expertise' with complementary knowledge (Brand and Karnoven 2007) in which some engaged directly with stakeholders, others pursued interdisciplinary research and others remained specialists who engaged with us as facilitators. Interdisciplinary research can cause researchers to gain new skills but also to gain satisfaction from addressing a real-world problem in a holistic manner (O'Brien et al. 2013), and we saw evidence of both in our project.

Because the emphasis of our project was on the development of our five technologies, our research was constrained by the project boundary. However, several times participants indicated that it normally takes 10–15 years to bring a concept to use, yet project funding cycles are typically only 2–4 years. This means that technology innovation requires an application for additional funding to test specific elements and aspects of emerging technologies. These projects are not independent but are nested within research programmes (such as THABPI), which are in turn funded by individual or coalitions between research councils, government or other organisations, including industry institutions. Particular forms of innovation were proposed for this project, largely supported for product (technology) innovation, but the THABPI focus demanded some system innovation recognition. Funders might consider in future a combination of shorter specific funds to support technology adaptation in later TRLs plus longer term emphasis on particular aspects of a system, such as tree health, that permits both the networking and strategising as well as specific advances.

6.3 Socio-technological Innovation or Socio-ecological Technological Innovation?

There are different approaches to analysing innovation systems. Hekkert et al. (2007) suggest exploring sub-functions such as knowledge development, knowledge diffusion, market formation and resource mobilisation. Analysis of these aspects in this study suggests that we have stimulated a new form of knowledge development and diffusion, because our stakeholders have an interest or responsibility in tree health that transcends a market consumer desire. Tree health and the wider field of biosecurity are issues of national and international concern, and fulfil a *required*, but not always *desired*, need for 'the public'. However, we have limited market formation because of the specific and constrained agency or government market for some of the technology 'products' that might emerge.

This study has also demonstrated that the notion of the Triple Helix of innovation systems (universities–industry–government) (Leydesdorff and Meyer 2006) may be naïve in a contemporary context with strong public interest and multi-stakeholder governance. Firstly, tree health technologies are not merely products for consumer consumption, although there are commercial applications for some detection technologies in some stakeholder contexts, such as horticultural nurseries. The need for new early detection technologies occurred across different parts of the system, involving a range of stakeholders, but the greatest need was identified by government and/or agency employees such as inspectors and foresters. Secondly, in our study, innovators were a combination and coalition of universities, research institutes and government agencies, including hybrid roles as described above; users included government, and to some extent universities and industry play a small but specific role in taking product designs to commercial output, sometimes for a limited market. Our findings thus concur with those of Smith et al. (2005) who described complex, non-homogenous systems. Thirdly, tree health policy is responding rapidly to social change (such as globalisation and trade) and environmental change (such as climate change and alien invasive species), shifting the context and needs for detection technologies. However, this Triple Helix model is still useful in highlighting how the lack of market potential for some of the early detection technologies might limit stimulus for innovation. We thus need to offer non-market-based stimulus, or alternative deployment, for public good technologies.

We agree with Mulder (2007) that technology development alone may not lead towards a sustainable future, but rather we need to consider wider socio-ecological consequences of technology deployment. Our study has identified a context where we need to draw on both understandings of social innovation (e.g. Mulgan 2007) and the more widely discussed product innovation for markets (e.g. Galanakis 2006) and associated socio-technical systems (e.g. Geels 2010). An additional novel contribution of our study to innovation studies is the focus on an environmental issue: tree health. We, therefore, have to consider not only the social and cultural context but also ecological complexity (White and van Koten 2016).

In this chapter, we have highlighted how the MLP offered by Geels (2010) can allow us to analyse innovation in environmental management. We have demonstrated how an innovation niche created by focused research funding catalysed progress across five different technologies. We disrupted and provoked more stakeholder interactions in order to offer new cognitive learning (Geels 2004), and we suggested how policy makers could offer new regimes to enable innovation to cascade across biosecurity measures. We saw evidence of some transformational change in innovation with some stakeholders thinking differently. For example, it was reported that, historically, inspectors would say they needed a particular product to measure a particular pathogen, but now they are asking more fundamental questions about the system of detection and roles and responsibilities. However, we did not clearly see a transition in the system. Transition is desirable to achieve long-term sustainable development (Mulder 2007), but conditions favourable to transition rarely emerge. Mulder (2007) describes the dilemma in which a new transitioned system has to establish against an already established system. It can take a significant event such as war, or natural disaster, to disrupt or enable such radical change. Whilst the recent outbreaks of tree disease in the UK have been perceived by some to be a natural disaster, it remains to be seen if this is sufficient to enable a transition in the system. Possibly this environmental change, combined with political change, can offer a sufficient external shock to the system to catalyse more rapid and radical change.

7 Conclusions

Whilst stakeholder engagement influenced socio-technological innovation in this project, it is difficult to assess *by how much* and hence to encourage prioritisation of engagement over other activities. Targeted, time-sensitive stakeholder engagement at critical TRLs is preferred by some stakeholders. However, the more diffuse benefits of broader social learning, mutual understanding and shared development of priorities and strategies remain critical.

Although collaborative approaches are essential in stimulating effective technology development, they can be costly (Davies and White 2012), and several stakeholders cited cost constraints already present in detection programmes. Future collaboration in this sector will thus have to consider the investment required for different forms and extents of engagement and their particular merits such as awareness raising for 'the public', specialist partnerships with inspectors, dissemination of information through government agencies and practitioner networks, as well as platforms for mixed groups and roles to debate the wider issues of biosecurity beyond specific technology requirements. If participation increases social learning and more democratic engagement in decision-making (Blackstock et al. 2007; Reed et al. 2009), what value do we put on this? We saw in this project that the interdisciplinarity and stakeholder engagement had some influences on specific technology (product) innovation (such as need for rugged mobile devices and devices to accommodate murky river water) and the wider system of tree health and biosecurity (such as highlighting operational contexts for technologies to scientists and the long-term and project-driven nature of science to practitioners). However, the cumulative serendipitous nature of engagement makes it difficult to defend a cost-benefit analysis that will omit some wider outcomes. We thus recommend that future projects and programmes maintain the commitment to an open, general Learning Platform, but we also suggest that specific interactions between scientists and practitioners (e.g. SLLs and partnerships), and other needs for engagement are planned within a strategy that recognises the wider socio-technological innovation system for tree health and biosecurity.

Tree health will require a suite of actions by a wide range of stakeholders, including changes in the behaviour of people visiting forests, consumers importing goods, producers and traders importing or selling plants in nurseries and those planting trees or forests. Early detection of tree pests and pathogens remains a critical aspect but will not resolve all challenges to the ecological integrity and social benefits of our forests and woodlands nor the economic strength of our forestry and horticultural industries. This study shows that there is wide interest in responding to tree pest and disease incursions and this interest should be harnessed across other areas that impact on biosecurity.

Acknowledgements We would like to thank all the project team members and stakeholders who generously gave their time to participate in the social research. The project 'New approaches for the early detection of tree health pests and pathogens' (http://protectingtreehealth.org.uk/) was supported by a grant funded jointly by the Biotechnology and Biological Sciences Research Council, the Department for Environment, Food and Rural Affairs, the Economic and Social Research Council, the Forestry Commission, the Natural Environment Research Council and the Scottish Government, under the Tree Health and Plant Biosecurity Initiative.

Notes

1. Project title: New approaches for the early detection of tree health pests and pathogens. http://protectingtreehealth.org.uk/. Funded in the UK through the LWEC Tree Health and Plant Biosecurity Initiative (THAPBI).
2. A simple presentation style where you show only images on 20 power point slides for 20 seconds each. The presentation is timed so slides move on automatically after 20 seconds (http://www.pechakucha.org).
3. The World Café approach involves a series of tables or settings with a host. Participants are broken up into groups, and each spends around 20 minutes at each table discussing a specific topic. The host of each table welcomes each group and fills them in on what happened in the previous round. Insights are shared at the end of the process (http://www.theworldcafe.com).
4. TRLs offer a more objective approach to assessing where technology sits in a deployment pipeline. They provide an easier approach for understanding the different resources or processes required for technologies at different levels of maturity. They also offer a means to assess the likelihood of success of a technology, which can help us prioritise the use and investment of limited resources. An outline of the historical development of TRLs and its relevance to our project can be found on http://protectingtreehealth.org.uk/documents/introduction.pdf.

References

Arnstein, S. R. (1969). A ladder of citizen participation. *Journal of the American Institute of Planners, 35*(4), 216–224. https://doi.org/10.1080/01944366908977225.

Bennett, N. J., Roth, R., Klain, S. C., Chan, K., Christie, P., Clark, D. A., et al. (2017). Conservation social science: Understanding and integrating human dimensions to improve conservation. *Biological Conservation, 205,* 93–108.

Blackstock, K. L., Kelly, G. J., & Horsey, B. L. (2007). Developing and applying a framework to evaluate participatory research for sustainability. *Ecological Economics, 60*(4), 726–742. https://doi.org/10.1016/j.ecolecon.2006.05.014.

Boyd, I. L., Freer-Smith, P. H., Gilligan C. A., & Godfray, H. C. J. (2013). The consequences of tree pests and diseases for ecosystem services. *Science, 342,* 1235773.

Brand, R., & Karnoven, A. (2007). The ecosystem of expertise: Complementary knowledges for sustainable development. *Sustainability: Science, Practice and Policy, 3*(1), 21–31.

Brasier, C. M. (2008). The biosecurity threat to the UK and global environment from international trade in plants. *Plant Pathology, 57,* 792–808.

Bryman, A. (2001). *Social research methods.* Oxford: Oxford University Press.

Dandy, N., Marzano, M., Porth, E., Urquhart, J., & Potter, C. (2017). *Who has a stake in ash dieback? A conceptual framework for the identification and categorisation of tree health stakeholders.* Special edition publication from COST Action Fraxback. http://www.slu.se/globalassets/ew/org/inst/mykopat/forskning/stenlid/dieback-of-european-ash.pdf.

Davies, A. L., & White, R. M. (2012). Collaboration in natural resource governance: Reconciling stakeholder expectations in deer management in Scotland. *Journal of Environmental Management, 112,* 160–169. https://doi.org/10.1016/j.jenvman.2012.07.032.

EARTO. (2014). *The TRL scale as a research and innovation policy tool, EARTO recommendations.* http://www.earto.eu/publications1.html.

Everett, R. A. (2000). Patterns and pathways of biological invasions. *Tree, 15*(5), 177–178.

Fagerberg, J., & Verspagen, B. (2009). Innovation studies—The emerging structure of a new scientific field. *Research Policy, 38*(2), 218–233.

Galanakis, K. (2006). Innovation process. Make sense using systems thinking. *Technovation, 26*, 1222–1232.

Geels, F. W. (2004). From sectoral systems of innovation to socio-technical systems. *Research Policy, 33*(6–7), 897–920. https://doi.org/10.1016/j.respol.2004.01.015.

Geels, F. W. (2010). Ontologies, socio-technical transitions (to sustainability), and the multi-level perspective. *Research Policy, 39*, 495–510.

Geels, F. W., & Kemp, R. (2007). Dynamics in socio-technical systems: Typology of change processes and contrasting case studies. *Technology in Society, 29*(4), 441–455. https://doi.org/10.1016/j.techsoc.2007.08.009.

Hekkert, M. P., Suurs, R. A. A., Negro, S. O., Kuhlmann, S., & Smits, R. E. H. M. (2007). Functions of innovation systems: A new approach for analysing technological change. *Technological Firecasting and Social Change, 74*, 413–432.

Leydesdorff, L., & Meyer, M. (2006). Triple Helix indicators of knowledge-based innovation systems: Introduction to the special issue. *Research Policy, 35*(10), 1441–1449.

Marzano, M., Dandy, N., Bayliss, H. R., Porth, E., & Potter, C. (2015). Part of the solution? Stakeholder awareness, information and engagement in tree health issues. *Biological Invasions, 17*(7), 1961–1977.

Mulder, K. F. (2007). Innovation for sustainable development: From environmental design to transition management. *Sustainability Science, 2*(2), 253–263. https://doi.org/10.1007/s11625-007-0036-7.

Mulgan, G. (2007). *Social innovation—What it is, why it matters and how it can be accelerated*. Oxford: Said Business School.

O'Brien, L., Marzano, M., & White, R. M. (2013). 'Participatory interdisciplinarity': Towards the integration of disciplinary diversity with stakeholder engagement for new models of knowledge production. *Science and Public Policy, 40*, 51–61.

Pahl-Wostl, C. (2006). The importance of social learning in restoring the multifunctionality of rivers and floodplains. *Ecology and Society, 11*(1), 10.

Perrings, C., Burgiel, S., Lonsdale, M., Mooney, H., & Williamson, M. (2010). International cooperation in the solution to trade-related invasive species risks. *Annals of the New York Academy of Sciences, 1195*, 198–212.

Prell, C., Hubacek, K., & Reed, M. (2009). Stakeholder analysis and social network analysis in natural resource management. *Society and Natural Resources, 22*(6), 501–518.

Reed, M. S., Graves, A., Dandy, N., Posthumus, H., Hubacek, K., Morris, J., et al. (2009). Who's in and why? A typology of stakeholder analysis methods for natural resource management. *Journal of Environmental Management, 90*, 1933–1949.

Smith, A., Stirling, A., & Berkhout, F. (2005). The governance of sustainable socio-technical transitions. *Research Policy, 34*, 1491–1510.

Sutherland, A., da Silva Wells, C., Darteh, B., & Butterworth, J. (2012). Researchers as actors in urban water governance? Perspectives on learning alliances as an innovative mechanism for change. *International Journal of Water, 6*(3/4), 311–329.

Webber, J. (2010). Pest risk analysis and invasion pathways for plant pathogens. *New Zealand Journal of Forest Science, 40*(Suppl.), 45–56.

White, R. M., & van Koten, H. (2016). Co-designing for sustainability: Strategising community carbon emission reduction through socio-ecological innovation. *The Design Journal, 19*(1), 25–46. http://www.tandfonline.com/doi/full/10.1080/14606925.2015.1064219.

Whitmarsh, L. (2012). How useful is the multi-level perspective for transport and sustainability research? *Journal of Transport Geography, 24*, 483–487.

13

Gaming with Deadwood: How to Better Teach Forest Protection When Bugs Are Lurking Everywhere

Marian Drăgoi

1 Introduction

Over the past several decades, a series of droughts in Central and South-East Europe have triggered die-off symptoms amongst some timber broadleaved species (Brasier and Scott 1994; Thomas et al. 2002; Borlea 2004; Pautasso et al. 2013; Nagel et al. 2014). Resinous species, like silver fir (*Abies alba,* Mill.) and Norway spruce (*Picea abies,* L.) have also been affected by bark beetles, especially in those stands outside of their natural range (Jonášová and Prach 2004; Stanovský 2002; Olenici et al. 2011). Even though regular silvicultural measures are unable to prevent affected trees from dying, maintaining a certain level of forest biodiversity and a closed forest canopy are important goals that forest management must fulfil. As such, standing and fallen deadwood are important contributors to forest biodiversity (Humphrey et al. 2005; Verkerk et al. 2011), particularly those that have a slow

---

M. Drăgoi (✉)
University of Suceava, Suceava, Romania

decay rate (Lassauce et al. 2012). Such trees are crucial for maintaining populations of bats (Lučan et al. 2009), birds (Drapeau et al. 2009; Joseph et al. 2011; Miles and Ricklefs 1984) and mammals (Radu 2006). In Central Europe, the legacy of the former Austro-Hungarian Empire means that upland areas consist of large forests with Norway spruce and Scots pine (*Pinus sylvestris,* L.) planted instead of the indigenous broadleaved species like beech, oak, ash, maple and hornbeam. Nowadays most of these resinous stands are severely affected by insect pests (Jonášová and Prach 2004; Olenici et al. 2011; Panayotov et al. 2015; Sproull et al. 2015), droughts (Anderegg et al. 2013), windfall and/or wildfires (Flannigan et al. 2000).

In Romania, after 1970, the communist regime resumed planting Norway spruce and pines beyond their natural habitats to increase the production of high-quality wood and resin. A tipping point of the forest policy was 1986 when a new set of technical standards came into force focused on preserving the forests' naturalness. Furthermore, while in 2000, only 5.3% of the forests had been restituted by the Government back to former landowners (Abrudan et al. 2009), by 2017, the same area is equally shared by the state (public forests) and private ownership. However, little has changed regarding the managerial options, and all forests are managed to produce logs for lumber or veneer, but not pulpwood or fuelwood. There is effectively no guidance for the private owner on what management objectives they are meant to achieve, nor the timber grade she or he might aim at. To confuse matters further, the restitution process was driven by three different laws while the forest inspectorates in charge of checking the lawfulness and quality of any harvest operations were barely organised in 2000 but subsequently reorganised in 2003, 2004, 2005 and 2016.

Without having a reliable and extended forest road network (Drăgoi et al. 2015), the National Forest Administration (NFA) could not harvest the allowable quota using environmentally friendly logging operations and shelterwood forest systems. With fierce competition for wood, caused by thousands of small logging companies (authorised to operate for the sake of free competition), the public authority's inspectors are not able to trace all timber theft, their job being especially difficult when the thief is the landowner. All these setbacks have been wrapped up in excessive bureaucracy, brought in by new institutions like environmental agencies, the Council of Competition, the Court of Accounts and many others.

However, the new institutions were not able to harmonise all the details of the forest policy, and a series of problems have occurred: sheer illegal logging, overharvesting through timber underestimation and different bureaucratic scams meant to get around legal obligations. Since 2010, the NFA has been continuously consolidating its position on the market by certifying its forest management according to Forest Stewardship Council (FSC®) standards. Hence, two divergent tendencies have occurred: on the one hand, the NFA staff pursued new environmentally friendly logging technologies and pest control in certified forests and, on the other hand, the forest rangers did their best to obtain more profit for their own account taking advantage of the weak control exerted by the forest inspectors. Cost-effective solutions are sought not only for economic reasons but also for simplifying the fieldworks; currently tagging trees for biodiversity (further referred to as TFB) and marking sanitation/salvage cuttings are two different tasks, carried out by the same people who have to wade through the forest twice: one time for sanitation fellings and the second time for tagging TFB. The order doesn't matter: usually, TFB are tagged prior to FSC audit, while sanitation/salvage fellings are stamped whenever is needed.

Where possible, forest rangers applied salvage cuttings in stands older than 60 years instead of regular harvesting operations in mature stands (for the sake of sustained yield principle the Forest Act allows this silvicultural swap). In addition to that, a systematic underestimation of the harvested volume was also an important scam as long as the amount of timber a logging company was charged for was not checked against the amount of timber transported by that company from the forest.[1]

Some illegal logging discovered in Retezat National Park in 2009 (Knorn et al. 2012) sparked media attention on harvesting operations generally whether they were legal or not legal. Later, in 2013, the Court of Accounts of Romania[2] published a retrospective report on the consequences of forestland restitution, focused mainly on illegal logging. Two years later, the Forest Code was amended and one important side effect of the public debates of that time was a sound involvement of NGOs in preventing all types of illegal fellings, even though many activists were unable to tell the difference between regular fellings, as prescribed by the forest plan, and the illegal ones. Since then there has been extensive public attention on forests and the wood industry, particularly on illegal logging (mitigated by the wood tracking system) and biodiversity conservation (management plans for Natura 2000 network and old-growth forests). These two areas of interest are intertwined in the forest certification process that has been triggered mainly by the NFA in 2010 and 2011. So far about 2.3 million hectares (two-thirds of the public forests according to the NFA site: www.rosilva.ro) of forests have been certified by the FSC® scheme. However, the demand for timber labelled with the FSC® logo fell behind the supply because many logging companies are not able to comply with the high-quality standards required for harvesting operations and the European requirements on timber traceability (Gavrilut et al. 2015; Hălălişan et al. 2012).

The FSC® standard brought to light the problem of sanitation and salvage fellings because it requires the presence of deadwood in the forest (TFB), without providing any rigourous threshold in terms of number of trees or volume of deadwood per hectare (Humphrey et al. 2005; Schroth and McNeely 2011; Johansson and Lidestav 2011). The differences between the two types of fellings are important for understanding which is the problem with selecting and maintaining a certain number of TFB. Sanitation felling involves harvesting dead trees up to one cubic metre per hectare per year without indicating the cause of death, while salvage fellings allow harvesting more than one cubic metre per hectare for specific biotic or abiotic reasons such as insect pests, wind, snow or whatever natural causes, including the wounds produced to remnant trees by prior harvesting operations. Bluntly speaking, sanitation fellings do what nature does, i.e. natural selection, while salvage fellings are intended to keep the pests out. But reckless sanitation fellings eventually bring about more salvage fellings, and this was a *modus operandi* for a long period of time.

Hence, in 2016, the public authority invested funding for crowdsourcing on issues with high social exposure, like nature conservation, preservation of old-growth forests, illegal fellings and timber traceability (Stanciu 2017). Nowadays NGOs and laymen can check whether or not a load of wood is legal or not by searching on the website www.inspectorulpadurii.ro.

As Romanian forests have been traditionally managed to provide timber and ecosystem services, the growing stock has been maintained at high levels, together with shelterwood systems. Leaving aside sheer illegal cuttings[3] and felonies like marking green trees for sanitation felling, two types of poor practice are still common, and both are loopholes in the technical standards and the Forest Code. The first one briefly explained here is the provision that says that any tract of sanitation or salvage fellings shall follow the same commercial procedures as any regular tract of wood sold on the stump. The procedures of marking and auctioning any tract of wood takes over 30 days due to the following operational requirements: (1) marking up and measuring the trees to be harvested in each area (also referred to as *timber cruising*); (2) assessing the volume and the value of each tract of wood; (3) organising the auction; and (4) issuing all required approvals to commence the harvesting operations (a special authorisation from the national protection agency is needed for Natura 2000 sites). In Norway spruce stands, seriously affected by bark beetles, Duduman et al. (2014) showed that harvesting the already dead trees did not stop the insects' propagation; on the contrary, the authors concluded that the delay in harvesting operations caused by the bureaucratic procedure helps insects' propagation. Subsequently, the gaps in the forest canopy allowed more sunlight to reach the trees' bark, thus speeding up the occurrence of a new generation of beetles. Two or three weeks after the initial attack, when the affected trees will have been harvested, the beetles will have already been boring adjacent trees.

The second issue stemming from the sustained yield principle is the provision that all salvage cuttings ranging from 1 to 5 m^3 yr^{-1} ha^{-1}, located in stands older than 60 years, shall be deducted from the main yield allowable cut, without any formal approval issued by the public authority (the public authority shall endorse tracts larger than

5 m³ yr⁻¹ ha⁻¹). This provision is very misleading because the Forest Code says that the amount of wood harvested from a certain forest unit cannot exceed the annual allowable cut prescribed by the forest management plan. Hence, the more scattered salvage fellings (up to 5 m³ yr⁻¹ ha⁻¹), the less "regular" harvesting operations will be carried out in mature (and often remote) stands. Because the many "tiny" tracts cannot be checked in the field by the forest authority, many weakened, but still alive trees, can be harvested. Replacing the main yield with salvage fellings provides another advantage to the forest manager, who pays less for the so-called regeneration fund that, according to the same Forest Code, is collected from the revenues brought by the main yield only. Thus, by adopting a strategy of "more salvage cuttings instead of main yield cuttings", the money that would otherwise go into the regeneration fund can be used to finance other activities, not precisely the ones envisaged by the regeneration fund (afforestation and thinnings).

The consequence of these poor practices, encouraged by the legal framework, is shown in Fig. 1: the gap between regular silvicultural

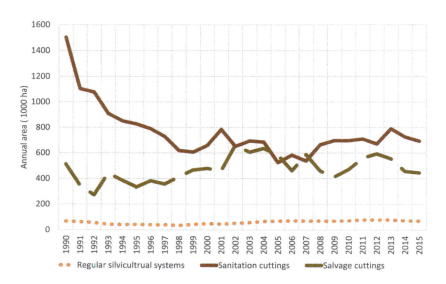

Fig. 1 Dynamics of the main types of fellings carried out in Romania since 1990 (*Source* Romanian national yearbooks)

systems and salvage fellings was still large in 2014, meaning that salvage cuttings replaced regular systems on large areas. The large share of sanitation cuttings is not a problem because the amount of wood per hectare and year is less than one cubic metre.

For two-thirds of the public forests managed by NFA, a feedback loop has been produced by the FSC® certification procedures in the sense that some TFB are left uncut in the forests. For the forests out of the scope of FSC® certification, maintaining a certain number of TFB is optional, but these trees must be properly labelled in the forest.

Romanian forestry has been confronted not only with *illegal logging* (Bouriaud and Marzano 2016) but also with the erosion of foresters' professional prestige (Lawrence 2009), undermined by an unsteady institutional and legal framework (Abrudan 2007; Knorn et al. 2012). Pursuing the same technical standards conceived as a command and control economy, forestry professionals face challenges in reconciling traditional forest management practices with the new socio-political context. The shift from the old paradigm which states that "all dead trees must be harvested", to a new one claiming that a certain number of dead trees must be spared for biodiversity purposes needs new procedures to train the forest rangers and the forest inspectors. Indeed, keeping TFB in order to maintain habitat for insectivorous birds doesn't help forest protection when the affected species are elm and ash as all affected trees must be harvested shortly after attack, without paying any attention to TFB. However, such situations are beyond the scope of this training scheme simply because such a process is even harder than one would expect because the professional responsibility of foresters has been eroded by the long and confusing process of land restitution.

2   Goal of the Study

The current policy is that NFA professionals (forest rangers and engineers) must maintain a certain amount of TFB to comply with the FSC® requirements. This goal has far-reaching implications at the level of forestry culture. Thus, this chapter describes a training drill that was developed and tested to provide the necessary tools for forest rangers and inspectors as they fulfil their obligations for forest management under the FSC requirements. The process for finding dead and weakened trees, marking, measuring them and storing the data into a tablet requires good coordination across a team of 2–3 people.

When it comes to salvage cuttings, it is not only about poor practices; it is about adopting a different mindset about what a healthy forest should mean: a series of stands of perfect and healthy trees or a series of stands enriched in biodiversity? Although some insects or fungi diseases (like the ones affecting elm and ash trees) cannot be eliminated by other means than sanitation/salvage fellings (sometimes resembling clearcuttings on small areas), in much numerous cases the foresters have been applying small-scale salvage fellings just to avoid harvesting operations in remote compartments (i.e. not related to biotic or abiotic threats). At the same time, the challenge of sparing some standing dead or near dead trees for complying with FSC® certified forests compounds the fieldworks carried out by the professional foresters (rangers, technicians and engineers).

The traditional way of marking the tree for fellings has never required "undo" or "unmark" procedures, excepting forest offences, when trees are demarked, and a special procedure applies. Because demarking a single tree takes time, the foresters must keep track of all trees already marked in the same compartment and the "undo" decision should be made prior to stamping. Indeed, avoiding (i.e. undoing) wrong stampings can be better learned if both operations are carried out simultaneously. In so doing the felony of marking a supposedly dead tree for salvage fellings can be avoided by marking that tree as TFB.

Learning to balance the tendencies to mark too many TFB (just for getting rid of duty) or too many salvage trees (as most of the foresters are currently doing) requires a thorough understanding of the role played by TBF. When it comes to reaching a certain amount of deadwood per hectare things are more complicated for two reasons: (1) the alternative to TFB is salvage, which may produce some profit and (2) TFB shall be evenly spread throughout the forest area. Therefore, the fieldwork requires competence in assessing the health status, identifying the most contagious pest and insects and assessing the volume of any affected tree. All these activities have been carried out on regular basis except for the decision to stamp TFB. This new series of decisions makes the difference between the mechanistic approach and the new one for two reasons at least: (1) a negative feedback loop is being triggered by the simple fact that two options are at hand, not only one; (2) a glimpse of reflection prior to stamping a tree is in place, in order to recall similar situations encountered in the past. Adding to these two mental processes, a more profound understanding of the forest ecosystem functions and boundaries, a keen sense of negotiation between environmental and economic goals and the ability to integrate newly acquired knowledge in everyday life, we have four out of the five strands of social learning, identified by Keen et al. (2005).

Assuming that none of the forest professionals wants to break the law by cutting healthy trees, we tried to conceive a sort of game with rules inspired not only by the legal framework but also nurtured by the belief that a certain amount of deadwood is welcome in any mature forest. It also provides a compromise between harvesting all dead trees and letting some TFB remain. Our game was inspired by the Operant Learning Theory (OLT), also known as Operant Conditioning Theory (OCT).

Burrhus Frederic Skinner, the American psychologist who developed OL/CT in the late 1930s, defined the goal of any learning process as changing the probability of having a certain response, under specific conditions (Skinner 1938; Thyer et al. 2012). In this regard, he suggested that learning "*is a series of discriminative stimuli and hence a series of reinforcers. It reinforces the act of blazing or otherwise marking the trail. Marking a path is, technically speaking, constructing a discriminative stimulus. The act of blazing or otherwise marking a trail thus has reinforcing consequences*" (Skinner 1988, 221). He also hypothesised that quite a large proportion of human behaviour is controlled by rules rather than by direct reinforcers. From his point of view, the outcome of applying a rule is a consequence of a particular response to a particular stimulus (Skinner 1969). In the context of salvage cuttings, we had to consider the real reinforcements and penalties brought about by the legal framework that refers to timber cruising and the FSC® standards and that influence the decision to "mark it as salvage timber" or "tag it as TFB".

Despite the fact that OL/CT oversimplifies the learning process and does not seem suitable for more complex learning situations, in this very particular case, where real penalties may apply as fines or may bring about major conditions,[4] according to FSC® procedures, we developed a training scheme inspired by OL/CT. Forest rangers, forest engineers and forest inspectors who are responsible for monitoring timber cruising for salvage and sanitation fellings could be better trained for tagging the TFB required by FSC® standards. It does not mean that all trees affected by pests should be kept uncut for the sake of biodiversity; it only implies that in the healthy forest a certain number of TFB shall be maintained. That being said, in "hotbed" areas, where infestation rates are very high, it is likely that the best approach to control the outbreak is for sanitation felling, maintaining no TFB in this instance.

3 Methodology

3.1 Operant Learning/Conditioning Theory

Basically, OL/CT assumes that behaviours are driven by reinforcements and punishments. Reinforcement occurs whenever an intensifying stimulus increases the likelihood to reproduce that behaviour—this is positive reinforcement; negative reinforcement is associated with a higher probability to maintain a given behaviour under decreasing stimuli. A punishment is a stimulus that reduces behaviour likelihood, and the same dichotomy applies; positive punishment—more stimulus, greater likelihood to resume the behaviour, and negative punishment—less stimulus, the lesser likelihood of maintaining that specific behaviour.

Two principles apply to OL/CT: (1) *Immediate consequences* (reinforcements or punishments) exert a stronger influence on behaviour than delayed consequences, and (2) behaviours already established can be maintained but with less effort (either slim rewards or penalties).

Apart from a long series of clinical studies focused on child behaviour therapy, summarised by Carr and Durand (1985), only one paper is relevant to our approach and refers to financial incentives for weight control (Jeffery 2012).

In our study, we have identified the optimum management option as maintaining a given number of standing dead trees per hectare with less sanitation fellings applied than is currently practised. A stimulus is the occurrence of a new "candidate" tree, physiologically weak and/or unsuitable for being felled for lumber or firewood: this is the "perfect" TFB. Further, on observing the methodical pattern of OL/CT, the reinforcements and punishments have been defined as follows:

(1) Positive reinforcements: tag a dead tree with "B" (for TFB);
(2) Negative reinforcement: tag a dead tree with "S" (sanitation cutting);
(a) Positive punishment: change the tag from "S" to "B";
(b) Negative punishment: change the tag from "B" to "S".

Tagging a TFB is the positive reinforcement because each new dead tree the operator comes across is a stimulus to look for another one, which can also be a TFB or a salvage tree. Conversely, marking a tree for salvage felling is negative reinforcement because it may be a strategy to harvest more wood in the most convenient way, as is happening now.

Demarking a tree from salvage to TFB is a positive punishment because that tree must be erased from the records and the effort and time taken to gauge its diameter, height and quality class is a waste of time.[5] The opposite action is a negative punishment because the field team must go back to a tree that has just been analysed, maybe a few minutes before. Erasing the letter "B" painted on its bark and resuming the timber cruising operations is obviously less costly than the previous operation.

Assessing the most appropriate number of TFB and salvage trees in any given stand is difficult. For this study, the "optimal" B/S ratio was estimated as 0.3–0.75. This was calculated by drawing on crowdsourcing data produced by a group of volunteers for the Romanian Ministry of Water and Forests in 2016 along with data taken from the evaluation forms issued in the last two years by Suceava branch of NFA relating to the average number of trees harvested per hectare as salvage cuttings. By combining the two datasets, a series of ratios were produced between the number of trees marked for the two types of sanitation fellings (S) and standing deadwood trees (B), per hectare.

3.2 Rules of the Game

Drawing on OL/CT, a field game for two teams of forestry students was designed to simulate the timber surveys in a mixed forest with Norway spruce and beech. The game involved the following rules:

(1) The final B/S ratio reached by each team after six hours of fieldwork should fall between 0.3 and 0.75;
(2) The positive punishment (F_{p+}) should be higher than the negative punishment (F_{p-}) for the reasons already explained. Both teams were encouraged to avoid as much as possible demarking salvage trees, once they have been impressed with the hammer, measured and recorded into the field evaluation form.
(3) An additional penalty per cubic metre was applied whenever a healthy tree was marked as salvage (according to Romanian legislation this is illegal);
(4) Each TFB is marked with a yellow fabric strip and each salvage tree with a red fabric strip.
(5) The total worth of the salvage trees is estimated according to the official rules, and the lump sum of all punishments are deducted from this value.

The same portion of natural forest was surveyed by two teams, each team consisting of three students; the scores of each team were updated each hour. In addition, there was a qualitative indicator of the work done by each team, which recorded how many times each team went "outside the box" of the B/S optimal range, and the amount of money "earned" by each team after six hours of fieldwork.

3.3 Timber Survey Location

The methodological framework was tested in September 2016 over two days, with two crews of three students each in Rarau-Giumalau natural reserve; this reserve harbours two Natura 2000 sites (see Fig. 2). The protected area is covered with old-growth forests of beech and Norway

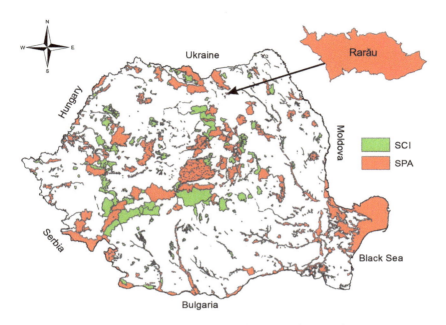

Fig. 2 Location of Rarau-Giumalau natural reserve (*Source* http://natura2000.eea.europa.eu)

spruce, and the natural selection is very intense. Hence, dead trees smaller than 20 cm in diameter were not taken into consideration either for salvage cuttings or deadwood because they do not occur very often in managed forests.

We chose a natural reserve because the density of dead and dying trees is much higher than in a managed forest; thus, it was less time consuming for the fieldwork carried out by the students.

The students were instructed to select trees larger than 30 cm in diameter as standing deadwood, observing the recommendation found in the literature (Dudley and Vallauri 2004). The positive punishment was set to 10 €/m^3, and the negative punishment to 3 €/m^3. The effective location of the six compartments where the timber cruising drill took place is presented in Fig. 3, indicated by the red line.

Both teams were organised in the same way: one student searched for the dead trees, while others measured the selected trees: species,

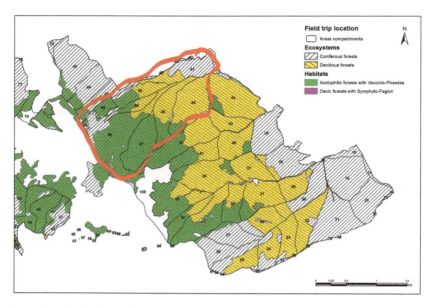

Fig. 3 Precise field-trip location in Rarau-Giumalau natural reserve (*Source* the management plan)

diameter, height and the wood quality (quality grade for salvage products, or decaying level for deadwood).

4 Results

The main outcome, in terms of B/S ratio per hour, is presented in Fig. 4. Because the sanitation tracts were not confined to a certain compartment or sub-compartment, the crews were advised to zigzag (uphill) within all compartments planned to be surveyed in a working day.

The penalties per hour (positive and negative punishments) are summarised in Fig. 5. All in all, the second team was penalised with 23.6€, while the first team, allegedly more efficient, was penalised with 14€. During the first hour, the first team got two negative punishments for swapping two TFB for sanitation cuttings, while the second team started a little bit awkwardly and got a positive punishment for demarking a salvage cutting (being afraid of not having enough TFB).

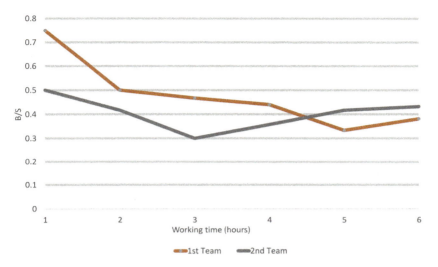

Fig. 4 Learning progress by working hours

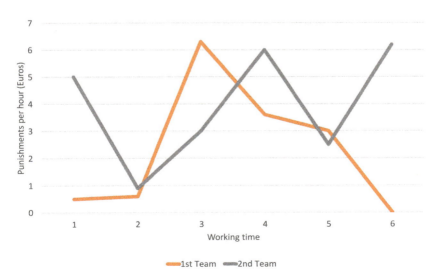

Fig. 5 Penalties recorded by the two teams

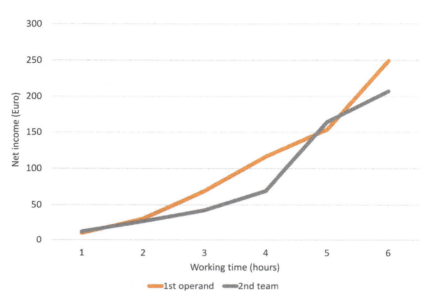

Fig. 6 Net cumulative incomes per hour

The dynamics of the net cumulative revenues gained by the two teams (income from salvage timber minus penalties) are presented in Fig. 6. The average number of trees marked for sanitation cuttings was seven trees per hour, for both teams. Two different strategies for tagging the trees were identified after the first two hours: the first team plunged from a high B/S ratio of 0.75 at the end of the first hour to nearly 0.33 at the end of the fifth hour; the second team worked steadily, keeping the B/S ratio near 0.4, which is close to the lower limit.

5 Summary

The first team tagged many TFB shortly after commencing the field-work (see Fig. 4), and more sanitation fellings afterwards; therefore, the "revenue" (the estimated worth of the trees to be harvested) went up faster in the last two hours of the working day, compared with the "revenue" gained by the second team, as shown in Fig. 6.

Did the first team do a better job, marking more trees for harvesting and lesser TFB at the end of the day? It is hard to tell because the quality of the work done depends to a large extent on the harvesting conditions for each tree or bunch of trees marked for salvage or sanitation cuttings. If the sanitation cuttings are dispersed in remote areas (and riparian, in most of the cases), it is better to tag those trees as TFB, because the cost of collateral damage brought about by harvesting operations is higher than the expected revenue of sanitation/salvage cuttings. These damages refer to wounds produced to other remnant trees and topsoil removal because each log needs to be towed for long distances.

So far these issues have never been contemplated by the professional foresters because they have had no other option than sanitation or salvage fellings (as already mentioned, the difference between sanitation and salvage is the amount of harvestable wood per year and hectare). Moreover, pursuing that threshold on one cubic metre per hectare per year is technically difficult when the whole forest is healthy, and trees are older than 60 years. Since the harvestable trees are rare, it is easier and cost-effective to mark salvage cuttings instead of sanitation, having the additional benefits already mentioned (the possibility to harvest less mature and over-mature stands, and less money paid to the regeneration fund).

Tagging TFB at a constant pace (as the second team did) is the best option in any situation, but the opposite strategy, chosen by the first team, could also be optimal if the fieldwork started from the top, not from the valley. Yet if happens that TFB are not evenly spared, it is better to have higher concentration of TFB uphill than downhill simply because the habitats are less disturbed uphill by anthropogenic factors, like illegal harvesting operations or poaching. Even though the second team earned less than the first team (see Fig. 6), its strategy of maintaining a constant trade-off between preserving biodiversity and salvage cuttings is recommended in any situation.

We confined our training scheme to maintaining a certain ratio between the two cumulated numbers of trees (S/B), and not to pursuing a certain amount of deadwood per hectare (as literature recommends) because, in the latter case, TFB refers to all types of deadwood (including snags laid on the ground), not only to the standing dead trees. However, once the stratum of TFB has been settled, a thorough monitoring of the decaying process shall be pursued afterwards.

6 Discussion

Through this small-scale training project, we tried to develop a training framework for students and professional foresters to encourage them to behave as information processors rather than simply acting. Even though saving a certain number of TFB will not substantially improve the forest health, it is a good premise for managing deadwood. Managing the deadwood involves a fairly complex screening process but getting enough forests managed in this way is the first condition of having a biodiversity monitoring system implemented, as FSC® and Natura 2000 management plans compel. That being said, for the purpose of this training programme, we did not consider the amount of deadwood but rather focused on the spatial distribution and the balance between salvage cuttings and TFB.

We conclude that by applying the new training scheme, foresters will be deterred from marking all big dying trees for salvage or sanitation felling and small trees as TFB, as they will realise that keeping small-size trees as TFB is not a long-term solution. If small trees are tagged as TFB, they eventually will be blown down and must be replaced by identifying other TFB, which requires additional effort next year. On the other hand, an old tree, not yet dead but physiologically very weak, can be confidently tagged as TFB. This type of conduct is encouraged by the new training system, which is unparalleled by any drilling scheme based on the technical standards only.

A similar game can be designed for the first thinnings when a certain set of "trees for the future" must be tagged, while another set of trees are to be harvested; such a drill is extremely important in mixed forests, where different species have different commercial and ecological values. However, it would be quite a challenge to design a drill for selecting the trees to harvest from mixed high forests when the group system is applied, as this would involve considering the ratio between shade-tolerant and light demanding species, the terrain aspect and the desired composition of the future generation of trees.

The training scheme presented in this chapter is the first attempt at solving the TFB issue. No other alternative exists, except for a simple checklist with criteria for salvage cuttings *and* TFB, used by different people, at different times. However, instead of training the foresters to go through the same area twice, firstly for salvage cuttings and secondly for a thorough selection of TFB, we came up with a training scheme that helps people address the two issues simultaneously.

Applied to forestry, the method presented in this chapter is not only about training, it is about changing the professional culture, in the sense that foresters should account for biodiversity issues on a regular basis. FSC® standards, embraced by the NFA require an integrated approach to pest control, subject to a regular audit, carried out by a different auditing company. Selecting TFB is just the first step towards having implemented the biodiversity management system.

Such a new approach to supporting professional training could make all the difference between the current behaviour of foresters, which is a mixture of rent-seeking practices (Nichiforel and Schanz 2011) and the desired behaviour, based on rapid and cost-effective field assessments. Keeping some standing TFB is just a part of the solution to the very complex problem of forest health under climate change. However, without a clear methodology properly designed for selecting TFB, all discussions around the biodiversity topics were somehow futile as long as professional foresters could not learn new practicalities, starting with a ratio between TFB and trees for salvage fellings. For the time being, this figure shall not be debated too much; rather, it might be regarded as a simple hint towards getting a trade-off between social aspects of forestry (like firewood provision) and biodiversity goals. Linking the two issues in a single training scheme is also important for getting to terms with local communities, who perceive biodiversity as a threat and forest diseases as opportunities for having cheap fuelwood. Instead of making the worst of them, we tried to make the best of the two worlds by designing this training scheme.

Maintaining forest health has many dimensions because each pest or disease outbreak needs to be treated individually, taking into account the weather conditions, the aspect, the stand density, the magnitude of potential damages and the biotic and abiotic propagation factors. There is no panacea in this respect, and the approach presented in this chapter offers an attempt to address just a small fraction of the whole problem.

Acknowledgements I am grateful to the two anonymous peer reviewers for their support in improving the first version of this chapter. I also acknowledge Mrs. Monica Vasile for her useful comments and feedback on the second draft of this paper, and to Mr. Dan Grigoroaea, ranger with the Muntii Calimani National Park for valuable hints about the ecologically optimal amount of deadwood per hectare in Norway spruce forests. Many thanks to my colleagues from Suceava office of the National Forest Administration for supporting our research with field data concerning sanitation cuttings.

Notes

1. This type of scam is no longer possible now due to the wood tracking system implemented after 2015.
2. Court of Accounts is the central authority in financial matters and public fund and public assets.
3. No evaluation form based on legal measurements, no marks applied to the tree prior to harvesting operations.
4. In FSC terminology, a major condition is a bunch of actions or misconducts that must be corrected within three months if the certificate was issued or within a year if the certification process is ongoing.
5. The two stamp impresses made with a special hummer (one on the stump and one on the trunk) must be taken away and destroyed according to a special procedure, which makes the process expensive.

References

Abrudan, I. V. (2007). Cross-sectoral linkages between forestry and other sectors in Romania. In Y. C. Dubé & F. Schmithüsen (Eds.), *Cross-sectoral policy developments in forestry* (pp. 183–189). Rome, Italy: CABI.

Abrudan, I. V., Marinescu, V., Ionescu, O., Ioras, F., Horodnic, S. A., & Sestras, R. (2009). Developments in the Romanian forestry and its linkages

with other sectors. *Notulae Botanicae Horti Agrobotanici Cluj-Napoca, 37,* 14–21.
Anderegg, W. R. L., Plavcová, L., Anderegg, L. D. L., Hacke, U. G., Berry, J. A., & Field, C. B. (2013). Drought's legacy: Multiyear hydraulic deterioration underlies widespread aspen forest die-off and portends increased future risk. *Global Change Biology, 19,* 1188–1196.
Anonymous. (2016). Ministry orders published in official. *Journal of Romania,* vol. 539, 551.
Borlea, G. F. (2004). Ecology of elms in Romania. *Forest Systems, 13,* 29–35.
Bouriaud, L., & Marzano, M. (2016). Conservation, extraction and corruption: Is sustainable forest management possible in Romania? In E. Gilberthorpe & G. Hilson (Eds.), *Natural resource extraction and indigenous livelihoods: Development challenges in an era of globalization* (pp. 221–240). London and New York: Routledge, Taylor & Francis.
Brasier, C. M., & Scott, J. K. (1994). European oak declines and global warming: A theoretical assessment with special reference to the activity of *Phytophthora cinnamomi. EPPO Bulletin, 24,* 221–232.
Carr, E. G., & Durand, V. M. (1985). Reducing behaviour problems through functional communication training. *Journal of Applied Behaviour Analysis, 18,* 111–126.
Coleman, D., Georgiadou, Y., & Labonte, J. (2009). Volunteered geographic information: The nature and motivation of producers. *International Journal of Spatial Data Infrastructures Research, 4,* 332–358.
Drăgoi, M., Palagianu, C., & Miron-Onciul, M. (2015). Benefit, cost and risk analysis on extending the forest roads network: A case study in Crasna Valley (Romania). *Annals of Forest Research, 58*(2), 333–345. https://doi.org/10.15287/afr.2015.366.
Drapeau, P., Nappi, A., Imbeau, L., & Saint-Germain, M. (2009). Standing deadwood for keystone bird species in the eastern boreal forest: Managing for snag dynamics. *The Forestry Chronicle, 85*(2), 227–234. https://doi.org/10.5558/tfc85227-2.
Dudley, N., & Vallauri, D. (2004). *Deadwood–Living forests.* WWF report, October 2004. Gland, Switzerland: World Wildlife Fund for Nature. http://wwwpanda.org/downloads/forests/deadwoodwithnotes.pdf.
Duduman, M. L., Olenici, N., Olenici, V., & Bouriaud, L. (2014). The impact of natural disturbances on the *Norway spruce* special cultures situated in Nord Eastern Romania, in relation with management type. In E. Scrieberna & M. Stark (Eds.), *Adaptation in forest management under*

changing framework conditions. Proceedings of IUFRO Symposium, 19–23 May 2014 (Groups 3.08.00 and 4.05.00) (pp. 44–45). Sopron, Hungary.

Estellés-Arolas, E., & González-Ladrón-De-Guevara, F. (2012). Towards an integrated crowdsourcing definition. *Journal of Information Science, 38*, 189–200.

Flannigan, M. D., Stocks, B. J., & Wotton, B. M. (2000). Climate change and forest fires. *Science of Total Environment, 262*(3), 221–229.

Gavrilut, I., Halalisan, A.-F., Giurca, A., & Sotirov, M. (2015). The interaction between FSC certification and the implementation of the EU timber regulation in Romania. *Forests, 7*, 3.

Hălălișan, A. F., Marinchescu, M., & Abrudan, I. V. (2012). The evolution of forest certification: A short review. *Bulletin of the Transylvania University of Brasov, Series II. Forestry, Wood Industry, Agricultural Food Engineering.*

Humphrey, J. W., Sippola, A. L., Lempérière, G., Dodelin, B., Alexander, K. N., & Butler, J. E. (2005). Deadwood as an indicator of biodiversity in European forests: From theory to operational guidance. *Monitoring and Indicators of Forest Biodiversity in Europe—From Ideas to Operationality, 51*, 193–206.

Jeffery, R. W. (2012). Financial incentives and weight control. *Preventive Medicine, 55*, S61–S67.

Johansson, J., & Lidestav, G. (2011). Can voluntary standards regulate forestry? Assessing the environmental impacts of forest certification in Sweden. *Forest Policy and Economics, 13*(3), 191–198.

Jonášová, M., & Prach, K. (2004). Central-European mountain spruce (*Picea abies* (L.) Karst.) forests: Regeneration of tree species after a bark beetle outbreak. *Ecological Engineering, 23*, 15–27.

Joseph, G. S., Cumming, G. S., Cumming, D. H., Mahlangu, Z., Altwegg, R., & Seymour, C. L. (2011). Large termitaria act as refugia for tall trees, deadwood and cavity-using birds in a miombo woodland. *Landscape Ecology, 26*, 439–448.

Keen, M., Brown, V. A., & Dyball, R. (Eds.). (2005). *Social learning in environmental management: Towards a sustainable future.* New York: Routledge.

Knorn, J., Kuemmerle, T., Radeloff, V. C., Szabo, A., Mindrescu, M., Keeton, W. S., et al. (2012). Forest restitution and protected area effectiveness in post-socialist Romania. *Biological Conservation, 146*, 204–212.

Lassauce, A., Lieutier, F., & Bouget, C. (2012). Woodfuel harvesting and biodiversity conservation in temperate forests: Effects of logging residue characteristics on saproxylic beetle assemblages. *Biological Conservation, 147*, 204–212.

Lawrence, A. (2009). Forestry in transition: Imperial legacy and negotiated expertise in Romania and Poland. *Forest Policy and Economics, 11*, 429–436.

Lučan, R. K., Hanák, V., & Horáček, I. (2009). Long-term re-use of tree roots by European forest bats. *Forest Ecology and Management, 258*(7), 1301–1306.

Miles, D. B., & Ricklefs, R. E. (1984). The correlation between ecology and morphology in deciduous forest passerine birds. *Ecology, 65*, 1629–1640.

Nagel, T. A., Diaci, J., Jerina, K., Kobal, M., & Rozenbergar, D. (2014). Simultaneous influence of canopy decline and deer herbivory on regeneration in a conifer–broadleaf forest. *Canadian Journal of Forest Research, 45*, 266–275.

Nichiforel, L., & Schanz, H. (2011). Property rights distribution and entrepreneurial rent-seeking in Romanian forestry: A perspective of private forest owners. *European Journal of Forest Research, 130*, 369–381.

Olenici, N., Duduman, M.-L., Olenici, V., Bouriaud, O., Tomescu, R., & Rotariu, C. (2011). The first outbreak of *Ips duplicatus* in Romania. In H. Delb & S. Pontuali (Eds.), *Proceedings of the Working Party 7.03.10 Methodology of Forest Insect and Disease Survey in Central Europe* (pp. 135–140), 10th Workshop, Vol. 89, 20–23 September 2010, Freiburg, Germany. Freiburg: FVA.

Panayotov, M., Bebi, P., Tsvetanov, N., Alexandrov, N., Laranjeiro, L., & Kulakowski, D. (2015). The disturbance regime of Norway spruce forests in Bulgaria. *Canadian Journal of Forest Research, 45*, 1143–1153.

Pautasso, M., Aas, G., Queloz, V., & Holdenrieder, O. (2013). European ash (*Fraxinus excelsior*) dieback–A conservation biology challenge. *Biological Conservation, 158*, 37–49.

Radu, S. (2006). The ecological role of deadwood in natural forests. In D. Gafta & J. Akeroyds (Eds.), *Nature conservation, environmental science and engineering (environmental science)* (pp. 137–141). Berlin, Heidelberg: Springer.

Schroth, G., & McNeely, J. A. (2011). Biodiversity conservation, ecosystem services and livelihoods in tropical landscapes: Towards a common agenda. *Environmental Management, 48*(2), 229–236.

Skinner, B. F. (1938). *The behavior of organisms*. New York: Appleton-Century-Croft.

Skinner, B. F. (1969). *Contingencies of reinforcement: A theoretical analysis*. Englewood Cliffs, NJ: Prentice Hall.

Skinner, B. F. (1988). An operant analysis of problem solving. In A. C. Catania & S. Harnad (Eds.), *The selection of behavior the operant behaviorism of B. F. Skinner: Comments and consequences* (pp. 221–222). Cambridge: Cambridge University Press.

Sproull, G. J., Adamus, M., Bukowski, M., Krzyżanowski, T., Szewczyk, J., Statwick, J., et al. (2015). Tree and stand-level patterns and predictors of Norway spruce mortality caused by bark beetle infestation in the tatra mountains. *Forest Ecology and Management, 354*, 261–271.

Stanciu, E. (2017). Secretary of state report at the end of the mandate. Manuscript (Unpublished).

Stanovský, J. (2002). The influence of climatic factors on the health condition of forests in the Silesian Lowland. *Journal of Forest Science, 48*, 451–458.

Thomas, F. M., Blank, R., & Hartmann, G. (2002). Abiotic and biotic factors and their interactions as causes of oak decline in Central Europe. *Forest Pathology, 32*, 277–307.

Thyer, B. A., Dulmus, C. N., & Sowers, K. M. (2012). *Human behaviour in the social environment: Theories for social work practice*. New Jersey: Wiley.

Verkerk, P. J., Lindner, M., Zanchi, G., & Zudin, S. (2011). Assessing impacts of intensified biomass removal on deadwood in European forests. *Ecological Indicators, 11*, 27–35.

14

The Effects of Mountain Pine Beetle on Drinking Water Quality: Assessing Communication Strategies and Knowledge Levels in the Rocky Mountain Region

Katherine M. Mattor, Stuart P. Cottrell, Michael R. Czaja, John D. Stednick and Eric R. V. Dickenson

1 Introduction

Climate change disturbances are recognized as a threat to water resources worldwide (Kiparsky et al. 2012; Kundzewicz et al. 2008; Pahl-Wostl 2007). The occurrence of increased salinization of coastal aquifers (Mazi et al. 2013), global flooding events resulting in contamination of drinking water (Cann et al. 2013), and ongoing drought across the globe are known to be tied to climate change (Kundzewicz et al. 2008). The responsiveness of water resources management to the complexities of climate is typically based on historical probabilities

---

K. M. Mattor (✉) · J. D. Stednick
Department of Forest and Rangeland Stewardship,
Colorado State University, Fort Collins, CO, USA

S. P. Cottrell · M. R. Czaja
Department of Human Dimensions of Natural Resources,
Colorado State University, Fort Collins, CO, USA

used to determine appropriate actions to minimize risk, i.e., through such measures as reservoirs for drought conditions. However, the ability to predict weather events and related disturbances is greatly reduced by climate change and introduces a need for increased adaptation in water resource decision-making (Pahl-Wostl 2007). Communication leading to knowledge exchange between scientists and water managers is necessary for successful adaptation in the face of climate-induced disturbances (Kiparsky et al. 2012). An improved understanding of the effects of ecological disturbances will enable those with responsibility for supplying drinking water to prepare for potential changes to drinking water quality, if necessary, or prevent unnecessary actions from being taken if limited effects are identified. Knowledge exchange also provides an improved opportunity for scientists to understand the challenges, concerns, and first-hand experience of drinking water providers and to apply this information to the formation and focus of their research.

The mountain pine beetle (MPB; *Dendroctonus ponderosae*) epidemic across western North America has raised concerns regarding its potential effects on forests and water resources (water quality and quantity) (Mikkelson et al. 2013b). Thus, the potential effects of ecological disturbances, such as MPB, require clear communication based on effective knowledge exchange between research scientists and drinking water providers (Anderson and Woosley 2005). In this context, sustainable natural resource management relies on the integration of knowledge across scientists and managers to develop a shared understanding of conditions and identify appropriate actions through

---

E. R. V. Dickenson
Water Quality Research and Development Division,
Southern Nevada Water Authority, Henderson, NV, USA

E. R. V. Dickenson
Department of Civil and Environmental Engineering,
Colorado School of Mines, Golden, CO, USA

bilateral communication between each group. The study reported here examines the existing levels of knowledge of the potential MPB effects on drinking water and associated concerns and communication needs of drinking water providers and drinking water professionals. Examining the current levels of communication associated and future needs can be used to inform adaptive capacity and management responses (Chapin et al. 2009; Kiparsky et al. 2012).

This chapter reports social science findings from an initial assessment as the first step in a larger five-year social-biophysical research effort (see http://igwmc.mines.edu/Research/WSC.html). The biophysical research conducted by scientists from the Colorado School of Mines examines the hydrological and biogeochemical effects of MPB to drinking water quality, while the social science component examines methods for improving scientist–manager communication and public outreach to integrate a collaborative learning process and effectively communicate project findings and management needs. An elicitation survey as this first step was used to identify the research information needs of drinking water providers and associated drinking water professionals, the perceived challenges associated with MPB and its effects on drinking water quality, as well as recommendations for communicating important water-related MPB issues to the general public in northern Colorado and southern Wyoming, USA. This assessment of current communication levels provides an important foundation for subsequent analysis of knowledge exchange across these constituents. These results will guide the development of learning-based approaches to improve engagement and knowledge exchange across scientists and drinking water providers.

In the following section, we provide an overview of the MPB epidemic in the US Rocky Mountain Region and examine barriers to communication and knowledge exchange between scientists and drinking water providers in attempts to manage it. The chapter concludes with a discussion of these findings in relation to the broader requirements of knowledge exchange and adaptation to climate-induced disturbances.

2 Mountain Pine Beetle Epidemic

Large forested areas of North America have been affected by insect outbreaks in recent decades. These beetle infestations are natural ecological processes. One of the primary insects is the MPB, endemic to the forests of western North America (Leatherman et al. 2007). The bark beetle typically infests and kills a small percentage of lodgepole pine trees within forested watersheds. However, more recently there has been a severe and widespread epidemic of the MPB in the region (CSFS 2016; USFS 2017). The MPB, along with several other species of bark beetles, have severely damaged coniferous forests in the western USA and Canada. Since 2002, unprecedented tree mortality has occurred across North America and is attributed to warmer air temperatures leading to increased MPB reproduction and greater susceptibility of trees to MPB because of drought stress (Hart et al. 2015; Hicke et al. 2012). This latest MPB infestation killed more than 3.4 million acres of Colorado lodgepole pine forest between 1996 and 2013 (CSFS 2016; USFS 2017).

The natural loss of trees in beetle-killed forests causes concern in terms of the epidemic's effect on streamflow generation mechanisms, water yield, and water quality (Bearup et al. 2014; Mikkelson et al. 2012; Mikkelson et al. 2013a, b; USFS 2017). Potential effects on water quality include increases in nitrogen and phosphorous concentrations, dissolved and total organic carbon (DOC and TOC, respectively), increased concentrations of heavy metals, natural organic matter (NOM), and disinfectant by-products (DBPs) precursors (Mikkelson et al. 2013a; Rhoades et al. 2013), in other words, the potential for unsafe drinking water.

Research by Mikkelson et al. (2012) using quarterly reports from drinking water providers showed higher TOC in MPB-impacted watersheds with respect to both mean and maximum concentrations compared to uninfected water supplies. Various mechanisms have been proposed to explain the increased TOC, including changes in acid deposition, variability in climate, and land use changes, or the MPB. Mikkelson et al. (2012) propose that the MPB infestation is another mechanism altering TOC loading and composition in surface and groundwater (Mikkelson et al. 2012). In parallel with increased TOC, Mikkelson et al. (2012) observed higher (DBP)[1] precursor concentrations in MPB-impacted watersheds. Combined, these were the first impacts on drinking water quality, due to climate change, observed during the 2004–2011 period (Mikkelson et al. 2012).

3   Communication and Knowledge Exchange Challenges

Climate change-induced ecological disturbances are increasingly recognized as one of the burgeoning challenges of the twenty-first century (Chapin et al. 2009). The ability of a social-ecological system to remain within a functioning state while adapting to disturbances has been tied to the system's level of adaptive capacity or resilience (Folke 2006). One of the several necessary preconditions for social adaptive capacity is knowledge transfer, defined as bilateral communication exchange leading to informed actions, with current literature emphasizing the need for increased knowledge transfer and communication to prepare for climate change-induced ecological disturbances (Dilling et al. 2015; Engle 2011; Folke et al. 2009; Walker and Salt 2006). This is especially true with regard to drinking water resources (Kiparsky et al. 2012). Knowledge consists of scientific and contextual information, values, and experience, all of which provide an individual with the basis for applying new information to guide their actions (Roux et al. 2006). The integration of knowledge across entities is defined as knowledge exchange and is considered successful when new information from one party is adopted by another (Roux et al. 2006). Numerous factors can influence the adoption of information, which results in changed behavior or actions at the individual and/or organizational levels.

Successful knowledge exchange between scientists and resource managers is often limited, and this decreases the efficiency and effectiveness of scientists' research and managers' actions (Addison et al. 2013; Cook et al. 2009; Hulme 2014). Several challenges to communication and resulting knowledge exchange across scientists and resource managers have been identified in previous studies. First, there are cultural differences between scientists and managers (Roux et al. 2006). These are two distinctly different populations with diverse educational backgrounds, motivations, and goals for their respective jobs, resulting in misunderstandings of the fundamental activities of each population (Borowski and Hare 2007; Gibbons et al. 2008; Timmerman et al. 2010). Such differences may lead to misunderstanding and confusion. Scientists believe managers do not understand the scientific process, are unable to convey their needs, and are not incentivized to address broader ecological issues (Roux et al. 2006). At the same time, managers believe scientists are detached from the issues occurring on the ground and are located within an insular system which does not produce information relevant to their needs (Roux et al. 2006).

The incentives scientists and managers encounter in their work are disparate; scientists are rewarded for novel findings, while managers are pressured to produce successful outcomes on the ground (Gibbons et al. 2008; Roux et al. 2006). To meet these objectives, managers tend to be more conservative and risk averse with regard to their management decisions and the information utilized to reach such decisions (Borowski and Hare 2007; Lemos 2008). Managers also tend to rely more heavily on past experience and traditional approaches than evidence-based knowledge reported by scientists when making management decisions (Cook et al. 2009; Hulme 2014). In contrast, scientists rely on model-based projections of anticipated disturbances even when the specifics of these changes cannot be specified (Addison et al. 2013; Kiparsky et al. 2012). These scientific findings often do not account for the tacit knowledge managers often rely on, resulting in limited implementation of the research findings (Hulme 2014).

The communication avenues between scientists and managers are key to knowledge exchange. These include limited opportunities for interaction between scientists and managers and differing sources of information (e.g., academic journals catered to research and association journals focused on implementation) (Gibbons et al. 2008). Opportunities for interaction are limited as scientists (e.g., university) are often too far removed from what is practiced in the field to fully understand the complexity of what is being practiced by managers (Borowski and Hare 2007; Dilling et al. 2015; Hulme 2014). Scientific papers catered to a specific audience or with increased technicality and presence of scientific jargon often lose relevancy to managers and are typically presented in journals specific to research audiences, rather than managers (Driscoll and Lindenmayer 2012; Hulme 2014; Medema et al. 2008). It is difficult to bridge this knowledge gap simply because the language and logic used within the two groups do not readily transfer among populations (Kieser and Leiner 2009). Differing information sources for these groups prevent scientists from understanding the site-specific information needs from the field and prevent managers from readily attaining up-to-date research findings (Mostert and Raadgever 2008).

These barriers to communication lead to gaps between the stated purpose of the scientific research and the managers' information needs (Dilling et al. 2015; Kiparsky et al. 2012; Medema et al. 2008). The mismatch between research purpose and management needs also results from inconsistent definitions of the problem, where researchers and managers may define or construe the issue differently in interdisciplinary work (Pahl-Wostl et al. 2007). With differing definitions of the issues, the identification of useful information across researchers and managers will be distinct (Timmerman et al. 2010). For example, in a study of the application of scientific-based water models by drinking water providers, Borowski and Hare (2007) found that this mismatch of purpose and need resulted in a lack of relevancy for the managers and rendered the models and exchange of knowledge unsuccessful.

Communication and associated knowledge exchange between scientists and drinking water providers can facilitate adaptation of water management systems to climate-induced disturbances. Contextual assessments of knowledge exchange identify the ability of a population or system to effectively respond to climate change-induced disturbances; yet, such assessments of water management systems are rare (Kiparsky et al. 2012). A bark beetle epidemic across forested watersheds in the western USA provides an opportunity to evaluate knowledge exchange between scientists and drinking water providers. This paper reports an initial exploration of knowledge levels and communication needs of drinking water providers to set the stage for further assessments of knowledge exchange across these entities in this region. The goal of this study was to increase learning-based engagement and subsequent knowledge exchange across scientists and drinking water managers.

4 Methods

The qualitative research reported here comprises the first phase of the social science component of a larger project funded by the National Science Foundation–Water Sustainability Climate Program (NSF 2015). The project investigates potential water resource changes resulting from the MPB epidemic and the associated drinking water management response to define feedbacks between climate change, insect-driven forest disturbance, biogeochemical processes, and management (e.g., forest and water treatment) practices. The study area is in northern Colorado and southern Wyoming of the western USA where over four million acres of forest were affected by this bark beetle (CSFS 2015; USFS 2017) (Fig. 1). This area includes the Platte River and Colorado River watersheds, which supply water to over 30 million residential users and 3.5 million acres of irrigated agricultural land (Colorado Watershed Assembly 2015).

The goal of this study was to identify the research information needs of drinking water providers and water-related professionals in Colorado and Wyoming regarding the MPB disturbance effects on drinking water resources. This study achieved this through three objectives: (1) explore

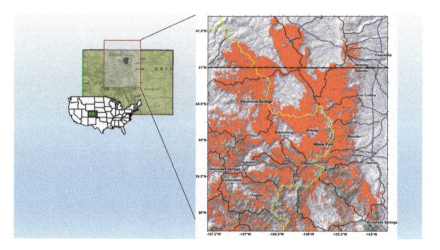

Fig. 1 Map of mountain pine beetle outbreak in northern Colorado and southern Wyoming, USA

drinking water provider and water-related professionals' knowledge levels regarding MPB effects on drinking water resources; (2) identify drinking water provider and professionals' perceptions of the potential effects and associated challenges from MPB affecting drinking water resources; and (3) determine important issues related to MPB and drinking water resources for future communication and knowledge exchange.

4.1 Stakeholder Survey

An Internet-based survey was developed to determine levels of knowledge and the information needs of drinking water providers and water-related professionals with respect to the MPB. The dispersed nature of the Internet-based survey approach reduces potential bias raised through group dynamics and provides respondents with the necessary flexibility and time to submit their responses (Doria et al. 2009). The survey draws on an expert elicitation approach. This technique is used to gather and summarize expert knowledge related to a specific topic of interest (Bosetti et al. 2012; Martin et al. 2012; Van Houtven et al. 2014). Expert elicitation in environmental and social sciences has increasingly been used to guide decision-making and inform future research endeavors (Donlan et al. 2010; Martin et al. 2012). For an exploratory analysis, this study did not utilize the complex expert elicitation approach formalized for quantitative prediction or uncertainty models (e.g., Donlan et al. 2010; Doria et al. 2009; Martin et al. 2012). A purposeful non-probability sampling of drinking water providers and associated professionals was used instead to provide foundational data for subsequent research on communication, knowledge exchange, and adaptive capacity in relation to ecological disturbances affecting drinking water resources (Creswell 2014).

4.2 Survey Development

A review of literature on public perceptions of MPB effects on forest health and water resources was conducted to help inform the development of the survey. The survey consisted of five open-ended questions concerning respondent knowledge of science and research on MPB, perceived drinking water issues and challenges, effective public communication and outreach about MPB effects on drinking water resources, and demographics (Table 1). Open-ended questions and qualitative analysis were used to obtain more detailed individualized responses and unexpected phenomena otherwise unattainable through closed-ended questions (Denzin and Lincoln 1998; Strauss and Corbin 1998). The online survey approach allowed for a much larger survey sample than face-to-face interviews, while the open questions provided respondents with the opportunity to provide detailed responses. A pilot test of the survey was conducted in spring 2013 with 14 colleagues, resulting in minor revisions to the survey, prior to its release.

4.3 Study Population and Sampling Procedures

Drinking water providers and professionals working with drinking water providers in northern Colorado and southern Wyoming were

Table 1 Survey objectives and question

Objective	Question
Knowledge levels	1. What do you already know about the science/research surrounding MPB impacts on water resources?
Information needs	2. Are there parts of the science/research surrounding MPB impacts on water resources that needs to be communicated to you more effectively? If so, please explain
Challenges of MPB epidemic	3. In general, what do you think are the biggest issues or challenges related to MPB affecting water resources? For each issue and challenge you list, please also indicate your recommendation(s) for best addressing that issue or challenge
	4. Do you feel that the quality of our drinking water has deteriorated or become unsafe to drink as a result of the MPB? In other words, are there any safety or health issues that downstream users should be concerned about?
Important issues to communicate with public	5. What are some key messages surrounding MPB impacts on water resources that need to be communicated to the public to improve their understanding?

identified as the focal population using a purposeful, non-probability sample (Creswell 2014). Professionals working with drinking water providers included research consultants, federal and state agency water experts, and academics. Respondents were identified through attendance lists from the June 2013 American Water Works Association (AWWA) Annual Conference and Exposition in Denver, Colorado, and the AWWA Rocky Mountain Region Section conference, held at Keystone, Colorado, in September 2013. Non-probability sampling, using inclusion and exclusion criteria, was used to identify the sample of 682 invited respondents from the conference attendees. Inclusion criteria included drinking water providers and water professionals within the Rocky Mountain Region that were involved in the provision of drinking water resources and provided their consent to participate in the research. The exclusion criteria included attendees working outside the study region and not involved in drinking water provision or management. Using the tailored design method, reminder emails were sent to non-respondents within two weeks of the initial invitation and the survey was available online for three months (September to November 2013) (Dillman et al. 2009).

4.4 Analysis

A qualitative thematic analysis was used via QSR NVivo (version 10) software to organize the data into key themes (Creswell 2014). These themes included information needs, watershed management issues, and drinking water challenges. We used open and axial coding to determine themes and sub-themes that emerged from the responses and that had not previously been identified (Creswell 2014; Denzin and Lincoln 1998; Strauss and Corbin 1998). Qualitative responses were categorized and independently coded by three of the authors to assure data validity. Percentages were used to describe the sample and the proportion of respondents with high versus low knowledge levels about the mountain beetle effects on water quality to better understand responses to the other questions in the survey.

5 Results

Of the 178 respondents who began the survey, a total of 96 respondents completed questions relevant to the analysis reported here. The majority (53%) of these respondents identified themselves as drinking water providers, while 27% were associated with consulting firms. Other respondents (20%) were associated with academic institutions, state or federal agencies, or community watershed groups.[2] The number of respondents with these affiliations corresponded with the proportions on the invitation list. On average, the respondents had held their current position for 13 years, ranging from 1 to 45 years of experience. The majority (79%) were working within drinking water system agencies serving over 10,000 people, and 73% used surface waters, rather than groundwater, as the water supply. Eighty-six percent of the respondents reported their associated water source originated from an MPB-affected watershed.

5.1 Knowledge Levels

One study objective was to identify the current levels of knowledge of water providers and other professionals in relation to potential MPB effects on drinking water (Table 1, Question 1). An analysis of these open-ended responses categorized respondents into four levels—none, little, moderate, and high—with those who stated they had such levels of knowledge cataloged into the relevant categories. For example, the respondents who stated they were very knowledgeable of the MPB impacts on water resources were placed within the high knowledge category. Overall, 55% of respondents reported having little to no knowledge about the MPB effects on drinking water resources, while 33% had moderate and 12% had high levels of knowledge. Taking a closer look, many drinking water providers reported having no or little knowledge of the effects of MPB on drinking water resources (57%), while only 19% reported moderate levels and 2% reported high levels of knowledge. In contrast, 48% of the remaining respondents reported little to no knowledge, 39% reported moderate levels, and 20% indicated high levels of knowledge of the MPB effects on drinking water. Where applicable the topics identified by the respondent were used to verify moderate and high knowledge levels. Open responses to Question 1 indicated self-reported knowledge levels and the topics the respondent was familiar with (Table 2). Topics identified by those with little reported levels of knowledge included forest management challenges, potential effects of wildfire on water quality, and soil erosion concerns. One respondent self-reported low knowledge but identified "forest fire, dead tree removal and soil stabilization in watershed area" as the biggest challenge of the MPB infestation. The most common areas of prior knowledge across respondents with moderate to high levels of knowledge pertained to potential effects on water quality, snowpack, and changes in runoff volume and timing. For example, one self-reported high knowledge respondent stated, "Type and quantity of TOC increases (if there are any); DBP formation potential; metals increases; length/severity of water quality changes." Respondents with moderate

Table 2 Topics identified by reported knowledge level

Knowledge level	Topics identified
None ($n = 26$) (28%)	Not applicable
Little ($n = 23$) (25%)	Forest management challenges
	Potential effects of wildfires on water quality
	Erosion and runoff impacts
	Limited information is available
Moderate ($n = 33$) (36%)	Water quality impacts (TOC, DBP precursors, organic matter, nutrients, color, metals)
	Increased risk of wildfire
	Increased runoff and associated erosion, turbidity, sediment
	Reduced snowpack
	Limited impact to water quality
	Read research papers
High ($n = 11$) (12%)	Water quality impacts (NOM, DBP precursors, TOC)
	Water yield, runoff
	Involved in mitigation, outreach, research efforts, statewide efforts
	Snowpack
	Streamflow timing
	Stay current through literature

to high levels of reported knowledge said they stayed current through published research. The respondents with high levels of reported knowledge identified being involved in cooperative mitigation, outreach, and research efforts. They also reported involvement with the Colorado Governor Ritter's Forest Health Advisory Council, the Colorado Bark Beetle Cooperative, Denver Water's "Forests to Faucets" program, and other smaller watershed outreach and mitigation programs in Colorado and Wyoming.

5.2 Research Communication

Respondents were asked if there was research about MPB effects on drinking water resources that needed to be communicated more effectively (Table 1, Question 2). The most common response was not specific to the types of research but focused on requests for scientists to disseminate research findings on MPB-related water management issues in more accessible formats (Table 3). Suggestions included the distribution of non-academic summaries, developing a centralized source of research findings, and advertising where relevant research summaries are located. As one drinking water provider explained, "Many of the published works are long and academic in nature. It would be nice to see a summary publication written at the practitioner level." Several respondents also requested that researchers share their preliminary findings and anticipated research direction through professional meetings, webinars, and practitioner publications (e.g., American Water Works Association's Opflow or Colorado Water).

The second most common response was that research was communicated effectively (Table 3). While limited knowledge was indicated by 53% of respondents in the previous question, many did not believe research needed to be communicated to them more effectively. The third topic was improved communication of research associated with the MPB effects on drinking water quality (Table 3). Respondents identified several water quality issues of concern, including DBP precursors and trihalomethanes (THM) formation potential, changes in TOC and NOM, and nutrient enrichment. Several other issues were

Table 3 Primary responses to survey questions by respondent type

	Primary responses
1. Research communication	• Improve accessibility of information (27%) • No parts of MPB research on water resources need to be communicated more effectively (23%) • Water quality effects of MPB (20%)
2. Identified challenges	• Addressing water quality effects of MPB (45%) • Increased fire danger and the resulting effects on water quality and quantity (37%) • Changes to water quantity as a result of MPB (27%)
3. Drinking water safety	• Drinking water has not deteriorated or become unsafe (67%) • Unsure or do not know (21%) • Yes—Deteriorated water quality; degraded enough to cause treatment issues (increasing chemical usage and cost) but not health or safety issues (10%)
4. Public outreach	• The effects of MPB on drinking water quantity and quality (25%) • The connection between forested watershed health and drinking water resources (17%) • General information on the MPB (current status, biological and ecological information) (14%)

identified but across fewer respondents and included requests for information on the MPB effects on water quantity (e.g., streamflow, groundwater recharge), recommended mitigation approaches for forests and/or drinking water management, the relation of MPB to wildfire occurrence, and the expected timeline for effects to occur after the beetle infestation.

5.3 Identifying Challenges

Respondents were asked to identify the biggest issues or challenges related to MPB affecting drinking water resources (Table 1, Question 3). The greatest concerns were in three thematic areas: water quality, water quantity, and wildfire potential (Table 3). The most identified challenge across all experts and knowledge levels was the treatment of potential water quality issues resulting from MPB-killed trees in forested watersheds. Increased turbidity, NOM, DBP precursors, TOC, DOC, and changes in taste and odor of water were identified as challenges. Many also indicated that the unknown duration of these effects was a considerable challenge, while others were concerned about the increased treatment costs associated with water quality changes and their ability to pass these costs on to consumers. The second most commonly identified challenge across respondents was concerns about increased wildfire danger due to more dead trees within MPB-associated forests and watersheds and the associated impacts on water quality and quantity. This was a relatively common concern of respondents with little to moderate levels of reported knowledge and less common among those with higher levels of reported knowledge (Table 2). Changes to water quantity were the next most commonly identified challenge across the percentage of respondents with little to high levels of reported knowledge. Specific concerns identified were the levels and timing of water runoff, associated erosion issues, and the effects on overall water yield and groundwater storage. As with the water quality issues, many respondents were concerned about increased treatment costs associated with water quantity issues, specifically sediment control across the watershed infrastructure. Additional challenges raised by respondents encompassed the uncertainty of the long-term impacts of the MPB infestation and effective communication of impacts to water users and the general public.

5.4 Limited Change in Source Waters Affected by MPB

When asked whether drinking water had deteriorated or become unsafe because of MPB, nearly all water providers and other respondents indicated no detrimental changes to water quality or safety (Table 3). This was an interesting finding given that prior scientific research has suggested that "disproportionate DBP increases and seasonal decoupling of peak DBP and TOC concentrations further suggest that the TOC composition is being altered" in drinking water systems (Mikkelson et al. 2012). This has the potential for making drinking water unsafe. Those who did state water quality had deteriorated, specifically due to increased DBP precursor formation, indicated these issues were effectively addressed through treatment and therefore did not present health or safety issues. These respondents did raise concern about the necessary increase in treatment costs at water treatment plants due to potential increases in TOC level reactions with chlorine and subsequent increases in THM from chlorination. The remaining respondents either stated they were unsure or did not know if drinking water had deteriorated or become unsafe.

5.5 Public Outreach Recommendations

The elicitation survey concluded with a request for key messages respondents thought should be communicated to the public to improve their understanding of MPB effects on drinking water. Overall, most respondents believed the general effects of MPB on drinking water quality and quantity were the most important messages to communicate to the public (Table 3). Specific messages associated with this topic included potential treatment cost increases, changes to water taste and odor, short- and long-term effects, as well as efforts being taken to address these changes. The second most commonly identified topic across respondents was the connection between watershed and forest health to drinking water resources as an important topic to convey to the public. Many believed it was important to communicate the forest and watershed management issues associated with MPB to the public, including whether it is linked to drought, wildfire potential, or climate change, while others recommended communicating methods being used to mitigate MPB effects on forests and drinking water resources. For example, one respondent highlighted the "state of the MPB infestation in the west, and what are the impacts to quality and quantity of water into the watershed" should be communicated to the public. Lastly, respondents recommended communicating general information about the MPB to the public, including that it is an endemic species, the current status of the outbreak and general biological and ecological information. Additional topic areas raised specifically by drinking water providers included the possibility of increased costs to water users because of additional treatments necessary to treat the effects of MPB, as well as the potential for wildfire (due to increased biomass on the forest floor from fallen trees) and the associated negative effects on drinking water quality and quantity.

6   Discussion

This chapter has presented the findings of a survey of access to, and understanding of, research information by drinking water providers and water-related professionals, looking specifically at the effects of a MPB epidemic on drinking water resources in the Rocky Mountain region. An elicitation survey revealed knowledge levels regarding MPB effects on drinking water resources, assessed respondents' perceptions of the primary challenges resulting from MPB effects on drinking water resources, and identified topic areas these professionals deemed important to communicate.

The limited knowledge of the impact of MPB on drinking water among a majority of the drinking water providers and other water professionals indicates a need for increased communication and improved knowledge exchange between scientists and managers. The difference in reported knowledge levels across populations with the drinking water providers reporting lower levels of knowledge than water professionals reveals disproportionate levels of information uptake across the sample. This could be attributed to the use of different sources of information for each population, such as peer-reviewed articles versus practitioner publications (Gibbons et al. 2008; Mostert and Raadgever 2008). While reported levels of knowledge across drinking water provider and water professional populations varied, the topics of concern reported by each population were similar. The topics identified need to be considered by scientists as they develop new research with applicability to drinking water providers. Topics included forest management challenges, soil erosion concerns, potential effects on water quality, and changes in water runoff. Further evaluation of the differences across these populations would provide greater insight toward the levels and areas of knowledge associated with the MPB effects on drinking water.

Another key finding was the call for improved and increased dissemination of the MPB disturbance research findings (Table 3). Many indicated a strong interest in learning about current research and how to better access findings. This confirms what has been identified in other studies where the avenues of communication between scientists and managers are inconsistent and pose barriers to effective knowledge exchange (Kiparsky et al. 2012; Medema et al. 2008; Mostert and Raadgever 2008). However, many respondents reported that MPB research did not need to be communicated to them more effectively, despite admitting to having little to no knowledge of the MPB effects on drinking water. This difference indicates that many drinking water providers and professionals do not realize the potential impact MPB-disturbed forests could have on drinking water resources. Findings may also indicate managers believe scientific reports lack relevance to the decisions made on the ground as the logic and terminology utilized by managers and scientists can differ immensely (Driscoll et al. 2011; Hulme 2014; Kieser and Leiner 2009; Medema et al. 2008).

The effect of MPB disturbance on water quality was most commonly identified as the topic needing to be communicated more effectively, as well as one of the greatest challenges. Effects on water quantity and potential increases in fire danger resulting from MPB forest disturbances were also identified as challenges. Although it is a common public perception, an increased level of fire risk in MPB-affected forests has not been scientifically established (Hicke et al. 2012). The prevalence of this concern may be attributed to the increased occurrence of wildfire throughout the study area in the past two decades, with substantial effects and associated costs to drinking water quality, infrastructure improvements, and watershed restoration. These identified challenges indicate key areas of concern for researchers to address.

Drinking water providers also reported a need for more information about the extent to which drinking water safety and quality has actually deteriorated. Identifying and communicating the challenges faced by drinking water providers offer an opportunity for scientists to address relevant management issues and better inform drinking water management. The response data may be limited by the potential reluctance of drinking water providers to admit that water quality changes may exist in local drinking water watersheds. These findings suggest a concurrent exchange of management field reports, and relevant scientific research would provide a venue for better-informed scientific research and management decisions. Such a venue would allow water providers to report drinking water changes for researchers to examine and possibly provide managers with information necessary to effectively address such changes.

Drinking water providers identified the importance of outreach to the general public regarding the effects of the MPB forest disturbance on drinking water. The responses highlighted areas of concern and interest and also provided insightful recommendations for communicating with the public such as general information on the MPB—current status, biological and ecological information, the connection between watershed (forest) health and drinking water resources, and the effects of MPB on drinking water quantity and quality. Several educational efforts occur across northern Colorado and southern Wyoming, through community watershed groups and statewide efforts. Further evaluation is necessary to determine the levels of communication scientists and water providers have with these outreach efforts. Some of the research conducted since the elicitation survey include MPB effects on streamflow and nutrients (Mikkelson et al. 2013b), hydrologic modeling of water supply impacts (Bearup et al. 2014), biogeochemical processes and water quality impacts (Brouillard et al. 2016), reactive transport modeling of metals (Mikkelson et al. 2014), and public perceptions of MPB effects on natural resources (McGrady et al. 2016) and the recreational experience (Wynveen et al. 2017).

Study results informed focus group attendees at the Colorado Watershed Assembly's Annual Conference, October 2014, representing forest and watershed health professionals in Colorado and two years later at panel discussions entitled "Hydrology and Water Quality Impacts from the MPB Infestation in the Rocky Mountain West" at the Joint Annual Conference of the Rocky Mountain Section of the American Water Works Association (RMsAWWA) and Rocky Mountain Water Environment Association (RMWEA) in September 2015 and the American Water Resources Association (AWRA) Annual Conference on Water Resources in Denver, CO, in October 2015. A special university honors seminar was also developed entitled "Naked Trees, Killer Beetles, and Dirty Water: Local Applications of Science & Outreach." The course was a collaborative learning effort between honors students and faculty at Colorado State University (CSU) and Colorado School of Mines (CSM) with Web-based classroom linkages for presentations. Results include outreach materials presented for K-12 students at the Windy Peaks Outdoor School in Jefferson County, Colorado, and a high school biology class in Fort Collins, Colorado, USA.

7 Concluding Remarks and Recommendations

Knowledge of climate-induced disturbances such as the MPB takes many forms and is inherently uncertain. Communicating scientific research and exchanging knowledge of the MPB effects on drinking water quality are necessary to inform management responses, but present numerous challenges. Assessing the levels of knowledge among drinking water providers and related professionals sets the stage to identify the extent of knowledge exchange needed to facilitate a communities' ability to effectively respond to ecological disturbances. The expert elicitation survey approach provided the means to examine general knowledge levels, identify information needs and primary concerns and current challenges of drinking water providers, and to explore potential outreach mechanisms to share research findings and improve communication with drinking water providers and the public overall. This information may enhance the development of mechanisms to improve communication of MPB information between scientists, drinking water providers, and the public overall.

This study provides important guidance for subsequent research on communication, knowledge exchange, and adaptive capacity associated with forest watershed disturbances. Identifying necessary adaptations to address the effects of ecological disturbances requires knowledge transfer and communication exchange between scientists, resource managers, and the public overall. The findings from this analysis set the stage for subsequent assessments of communication and knowledge exchange among other populations. Further research will build on these findings to assess knowledge levels, communication needs, and concerns of watershed groups, natural resource managers, and the general public, in addition to optimal venues for the exchange of field reports and scientific research. The development of such mechanisms can improve knowledge exchange and communication of climate change-related disturbances between scientists and managers across other fields and regions.

Acknowledgements The National Science Foundation Water Sustainability and Climate Program Project Grants 1204460 and 1204787 supported this work. The authors wish to thank A. Mitchell for assistance with developing the survey instrument, B. Brouillard for gathering contact information, and S. Brooker for assisting with a literature review.

Notes

1. Disinfection by-products are chemical, organic, and inorganic substances that can form during a reaction of a disinfectant (in this case chlorine in water treatment facilities) with naturally present organic matter in the water, in essence potentially harmful drinking water quality (Mikkelson et al. 2013a, b).
2. The consulting firms and other respondents work closely with drinking water providers and are assumed to share information and resources, and they were therefore included in the survey. We provide findings from across all respondents.

References

Addison, P. F., Rumpff, L., Bau, S. S., Carey, J. M., Chee, Y. E., Jarrad, F. C., et al. (2013). Practical solutions for making models indispensable in conservation decision-making. *Diversity and Distributions, 19,* 490–502.

Anderson, M. T., & Woosley, L. H., Jr. (2005). *Water availability for the western United States-Key scientific challenges: U.S. Geological Survey Circular 1261*. Washington, DC: US Department of Interior, US Geological Survey.

Bearup, L. A., Maxwell, R. M., Clow, D. W., & McCray, J. E. (2014). Hydrological effects of forest transpiration loss in bark beetle-impacted watersheds. *Nature Climate Change, 4,* 481–486. https://doi.org/10.1038/NCLIMATE2198.

Borowski, I., & Hare, M. (2007). Exploring the gap between water managers and researchers: Difficulties of model-based tools to support practical water management. *Water Resources Management, 21,* 1049–1074. https://doi.org/10.1007/s11269-006-9098-z.

Bosetti, V., Catenacci, M., Fiorese, G., & Verdolini, E. (2012). The future prospect of PV and CSP solar technologies: An expert elicitation survey. *Energy Policy, 49,* 308–317. https://doi.org/10.1016/j.enpol.2012.06.024.

Brouillard, B. M., Dickenson, E. R. V., Mikkelson, K. M., & Sharp, J. O. (2016). Water quality following extensive beetle-induced tree mortality: Interplay of aromatic carbon loading, disinfection byproducts, and hydrologic drivers. *Science of the Total Environment, 572,* 649–659.

Cann, K. F., Thomas, D. R., Salmon, R. L., Wyn-Jones, A. P., & Kay, D. (2013). Extreme water-related weather events and waterborne disease. *Epidemiology and Infection, 141,* 671–686. https://doi.org/10.1017/S0950268812001653.

Chapin, F. S., Kofinas, G. P., Folke, C., & Chapin, M. C. (2009). *Principles of ecosystem stewardship: Resilience-based natural resource management in a changing world.* New York: Springer.

Colorado State Forest Service (CSFS). (2015). *2014 Report on the health of Colorado's forests.* Fort Collins, CO: Colorado State Forest Service. Available from http://csfs.colostate.edu/forest-management/common-forest-insects-diseases/mountain-pine-beetle (Accessed June 2, 2016).

Colorado State Forest Service (CSFS). (2016). *2015 Report on the health of Colorado's forests: 15 years of change.* Fort Collins, CO: Colorado State Forest Service.

Colorado Watershed Assembly. (2015). *Water facts.* Available from http://www.coloradowater.org/Colorado%20Water%20Facts (Accessed June 2, 2015).

Cook, C. N., Hockings, M., & Carter, R. W. (2009). Conservation in the dark? The information used to support management decisions. *Frontiers in Ecology & Environments, 8,* 181–186. https://doi.org/10.1890/090020.

Creswell, J. W. (2014). *Research design: Qualitative, quantitative, and mixed method approaches* (4th ed.). Thousand Oaks, CA: Sage.

Denzin, N. K., & Lincoln, Y. S. (Eds.). (1998). *The landscape of qualitative research: Theories and issues.* Thousand Oaks, CA: Sage.

Dilling, L., Lackstrom, K., Haywood, B., Dow, K., Lemos, M. C., Berggren, J., et al. (2015). What stakeholder needs tell us about enabling adaptive capacity: The intersection of context and information provision across regions in the United States. *Weather Climate & Society, 7,* 5–17. https://doi.org/10.1175/WCAS-D-14-00001.1.

Dillman, D. A., Smyth, J. D., & Christian, L. M. (2009). *Internet, mail, and mixed-mode surveys: The tailored design method* (3rd ed.). Hoboken, NJ: Wiley

Donlan, C. J., Wingfield, D. K., Crowder, L. B., & Wilcox, C. (2010). Using expert opinion surveys to rank threats to endangered species: A case study with sea turtles. *Conservation Biology, 24,* 1586–1595. https://doi.org/10.1111/j.1523-1739.2010.01541.x.

Doria, M. D., Boyd, E., Tompkins, E. L., & Adger, W. N. (2009). Using expert elicitation to define successful adaptation to climate change. *Environmental Science & Policy, 12,* 810–819. https://doi.org/10.1016/j.envsci.2009.04.001.

Driscoll, C. T., Lambert, K. F., & Weathers, K. C. (2011). Integrating science and policy: A case study of the Hubbard Brook Research Foundation Science Links Program. *BioScience, 61,* 791–801. https://doi.org/10.1525/bio.2011.61.10.9.

Driscoll, D. A., & Lindenmayer, D. B. (2012). Framework to improve the application of theory in ecology and conservation. *Ecological Monographs, 82,* 129–147.

Engle, N. L. (2011). Adaptive capacity and its assessment. *Global Environmental Change, 21,* 647–656. https://doi.org/10.1016/j.gloenvcha.2011.01.019.

Folke, C. (2006). Resilience: The emergence of a perspective for social-ecological systems analyses. *Global Environmental Change, 16,* 253–267. https://doi.org/10.1016/j.gloenvcha.2006.04.002.

Folke, C., Colding, J., & Berkes, F. (2009). Synthesis: Building resilience and adaptive capacity in social-ecological systems. In F. Berkes, J. Colding, & C. Folke (Eds.), *Navigating social-ecological systems: Building resilience for complexity and change* (pp. 352–387). Cambridge, UK: Cambridge University Press.

Gibbons, P., Zammit, C., Youngentob, K., Youngentob, K., Possingham, H. P., Lindenmayer, D. B., et al. (2008). Some practical suggestions for improving engagement between researchers and policy-makers in natural resource management. *Ecological Management & Restoration, 9,* 182–186. https://doi.org/10.1111/j.1442-8903.2008.00416.x.

Hart, S. J., Schoennagel, T., Veblen, T. T., & Chapman, T. B. (2015). Area burned in the western United States is unaffected by recent mountain pine beetle outbreaks. *Proceedings of National Academy of Sciences, 112,* 4375–4380. https://doi.org/10.1073/pnas.1424037112.

Hicke, J., Johnson, M., Hayes, J., & Preisler, H. (2012). Effects of bark beetle-caused tree mortality on wildfire. *Forest Ecology and Management, 271,* 81–90. https://doi.org/10.1016/j.foreco.2012.02.005.

Hulme, P. E. (2014). Bridging the knowing-doing gap: Know-who, know-what, know-why, know-how and know-when. *Journal of Applied Ecology, 51,* 1131–1136. https://doi.org/10.1111/1365-2664.12321.

Kieser, A., & Leiner, L. (2009). Why the rigour–relevance gap in management research is unbridgeable. *Journal of Management Studies, 46,* 516–533. https://doi.org/10.1111/j.1467-6486.2009.00831.x.

Kiparsky, M., Milman, A., & Vicuña, S. (2012). Climate and water: Knowledge of impacts to action on adaptation. *Annual Review of Environment and Resources, 37,* 163–194. https://doi.org/10.1146/annurev-environ-050311-093931.

Kundzewicz, Z. W., Mata, L. J., Arnell, N. W., Döll, P., Jimenez, B., Miller, K., et al. (2008). The implications of projected climate change for freshwater resources and their management. *Hydrological Sciences Journal, 53,* 3–10.

Leatherman, D., Aguayo, I., & Mehall, T. (2007). *Trees and shrubs: Mountain pine beetle* (Colorado State University Extension Service Fact Sheet No. 5, 528). Fort Collins, CO: Colorado State University.

Lemos, M. C. (2008). What influences innovation adoption by water managers? Climate information use in Brazil and the United States. *Journal of American Water Resources Association, 44,* 1388–1396. https://doi.org/10.1111/j.1752-1688.2008.00231.x.

Martin, T. G., Burgman, M. A., Fidler, F., Kuhnert, P. M., Low-Choy, S., Mcbride, M., et al. (2012). Eliciting expert knowledge in conservation science. *Conservation Biology, 26,* 29–38. https://doi.org/10.1111/j.1523-1739.2011.01806.x.

Mazi, K., Koussis, A. D., & Destouni, G. (2013). Tipping points for seawater intrusion in coastal aquifers under rising sea level. *Environmental Research Letters, 8,* 014001. https://doi.org/10.1088/1748-9326/8/1/014001.

McGrady, P., Cottrell, S., Raadik Cottrell, J., Clement, J., & Czaja, M. (2016). Local perceptions of mountain pine beetle infestation, forest management, and connection to national forests in Colorado and Wyoming. *Human Ecology, 44*(2), 185–196. https://doi.org/10.1007/s10745-016-9816-y.

Medema, W., McIntosh, B. S., & Jeffrey, P. J. (2008). From premise to practice: A critical assessment of integrated water resources management and adaptive management approaches in the water sector. *Ecology and Society, 13*(29). http://www.ecologyandsociety.org/vol13/iss2/art29/.

Mikkelson, K. M., Dickenson, E. R., Maxwell, R. M., McCray, J. E., & Sharp, J. O. (2012). Water-quality impacts from climate-induced forest die-off. *Nature Climate Change, 3,* 218–222. https://doi.org/10.1038/nclimate1724.

Mikkelson, K. M., Bearup, L. A., Maxwell, R. M., Stednick, J. D., McCray, J. E., & Sharp, J. O. (2013a). Bark beetle infestation impacts on nutrient cycling, water quality and interdependent hydrological effects. *Biogeochemistry, 115,* 1–21. https://doi.org/10.1007/s10533-013-9875-8.

Mikkelson, K. M., Maxwell, R. M., Ferguson, I., Stednick, J. D., McCray, J. E., & Sharp, J. O. (2013b). Mountain pine beetle infestation impacts: Modeling water and energy budgets at the hill-slope scale. *Ecohydrology, 6,* 64–72. https://doi.org/10.1002/eco.278.

Mikkelson, K. M., Bearup, L. A., Navarre-Sitchler, A. K., McCray, J. E., & Sharp, J. O. (2014). Changes in metal mobility associated with bark beetle-induced tree mortality. *Environmental Sciences; Processes Impacts, 16,* 1318–1327.

Mostert, E., & Raadgever, G. T. (2008). Seven rules for researchers to increase their impact on the policy process. *Hydrology and Earth System Sciences, 12,* 1087–1096. https://www.hydrol-earth-syst-sci.net/12/1087/2008/.

National Science Foundation (NSF). (2015). *Water sustainability climate.* Available from http://www.nsf.gov/funding/pgm_summ.jsp?pims_id=503452 (Accessed June 5, 2015).

Pahl-Wostl, C. (2007). Transitions towards adaptive management of water facing climate and global change. *Water Resources Management, 21,* 49–62. https://doi.org/10.1007/s11269-006-9040-4.

Pahl-Wostl, C., Craps, M., Dewulf, A., Mostert, E., Tabara, D., & Taillieu, T. (2007). Social learning and water resources management. *Ecology and Society, 12,* 5. https://www.ecologyandsociety.org/vol12/iss2/art5/.

Rhoades, C. C., McCutchan, J. H., Jr., Cooper, L. A., Clow, D., Detmer, T. M., Briggs, J. S., et al. (2013). Biogeochemistry of beetle-killed forests: Explaining a weak nitrate response. *Proceedings of National Academy of Sciences, 110,* 1756–1760. https://doi.org/10.1073/pnas.1221029110.

Roux, D. J., Rogers, K. H., Biggs, H. C., Ashton, P. J., & Sergeant, A. (2006). Bridging the science-management divide: Moving from unidirectional knowledge transfer to knowledge interfacing and sharing. *Ecology and Society, 11,* 4. https://www.ecologyandsociety.org/vol11/iss1/art4/.

Strauss, A., & Corbin, J. (1998). *Basics of qualitative research: Techniques and procedures for developing grounded theory* (2nd ed.). Thousand Oaks, CA: Sage.

Timmerman, J. G., Beinat, E., Termeer, K., & Cofino, W. (2010). Analyzing the data-rich-but-information-poor syndrome in Dutch water management in historical perspective. *Environmental Management, 45,* 1231–1242. https://doi.org/10.1007/s00267-010-9459-5.

United States Department of Agriculture Forest Service (USFS). (2017). *Rocky mountain bark beetle: More than 4 million acres impacted.* Available from http://www.fs.usda.gov/main/barkbeetle/home (Accessed October 12, 2017).

Van Houtven, G., Mansfield, C., Phaneuf, D. J., Von Haefen, R., Milstead, B., Kenney, M. A., et al. (2014). Combining expert elicitation and stated preference methods to value ecosystem services from improved lake water quality. *Ecological Economics, 99,* 40–52. https://doi.org/10.1016/j.ecolecon.2013.12.018.

Walker, B. H., & Salt, D. (2006). *Resilience thinking sustaining ecosystems and people in a changing world.* Washington, DC: Island Press.

Wynveen, C., Schneider, I. E., Cottrell, S. P., & Arnberger, A. (2017). Assessing place attachment measurement: A cross-site comparison in the United States and Germany. *Society and Natural Resources.* https://doi.org/10.1080/08941920.2017.1295499.

15

Forest Collaborative Groups Engaged in Forest Health Issues in Eastern Oregon

Emily Jane Davis, Eric M. White, Meagan L. Nuss and Donald R. Ulrich

1 Introduction

In the western USA, there has been substantial political and social dialogue about forest health since the late 1990s. In this region, the federal government owns large percentages of the land base and is directed to manage it for the public good and resource conservation. There is often disagreement about the condition of federal forests and what

E. J. Davis (✉)
Department of Forest Ecosystems and Society, Oregon State University, Corvallis, OR, USA

E. M. White
Pacific Northwest Research Station of the USDA Forest Service, Olympia, WA, USA

M. L. Nuss
Philomath, USA

D. R. Ulrich
Los Alamos, USA

management they may need to improve forest health (Vaughn and Cortner 2005). Some members of the scientific community, forest industry, and others suggest that decades of past wildfire suppression have led to a generally unhealthy, "overstocked" forest prone to uncharacteristic wildfire, insects, and disease and in need of significant management intervention. Yet other scientists and environmental organizations often purport that claims of these forest health issues are exaggerated or that such disturbances are part of natural ecosystem process and that management to "correct" them is not warranted (Abrams et al. 2005). In short, there are conflicting values about how to manage forest health and social discord (Shindler et al. 2002). The US Forest Service, which manages many national forest lands, has had established processes for public participation and interagency consultation for decades. But conflict among stakeholders over forest health and other public land management issues has continued. Stakeholders hold differing viewpoints on many issues including logging, preservation, endangered species management, and public access. They may include environmental organizations, forest industry, local or county government, tribes, other government agencies, and private citizens and landowners.

Forest collaborative groups have become increasingly common in the western USA in an attempt to overcome this discord. In the words of Tom Tidwell, Chief of the US Forest Service, these groups are intended to help forest management move "toward a shared vision that allows environmentalists, forest industry, local communities, and other stakeholders to work collaboratively toward healthier forests and watersheds, safer communities, and more vibrant local economies" (2012). These collaborative groups meet regularly and have multi-stakeholder dialogue about their priorities and where they may have common ground agreement (Schultz et al. 2012; Cheng and Sturtevant 2012). They provide input, often written, to the Forest Service on planned management activities in a given area of national forest land. They also may engage in multi-party monitoring during and after implementation of management actions.

One region of particular note for collaboration is eastern Oregon, where there have been uniquely concerted efforts to restore forest health. A distinction is often made between the "Eastside" of the state, located east of the Cascade Range to the Idaho border, and the "Westside," located west of the Cascades to the Pacific Ocean (Thorson 2003). The Eastside, also informally known as the "dry-side," receives far less precipitation and has forest and range ecosystems adapted to frequent fire. Policy makers and the forest industry have expressed concern that Oregon's Eastside national forestlands are not sufficiently managed to mitigate these forest health risks and their subsequent effects on local communities and economies (White et al. 2015, 2016). This region's forests are subject to a range of disturbances including drought, wildfire, insects, and diseases.

The Forest Service's Pacific Northwest Regional Office (Region 6), which includes eastern Oregon, initiated an "Eastside Restoration Strategy" in 2012 to address forest health issues on national forests. The Strategy states that it requires "robust community collaborative groups to identify and overcome obstacles to restoration activity" to achieve these goals. In 2013, the state of Oregon created a state Federal Forest Health Program[1] to increase active restoration of the health of these national forests. This included grant support for collaborative groups. The assumption of these programs is that collaborative groups may reduce the burden of environmental planning of restoration projects by helping the agency understand where there may be social agreement, decreasing the potential of litigation or appeal from dissatisfied stakeholders, and speeding up planning timelines (Goldstein and Butler 2010).

Forest collaborative groups have thus been made central in the pursuit of improved forest health in eastern Oregon. Expectations and critiques of these groups and their role in federal forest health have subsequently grown. However, there is no single or mandated definition of what constitutes a forest collaborative group, what forest health issues it would address, and how it engages with the social dimensions of forest health. Policy makers and managers often make general statements about the importance of "building agreement" around forest health, but precisely what this looks like and how groups organize to do so are not well documented. Existing research on social dimensions of forest health tends to focus on individuals' values (e.g., Jenkins 1997; Fuller et al. 2016), but less is known about collective action and collaboration around forest health.

Given this current lack of definition and information about this form of collaboration on forest health, we undertook a mixed methods study of eastern Oregon's collaborative groups to document their basic characteristics and how they conceive of and work on specific forest health issues. We offer findings about how eastern Oregon forest collaborative groups are organized, processes used to build agreement on forest health, how stakeholders conceive of forest health, and participant perceptions of satisfaction and group success at achieving its desired outcomes. We also identify commonalities and differences across groups. This comparison could inform future policy by suggesting where there are issues of common interest that may warrant region-wide approaches, and other areas where concerns and needs may be more localized. Also, although the Oregon collaborative context is somewhat unique, this study may contribute some general insights applicable to other settings around the world where groups of stakeholders and land managers are attempting to act collectively on forest health issues.

2 Context: Public Lands Collaboration in the US West

Across the 11 states of the US West, the Forest Service owns and manages approximately half of the forested land base through a system of national forests and grasslands (Smith et al. 2009). This federal agency is responsible for managing these lands for the public good and for resource conservation, which includes multiple uses such as timber harvest, cattle grazing, ecosystem services, recreation opportunities, and wilderness. It also leads wildfire suppression efforts on its lands. In 1891, the US Forest Reserves Act was passed, which authorized the nation's president to designate lands as public "forest reserves." At the time, a primary motivation driving this Act was overharvesting by private industry interests. In 1905, these forest reserves were transferred to the US Forest Service for management.

Over the twentieth century, several policies directed the agency to produce sustained yields of timber and then to consider multiple uses in that yield (Multiple Use Sustained Yield Act of 1960/Public Law 86-517). By the 1970s, larger trends toward wildlife species protection and recognition of environmental issues began to call into question the rate of timber harvest on public lands and opened the door for increased public comment and participation in agency planning, through processes such as those required by the National Environmental Policy Act and the National Forest Management Act. By the 1980s, scientists, environmental organizations, and recreationists were raising concerns about the effects of harvesting on wildlife habitat in the northwest region of the USA, particularly for the northern spotted owl (*Strix occidentalis caurina*) and Chinook salmon (*Oncorhynchus tshawytscha*), which further led to a ruling that the Forest Service was failing to protect these and other species. The result was the 1994 Northwest Forest Plan for the northwest region and a broader shift to an "ecosystem management" paradigm across the West. Timber harvests declined and national forest management generally became more focused on ecological restoration, recreation, and other non-extractive values, as well as on fighting wildfires and reducing wildfire risk through mitigation activities (Abrams et al. 2015). For many rural communities adjacent to national forest lands, this contributed to challenges with unemployment and social issues as jobs in the forest industry were lost. It also fostered extensive social conflict over how national forests should be managed (Proctor 1998).

As a result of these substantive shifts in national forest policy and ongoing social conflict, national forest management in the USA in general has moved away from top to down, hierarchical decision-making toward more collaborative and participatory governance approaches over the past two decades (Wondolleck and Yaffee 2000). Collaboration is defined as problem-solving wherein a diverse group of interdependent stakeholders addresses common issues and resolves environmental disputes through deliberation, consensus building, co-learning, and generating solutions (Goldstein and Butler 2010; Margerum 2011). The impetus to collaborate comes from many drivers. For the Forest Service, there has been an explicit policy shift toward all-lands management and partnerships with stakeholders to address resource challenges that cross jurisdictional boundaries, while at local and regional levels, many community leaders have self-organized to address natural resource-related socioeconomic threats and challenges (Baker and Kusel 2003). In much of the interior northwest, funding for and emphasis on collaboration on national forests has largely come from policies and programs that focus on wildfire risk reduction and dry forest health restoration, including the National Fire Plan (2000), Healthy Forests Restoration Act (2003), Federal Land Assistance Management and Enhancement Act (2009), and Collaborative Forest Landscape Restoration Program (2010).

The forest health issues that collaboration attempts to address in the "Inland Northwest" of the USA (Hessburg et al. 2015)[2] largely pertain to two primary broad forest types: dry ponderosa pine (*Pinus ponderosa*) and mixed conifer, and moist mixed conifer. Mixed conifer species may include various true fir species (e.g., *Abies grandis* and *Abies concolor*) and others (e.g., *Pseudotsuga menziesii* and *Larix occidentalis* Nutt.) depending on the location, elevation, and moisture (Stine et al. 2014; Hessburg et al. 2005). Recent research on ecological restoration in the Inland Northwest has provided frameworks and concepts that forest collaborative groups attempt to use to create middle ground, such as *ecological restoration, landscapes*, and *resiliency*. "Wildfires and insect outbreaks are an inevitable part of future landscapes. Future management should aim to restore more resilient vegetation patterns that can help to realign the severity and patch sizes of these disturbances, promote natural post-disturbance recovery, reduce the need for expensive active management, and drastically reduce the role and need of fire suppression" (Hessburg et al. 2015, p. 1805). The stated mission of many collaborative groups across the Inland Northwest is to restore this resiliency and build agreement about what it means in practice.

3   Forest Collaborative Groups

Over time, these Forest Service policies and programs have demanded growing levels of stakeholder involvement because they encourage outcomes that necessitate sustained engagement beyond a limited number of public meetings or short-term projects, and include involvement in planning and implementation (Butler et al. 2015). Research has frequently noted that this level of collaborative engagement is typically more successful when stakeholders have a "collaborative body [with] the ability to organize and sustain itself as a group" (Cheng and Sturtevant 2012).

In the Inland Northwest, there are now numerous such collaborative groups. They tend to have defined missions, operating policies and procedures, and a process of regular meetings. They have dialogue about planned forest management activities, which may include the Forest Service sharing data about current conditions and restoration needs, and scientists providing insights into the impacts of potential management actions. Through this dialogue, they focus on reaching some type of group agreement (Brown 2012; Davis et al. 2015a, 2017). Collaborative processes vary, but many typically prepare written statements of input that they provide to the Forest Service. These statements are intended to capture the desired outcomes that collaborative members have, and the types of management activities they do or do not support to achieve them. Input may be provided for a specific planned project on the landscape, or it may be focused on a forest type or management issue in general. These statements may be called, for example, recommendations, common restoration principles, or zones of agreement, depending on the group. Importantly, the Forest Service is the manager of the land, and although collaborative groups provide input, there is currently no legal or otherwise mandated requirement that the agency uses it; there is often just a requirement "to collaborate" without specific direction (Monroe et al. 2016). Decisions about the planned activities and how to implement them are made by the agency. Thus, collaborative groups have a fair degree of social importance and authority, but no official legal authority or status relative to federal land.

There are also no official regulations or guidelines about what structures and processes a collaborative group must have, although some entities such as the National Forest Foundation provide suggested best practices. This has led to variation in the size, composition, organization, and funding base of groups across the entire West. Authors' experiences attending and studying these groups also suggest that there can be significant fluctuation in stakeholder attendance from meeting to meeting. This challenges generalizations and standard descriptions. However, most groups have a regular facilitator or coordinator and may have some grant funding from entities such as the National Forest Foundation, private foundations, or the Forest Service (Davis et al. 2015b). They are usually not officially incorporated as nonprofit organizations, and administrative support is provided by another organization. More discussion of group structure in eastern Oregon is shared in our findings.

4 Methods

In eastern Oregon, there are seven active forest collaborative groups that focus on restoring the health of federal forests (Fig. 1). We used a mixed methods approach to describe and analyze how these collaborative groups are organized, what they do, what forest health issues concern them, and participant perceptions of satisfaction and outcomes. First, we conducted a rapid assessment (Beebe 2001) to gather information about each group's basic organizational structure. We conducted the assessment using paper questionnaires at a statewide meeting of Oregon's collaborative groups in autumn 2014. Accessing respondents at a single point like this reduces burden, standardizes the approach, and allows data collection in a short period of time. One questionnaire per group was completed by each group's facilitator or coordinator. These facilitators and coordinators typically manage a collaborative's process as well as often its fiscal and other structures, and are a reputable source of this type of information about their groups. We used simple descriptive statistical techniques in Microsoft Excel to identify the similarities and differences in the organizational characteristics across groups.

Second, we collected available collaborative group documentation such as operational charters and Web sites for further details on groups' missions, forest health issues of focus, and activities. However, not all collaborative groups have this documentation, and there is a lack of consistency in the type and detail of information available on each group. Data used are documented in Appendix 1. Documentation primarily was from the period 2012 to 2015. We used basic content analysis and keyword searches to analyze all documents, identifying references to "forest health" in general and to any specific forest health issues such as uncharacteristic wildfire, insects, disease, or other disturbances. Since the quantity of documentation and level of detail available varies from group to group, quantitative measuring of keyword frequency and other variables would have possibly presented an inaccurate picture. We therefore took a more qualitative approach by looking at repeated themes, terms,

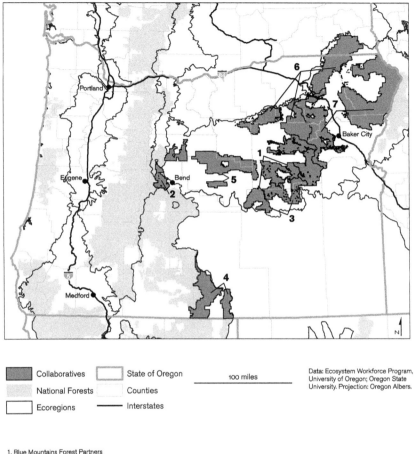

1. Blue Mountains Forest Partners
2. Deschutes Collaborative Forest Project
3. Harney County Restoration Collaborative
4. Lakeview Stewardship Group
5. Ochoco Forest Restoration Collaborative
6. Umatilla Forest Collaborative Group
7. Wallowa-Whitman Forest Collaborative

Fig. 1 Federal forest collaborative groups in eastern Oregon as of 2016

and word associations, and used an inductive approach to guide interpretation (Patton 2005). It should also be noted that meeting notes are not necessarily a complete transcription of all discussions and that not all issues may have been discussed and recorded consistently.

Third, we conducted an Internet survey of forest collaborative group stakeholders across Oregon and, for this chapter, analyzed a subset of the data from the seven eastern Oregon groups. Survey questions presented here focused on the perceived performance of various outcomes of the collaborative, and satisfaction with the collaborative's effectiveness at achieving its goals. The survey was administered using Oregon State University's Qualtrics license, an industry-standard online survey program. Collaborative groups use the Internet for communications, and with a few exceptions, participants are on email lists. We conducted a pilot test with five collaborative stakeholders representing diverse perspectives and geographical locations in Oregon. We then sent invitations to the coordinators/facilitators of each of the recognized forest collaborative groups, with a request that they send it to their participant email distribution lists. We followed up with repeat invitations until we ensured that the survey had reached all of these groups. Respondents were asked a series of filtering questions to determine if they did participate in a forest collaborative, identify the single collaborative group with which they were most active currently, and respond to the survey with that group in mind (Table 1). Since these groups vary in size, and in the nature of participation, single static group size is not easily identified. We received 97 usable responses, with the largest proportion

Table 1 Collaborative group survey respondent affiliations

Group affiliation	Frequency	Percent of respondents from
Blue Mountains Forest Partners (BMFP)	11	11.3
Deschutes Collaborative Forest Project (DCFP)	25	25.8
Harney County Restoration Collaborative (HCRC)	13	13.4
Lakeview Stewardship Group (LSG)	10	10.3
Ochoco Forest Restoration Collaborative (OFRC)	14	14.4
Umatilla Forest Collaborative Group (UFCG)	9	9.3
Wallowa-Whitman National Forest Collaborative (WWNFC)	15	15.5
Total	97	100

Table 2 Sectoral survey respondent affiliations

Participant affiliation	Frequency	Percent of respondents
No response	2	2
Municipal government	3	3
Conservation organization	4	4
Academic institution	5	5
Non-governmental organization	7	7
Private citizen	14	14
Multiple capacities	20	21
Federal agency	35	36
Total	97	

(36%) being federal agency respondents (Table 2). Given the limited number of respondents in most affiliation types, analysis by different stakeholder types was not statistically viable. Survey data were cleaned and analyzed using IBM SPSS Statistics 22, a statistical analysis software program commonly used for survey data.

Finally, we also drew on authors' experiences of working with these collaborative groups through several past research projects and in providing technical assistance to contribute additional insights and aid in interpretation of data. This afforded extensive participant observation, allowing the authors to build familiarity with the issues and activities of eastern Oregon collaborative groups. Previous activities include providing facilitation to two groups,[3] conducting research on the economic outcomes of the work of one group,[4] monitoring the impacts of Oregon's Federal Forest Health Program and the Forest Service's Eastside Strategy for five groups,[5] and case studies of the mechanics of collaborative processes of two groups.[6] We re-reviewed reports and results of these activities where applicable to identify relevant insights for this chapter, such as our own data and field notes.

5   Eastern Oregon's Forest Health "Crisis"

Eastern Oregon is characterized by a relatively dry climate as it lies in the rain shadow of the Cascade Mountain Range. The primary Level III ecoregions in eastern Oregon include the East Cascades Slopes,

Columbia Plateau, Blue Mountains, and Northern Basin and Range (Commission for Environmental Cooperation 1997). The majority of the region's forested land is found in the East Cascades Slopes and Blue Mountains ecoregions. Common tree species in this region include ponderosa pine, true firs, larch, and spruce, found in relatively dry, ponderosa pine-dominated plant association groups at lower elevations; and increasing in moisture and trending toward mixed conifer types at higher elevations and northern areas of the region (Hessburg et al. 2005).

There has been widespread political, social, and scientific dialogue in Oregon about the condition of eastern Oregon's forests (Langston 1995). Concerns about forest health were raised in the Blue Mountains area in particular:

> The fire-adapted forests of the Blue Mountains are suffering from a forest health problem of catastrophic proportions. Contributing to the decline of forest health are such factors as the extensive harvesting of the western larch and ponderosa pine overstory during the 1900s, attempted exclusion of fire from a fire-dependent ecosystem, and the continuing drought. The composition of the forest at lower elevations has shifted from historically open-grown stands primarily of ponderosa pine and western larch to stands with dense understories of Douglas-fir and grand fir. Epidemic levels of insect infestations and large wildfires now are causing widespread mortality that has a profound effect on forest health by adversely affecting visual quality, wildlife habitat, stream sedimentation, and timber values. (Mutch et al. 1993, 1)

Similar evidence of forest health issues has also been found in the East Cascades (e.g., Heyerdahl et al. 2014).[7] The primary concerns are uncharacteristically large and severe wildfires, tree mortality from a variety of causes, and specific insects and infestations such as pine beetles and mistletoe. But it also should be noted that some scientists and environmental organizations have argued that the evidence of a widespread forest health crisis is not sufficient, that active management is not appropriate, and/or that the issue has become overly politicized (e.g., DellaSalla et al. 1995; Willams and Baker 2012). In tandem with forest health issues, there has been long-standing concern about the employment and economic situation of the declining forest industry in the region and its necessity to make restoration of forest health economically feasible. The Forest Service has stated that: "Eastside communities need the raw material and jobs created by restoration work. Restoration work depends on a healthy forest products industry to provide labor, capital, and equipment, and robust community collaborative groups to identify and overcome obstacles to restoration activity. Therefore, we cannot afford to lose the forests, and we cannot afford to lose the mills" (US Forest Service Eastside Restoration Web Site 2016).

As a result of this combined ecological and economic crisis narrative, there has been significant political momentum leading to unique state and federal interventions to support more active forest restoration in eastern Oregon. There has been (1) legislated state-level investment in federal forest health, (2) focused efforts from the Forest Service's Regional Office, and (3) in both state and federal interventions, a strong emphasis on forest collaborative groups as a critical component to improving forest health. In 2006, the state's Board of Forestry created the Federal Forestland Advisory Committee (FFAC) to develop an ecologically, economically, and socially sustainable vision for federal forests in Oregon, given their extent and importance. The FFAC developed a recommendations report in 2009, stating that "The Governor and the State Legislature should assist federal agencies in providing administrative, financial, and technical resources to local collaborative partnerships to build trust and help identify scientifically informed and socially acceptable forest management," and that "Collaboration among diverse interests…must become the norm" (Oregon Board of Forestry 2009, 31–32). In 2013, the state of Oregon created a $2.9 million Federal Forest Health program for eastern Oregon (Oregon SB 5521), which provided grants to collaborative groups, as well as regional technical assistance and the capacity of state forestry staff to work directly on tasks such as timber marking on national forests. Grants to collaborative groups were titled "Federal Forest Health Collaborative Capacity Assistance Grants" and were intended to help these groups increase the pace and scale of their collaboration with the Forest Service on forest restoration activities to improve forest health. The Forest Service also has focused on forest health in eastern Oregon (and similar areas in

Washington) by developing a general "Eastside Restoration Strategy" that encourages national forests to readjust how they plan and implement work in order to accomplish more forest restoration activity at a faster pace than in the past. This includes a dedicated planning team to help augment existing national forest system capacity and plan large landscape projects in the Blue Mountains.

In summary, the ecological and economic need to restore forest health is a concern to many stakeholders in eastern Oregon. Organized forest collaborative stakeholder groups have been enshrined in state and federal programs and efforts as central to the solution by building increased social agreement and slowing potential legal resistance to forest restoration. Yet there are few descriptive studies to date of how these groups operate and the particular goals of forest health that they seek to address; across the West, scholars note that collaborative groups have proliferated yet perceive them as "highly idiosyncratic in detail" (Franklin et al. 2014) and not well documented.

6 Findings

6.1 Are Eastern Oregon's Forest Health Collaborative Groups Organized Similarly?

We conducted a basic characterization of how formal or informal collaborative groups were in their structure and processes including the following characteristics, which are based on other studies of collaboration and organizational factors in success (e.g., Ansell and Gash 2007; Cheng and Sturtevant 2012):

- *Registered nonprofit organization status*: Indicator of level of own fiscal and administrative capacities, ability to access and absorb resources.
- *Dedicated facilitator*: Indicator of stewardship of process and leadership.
- *Policies and procedures*: Indicator of governing rules and processes that can help ensure fairness and consistency.

- *Leadership committee*: Indicator of a stratified system for managing the group and taking action in smaller group settings as needed.
- *Shares input with Forest Service*: Indicator of how formally input is delivered—not at all, verbally, or written.

We used a three-point scale for each of these categories with the criteria (see footnotes, Table 3) from not having the specified characteristic (0) to some intermediate expression of it (1) to fully exhibiting it (2), then aggregated these rankings for each collaborative to provide an overall score of 0–10. Higher scores indicate that a group has more formal organizational characteristics that previous research has identified as important to collaborative capacity. Lower scores indicate a group is less formally organized.

Overall, these groups appear to be venues for dialogue with limited organizational infrastructure, rather than structured organizations with staff, executive leadership, programs, and resources (see also Davis et al. 2015b). We found that all of the collaborative groups in eastern Oregon have a dedicated facilitator and that almost all of them also had written procedures for sharing input with the Forest Service. Four of the seven groups had the same organizational features (score of 8; every feature except having their own registered nonprofit status, which means that they must rely on another organization with this status to manage any funds or administrative needs for them). This may suggest that collaborative groups have focused on process needs such as having a dedicated facilitator and ground rules, but have not necessarily emphasized structures for managing the collaborative as an organization, such as setting up administrative procedures and using committees to share workloads and leadership beyond the facilitator. One organization had all characteristics plus was a nonprofit, and two groups had aggregate scores of 4 and 5, suggesting less formal operations.

We drew on our previous experience with these groups as well as available documentation to hypothesize what may drive the different organizational forms collaborative groups take. The two least formal groups operated in Harney and Lake counties of the southeastern region of Oregon, where population density is the lowest in the state and communities are fairly remote. This informality may be a result of

Table 3 Organizational characteristics of eastern Oregon's forest collaborative groups

Group	Registered nonprofit organization status[a]	Dedicated facilitator[b]	Policies and procedures[c]	Leadership committed[d]	Shares input with Forest Service[e]	Aggregate score
Harney County Restoration Collaborative	0	2	1	0	1	4
Lakeview Stewardship Group	0	2	1	0	2	5
Deschutes Collaborative Forest Project	0	2	2	2	2	8
Ochoco Forest Restoration Collaborative	0	2	2	2	2	8
Umatilla Forest Collaborative Group	0	2	2	2	2	8
Wallowa-Whitman National Forest Collaborative	0	2	2	2	2	8
Blue Mountains Forest Partners	2	2	2	2	2	10

[a] 0 = no 501c3 status; 2 = yes
[b] 0 = none; 1 = rotating; 2 = regular/dedicated
[c] 0 = none; 1 = one; 2 = 2 or more
[d] 0 = none; 1 = meets infrequently; 2 = meets
[e] 0 = no procedures; 1 = verbal input only; 2 = written input

these conditions and local culture. For the four groups with the same characteristics, three started around the same time period (2011–2012). Authors' experience with these groups revealed that the Ochoco, Umatilla, and Wallowa-Whitman groups all hosted members of the more long-standing Blue Mountains Forest Partners (BMFP) to provide advice when they began, and carefully reviewed and used the BMFP's organizational documents as ideas and templates for their own. One group, the Deschutes group, started slightly earlier (2009) as a result of the opportunity provided by the federal Collaborative Forest Landscape Restoration Program to organize, and built on years of previous collaboration in its local area.

6.2 Do Eastern Oregon Collaborative Groups Use Similar Processes to Discuss Forest Health?

We also used documents and meeting notes to review the specifics of the collaborative process used in each group. We found that all of these collaborative groups had a fairly similar process wherein they would work with the Forest Service to understand the resource conditions and potential forest health issues on a specific area of land where the agency was planning future management action. This "planning area" was determined by the Forest Service and varied in size from 10,000 up to 100,000 acres. The duration of this process varied but was usually about one year per given area of land and groups typically met once a month (with the exception of the Lakeview Stewardship Group, which met two or three times a year). Often, agency staff appeared to have provided presentations to the groups about the forest health issues that were present in the area. In addition, scientists would give visiting presentations or field tours to look at forest health issues on the ground. In almost all cases, a few scientists either visited fairly regularly or were in fact considered "science advisors" to specific groups. With this knowledge, the stakeholders would discuss their desired outcomes and goals for the future of that area, often discussing competing views of what they each found to be an acceptable management action for their values. Acceptability appeared to consistently hinge on topics such as the amount and type of trees to be removed in a proposed restoration or thinning treatment, the intended positive effects for wildfire and insect/disease risk reduction, and the potential positive or negative effects of wildlife (primarily mule deer, elk, and several avian species). Stakeholders then would attempt to draw up some recommendations that were provided in written format to the Forest Service. These recommendations usually indicated where all stakeholders agreed and identified recommendations or issues on which a minority did not, and explained why.

We observed three important aspects of this collaborative process. First was that despite collaborative groups being considered independent bodies not convened by or responsible to the Forest Service, the Forest Service and scientists appeared to play a fairly substantial role in establishing what forest health issues needed to be addressed. As we have noted, the Forest Service is the manager of the land and collaborative groups have no official or legally mandated role. Forest Service personnel usually have collected data about resource conditions and shared it with the groups, which appeared to provide framing concepts and issues that shaped discussion. However, there were a few instances of stakeholders contributing their own data or local knowledge to the discussions. For example, local forest industry representatives or longtime citizens would express observations about what they had seen happening in the forest over time, or some specific history about the area being collaborated upon. In the Umatilla Forest Collaborative Group in 2013, meeting notes show that an environmental stakeholder shared results of their own data collection and there was some group discussion of these findings. We also saw a few instances of collaborative groups funding and supporting their own monitoring crews independent of the Forest Service (most notably, the BMFP).

Second, we found that level of specificity of management recommendations varied significantly. Some of the recommendations were very specific, e.g., they requested a certain residual basal area of volume to be left after treatment, or identified particular areas or conditions under which they did not want any treatments to occur. But many of them were more broadly stated as general principles or management directions, without specific prescriptions or quantifiable parameters. Several groups also focused on providing recommendations at a larger scale, without links to a specific planning area on the landscape. These attempted to capture general management needs and recommendations that stakeholders could agree on and that could potentially be applied across planning areas. These were variously called "zones of agreement" (described in Seager et al. 2015; Nuss and Davis 2015) or "restoration principles."

Third, it is clear that collaborative group recommendations typically do not represent 100% consensus of all stakeholders for a given forest. We found that minority reports, usually from environmental groups, seemed to be commonly made, showing deviation from what other stakeholders in a collaborative group were recommending on certain issues, or a group's recommendations might include a statement about where the group did not find consensus. We cannot calculate exactly what proportion of collaborative recommendations contain minority reports or lack of full consensus due to inconsistency in the availability of collaborative documents. Even further, meeting notes and Forest Service documents show that some environmental stakeholders did not consistently participate in collaborative groups, chose to deliver their recommendations through the standard public process, and expressed that they did not support what collaborative groups were recommending (e.g., Coulter et al. 2015). Therefore, collaborative group input does not necessarily or consistently capture the full spectrum of stakeholder values that may be present for a given area. They also are not a guarantee that there will not be public objection to a project. The recommendations of collaborative groups may be more appropriately seen as the result of dialogue among a subset of stakeholders and the Forest Service, rather than a comprehensive statement of all relevant stakeholder values. It also may not be appropriate to expect all collaborative groups to be able to reach or sustain 100% consensus on all forest health issues. Collaborative recommendations may be more usefully seen as a snapshot of different stakeholders' values and beliefs at the time of their preparation, rather than an enduring and broadly applicable statement of social agreement.

6.3 Do Collaborative Groups Focus on Similar Forest Health Issues, and What Are They?

We used a review of the mission documents and meeting notes of collaborative groups as well as representative project documentation from the Forest Service to identify forest health issues of focus and compare them across groups. Forest collaborative groups across eastern Oregon appear to focus on similar forest health issues, and to use forest health as a broad framing concept. Despite differences in location, specific national forest and communities involved, and exact forest types, we found that eastern Oregon's forest collaborative groups worked with markedly similar concepts and broad forest health concerns (Table 4).

First, we observed that the term "forest health" is not widespread in collaborative dialogue in comparison with "forest restoration." The most common unifying concept across all seven groups is the notion that eastern Oregon's forests are ecologically departed from their historic range of variability (HRV) and require active management to restore them to this range. HRV is an ecological concept used to describe reference or benchmark conditions of a fully functioning ecosystem (Keane et al. 2009). In eastern Oregon forest collaborative dialogue, this typically appears to refer to historical pre-European conditions when natural fires were not suppressed and anthropogenic fires were also used to control forest conditions. This broad term can signify departure in many aspects, such as fire return interval, tree species diversity, and stand structure. It is within the framework of HRV that collaborative groups appear to have their discussions about how to restore forest health. However, there was also some evidence of debates about HRV in collaborative meeting notes. For example, some environmental stakeholders and scientists have questioned the fire history of forests in the US West, arguing that current conditions are not as departed from historical condition as some studies suggest and some stakeholders believe (e.g., Williams and Baker 2012; discussions in the Deschutes Collaborative Forest Project). In a few instances, we also observed that some stakeholders questioned the use of HRV and preferred to

Table 4 Overview of missions and forest management[a] issues of focus

Collaborative group	Common forest management terms discussed repeatedly in collaborative documents	Collaborative mission statement or stated goals/purpose
Wallowa-Whitman Forest Collaborative	• Inventoried Roadless Areas, Inventoried Old Growth Management Areas • Removal of trees over 21 inches • Management in riparian areas	"We work to improve the social, economic, and ecological resiliency of the Wallowa-Whitman National Forest and local communities through collaboration by a diverse group of stakeholders"
Umatilla Forest Collaborative Group	• Management in riparian areas • Moist mixed conifer forests • Restoring composition and structure of stands to historic range of variability	"To develop and promote balanced solutions from a diverse group of stakeholders to improve and sustain ecological resiliency and local community socio-economic health in and near the Umatilla National Forest"
Blue Mountains Forest Partners	• Restoring composition and structure of stands to historic range of variability • Creating variety across the landscape ("patchy, gappy, clumpy") • Hazardous fuels reduction	"The BMFP is a diverse group of stakeholders who work together to create and implement a shared vision to improve the resilience and well-being of forests and communities in the Blue Mountains"
Harney County Restoration Collaborative	• Restoring composition and structure of stands to historic range of variability • Returning fire to the landscape	"Our goal is to restore healthy and resilient forests. Our projects provide social and economic benefits to the local community. We are continually learning and developing best practices that may be applied in other areas"

(continued)

Table 4 (continued)

Collaborative group	Common forest management terms discussed repeatedly in collaborative documents	Collaborative mission statement or stated goals/purpose
Ochoco Forest Restoration Collaborative	• Removal of trees over 21 inches • Management in riparian areas • Restoring composition and structure of stands to historic range of variability	"The Ochoco Forest Restoration Collaborative is a diverse group of stakeholders who work together to create and implement a shared vision to improve the resilience and well-being of forests and communities in the Ochoco Mountains"
Deschutes Collaborative Forest Project	• Restoring composition and structure of stands to historic range of variability • Creating variety across the landscape ("patchy, gappy, clumpy") • Addressing dwarf mistletoe infestation	"To restore our forests to a healthier, more resilient condition through balanced, science-driven restoration projects"
Lakeview Stewardship Group	• Restoring species, diversity, and structure of stands to range of historic variability • Returning fire to the landscape	"Sustain and restore a healthy, diverse, and resilient forest ecosystem that can accommodate human and natural disturbances. Sustain and restore the land's capacity to absorb, store, and distribute quality water. Provide opportunities for people to realize their material, spiritual, and recreational values and relationships with the forest"

[a]These groups also discuss other topics such as fish and wildlife habitat, recreation, and roads; this overview focuses solely on topics directly related to forest health/forest management

use "future range of variability" or resilience to future climate change as framing concepts, suggesting that forest restoration should focus on likely future rather than past conditions (e.g., discussions in the Lakeview Stewardship Group).

Second, collaborative dialogue typically tended to focus on examining the type and degree of forest restoration treatments intended to restore stand species diversity and structure within the framework of HRV, and depending on forest type. Dry ponderosa pine-dominated forests and dry mixed conifer forests were the most common types of forests discussed, although moist mixed conifer was also a topic for several groups. Thinning of trees and prescribed burning were the most common treatments, and discussion often focused on the extent, intensity, and after-effects of these treatments on various tree and other (wildlife, fish) species in eastern Oregon forests. Although reducing wildfire risk by reducing hazardous fuels (selective removal of forest vegetation to reduce the risk of uncharacteristic wildfire) was a frequent discussion topic, conversations appeared to often move beyond the effect of treatments on wildfire risk alone toward using treatments to create a more heterogeneous forest in terms of age, species composition, and structure. Two groups (Deschutes Collaborative Forest Project and Blue Mountains Forest Partners) frequently discussed their desired outcomes in terms of "patchy, gappy, clumpy" stands across the landscape, recommending treatments that would create a mosaic of variability (as described in Oregon Wild 2012). The resilience of such a mosaic pattern to insects and disease was not frequently discussed, however, and managing for values such as reduced wildfire risk and wildlife habitat was a much more common topic.

Finally, we found that wildfire risk was the most substantial forest health issue discussed across all groups, and that other concerns such as insects and disease generally were not as common in the recorded dialogue. The benefits and tradeoffs of thinning and prescribed burning treatments for hazardous fuels reduction were by far more consistently and universally discussed within and across all seven groups. Insects and disease were more frequently referred to in a general sense as disturbances that affected the forest. Specific insects and diseases were rarely mentioned in meeting notes, and did not appear regularly in collaborative recommendations. The exception was the Deschutes Collaborative Forest Project, where dialogue about the effects of dwarf mistletoe (*Arceuthobium* species) infestation on second-growth ponderosa pine occurred and was included in written collaborative recommendations on at least one occasion.

6.4 What Are the Perceived Outcomes and Satisfaction Levels of Collaborative Groups?

We used our Internet survey of collaborative group stakeholders to query their reported satisfaction with their group and their perceptions of its success at achieving various outcomes. Despite the limitations of this type of perceptual data, it offers one lens onto how well collaborative groups may or may not be working for their participants. First, we examined satisfaction by asking participants how satisfied they were with 24 aspects of their group and how satisfied they were overall with their group's ability to achieve desired outcomes, using a five-point Likert scale (Fig. 2). These aspects of collaborative process and structure were drawn from existing literature on collaborative capacity and from authors' experiences.

We found that a majority (over 59%) of respondents were satisfied or very satisfied overall, and about one-quarter were dissatisfied or very dissatisfied. On all aspects, respondents were much more likely to be satisfied or very satisfied than to be dissatisfied or very dissatisfied. The most satisfaction was with agency commitment to the group (nearly 90% satisfied or very satisfied, 64% of which were very satisfied). The most dissatisfaction was with shared mission/vision of the group, but this did not exceed one-quarter of respondents. These results demonstrate that participants who were willing to take the survey have fairly extensive satisfaction with different facets of their group; however, other participants who were not inclined to take the survey may have held different viewpoints.

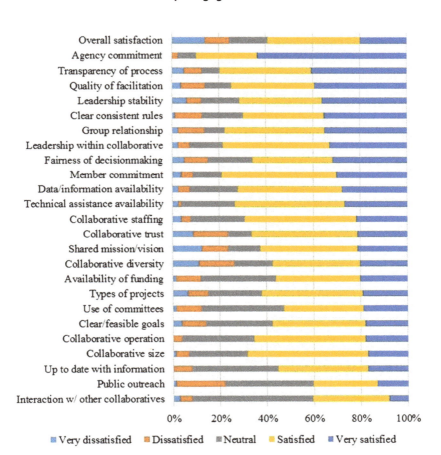

Fig. 2 Respondents' reported satisfaction with aspects of their collaborative groups; response for individual aspects ranged from 74 to 86 respondents (Overall survey *n* was 97)

We also asked respondents to report how successful they thought their group had been at achieving various outcomes, using a three-point Likert scale (Fig. 3). Most respondents indicated feeling that their group was moderately successful at achieving most of the outcomes. 45% of respondents stated that improved communication and trust within the group were very successful. The highest proportion of "not successful"

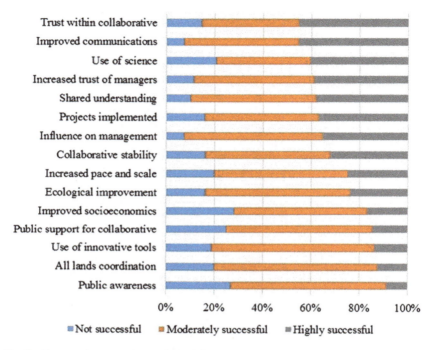

Fig. 3 Respondents' reported satisfaction with aspects of their collaborative groups; response for individual aspects ranged from 76 to 82 respondents (Overall survey *n* was 97)

responses, about a quarter each, were affiliated with producing socioeconomic improvements, garnering public support for the group itself, or garnering public awareness of forest management issues. Nearly a quarter reported that ecological improvement was highly successful and 60% reported that it was moderately successful, indicating that participants largely view their groups as effective at addressing forest health issues.

7   Conclusions

The state of Oregon and the US Forest Service have invested resources in addressing a perceived forest health crisis on federal lands; specifically, eastern Oregon's pine and mixed conifer forests. There is a social process around forest health in eastern Oregon wherein collaborative groups of stakeholders have regular dialogue about their desired outcomes and recommended treatments, and deliver these recommendations to the Forest Service. Collaborative groups are viewed as a potentially important mechanism for building social agreement around forest health, but there is little research or systematic documentation of what collaborative groups actually do and the forest health issues they address.

Our review of collaborative structure and documents, combined with our professional experiences, suggested that eastern Oregon's forest collaborative groups were fairly similar in how they were organized and with one exception, these collaborative groups were not formally established organizations with any kind of nonprofit, corporate, or other official status that would allow them to manage fiscal resources and staff directly; they were venues for dialogue. They relied on other organizations in a sponsor or fiscal agent type role. We also observed that "forest health" for all groups is more of an umbrella term encompassing many different issues. It is often used fairly vaguely, and to indicate a general and widespread need for forest restoration activities including thinning and prescribed burning. Ecological concepts such as historic range of variability and restoring diversity in forest structure and species were more commonly discussed than specific insects and diseases. Finally, surveyed participants generally indicated a high level of satisfaction with their collaborative group's ability to achieve its goals, and rated their groups as moderately or highly successful at a number of management outcomes.

Given these similarities, we would posit the following implications for future policy and practice. First, since collaborative groups are not typically organized as formal entities, they may face challenges with long-term durability and ability to attract, maintain, and independently manage their operations (alternately, this may allow them to focus on dialogue rather than management and administration). Policy makers and grantors may wish to consider that most collaborative groups are not organizations with extensive capacity for managing funds, staff, and project implementation; and establish fair expectations. Second, although localized dialogue specific to each collaborative would certainly remain important, there may be a need for more region-wide workshops and learning opportunities so that all collaborative groups can equitably access consistent scientific information about forest health. This may also improve efficiency for the scientists and organizations that assist collaborative groups with this information. Third, we found that collaborative groups did not necessarily represent all the stakeholder perspectives on each national forest and that they do not lead to 100% agreement about planned forest restoration actions. In particular, some environmental organizations were not consistently participating, or even actively stating that they did not view collaborative groups as an effective venue for true multi-stakeholder dialogue and consensus building. Yet the expectations and language about collaborative groups found in both state and federal policies and strategies continues to suggest that collaborative groups are valuable for reducing conflict and public objections. We propose that more research into this topic is necessary, as there is very little (but see Summers 2014); and that any evaluation or monitoring of collaborative groups for state or federal programs clarify the limitations of collaborative groups in this regard rather than sustain unrealistic expectations. Monitoring and evaluation could also explore if the outcomes of collaboration with which respondents expressed satisfaction are indeed occurring, and if these are suitable metrics to use for understanding the impacts and value of collaborative groups.

Despite our focus on eastern Oregon, there are potential implications for research on social dimensions of forest health worldwide. For example, eastern Oregon's forest collaborative groups operate in a context wherein government owns and manages the land. Observations about how they interact with government and may be moving toward co-management of forest health and forestlands (or not) may be applicable to other settings with similar landownership patterns and public–government relationships. Another consideration is that forest health issues are not merely biophysical; and they cross political and other boundaries, necessitating the engagement and cooperation of multiple stakeholders, land owners, scientists, managers, and policy makers. Previous studies of social dimensions have tended to emphasize surveying individual values or preferences, or impacts on populations affected by forest health. Examining collaborative groups and processes, particularly the themes of trust and communication that emerged as important in our survey, may help uncover new insights into the importance of social interaction and collective action. Therefore, we anticipate a need for future study of collaborative governance and how diverse stakeholder values for forest health are negotiated in such settings.

Appendix 1

Collaborative Documentation Reviewed

Where a link is not provided, documents came from authors' files and are not publicly available.

- Meeting notes and organizational documents (charter, ground rules, operations manual, etc.) from all groups as available, 2012–2015. Documentation is not consistently available or reported similarly.
- Long-Range Strategy for the Lakeview Federal Stewardship Unit. 2011 update. Available at: https://www.fs.usda.gov/Internet/FSE_DOCUMENTS/stelprdb5356799.pdf.
- Lakeview Landscape Stewardship Proposal. 2010. Available at: https://www.fs.fed.us/restoration/documents/cflrp/2011Proposals/Region6/FremontWinema/Lakeview.docx.
- Final Draft of Emerging Consensus on Lower Joseph Creek Project. 2015. Wallowa-Whitman Forest Collaborative Group. Available at: http://www.wallowawhitmancollaborative.org/s/WWFC_LoJo_DRAFT_Consensus_Position_April_8_2015.pdf.
- Zones of Agreement. Blue Mountains Forest Partners. 2013–present. Available at: http://www.bluemountainsforestpartners.org/work/zones-of-agreement/.
- Southern Blues Coalition Collaborative Landscape Restoration Proposal. 2010. Available at: https://www.fs.fed.us/restoration/documents/cflrp/2011Proposals/Region6/Malheur/2011SouthernBluesRestorationCoalitionCFLRPProposal.pdf.

- Restoration Recommendation Framework by Forest Type. 2012. Deschutes Collaborative Forest Project.
- Deschutes Collaborative Forest Project 2013 Addendum to CFLRP. Available at: https://www.fs.fed.us/restoration/documents/cflrp/2010 Proposals/Region6/Deschutes/DCFPAddendumApril2013.pdf.
- Collaborative Input to the Wolf Watershed Analysis. 2012. Ochoco Forest Restoration Collaborative.
- Collaborative Input Statement for the Kahler Basin Planning Area. 2014. Umatilla Forest Collaborative Group.
- Common Principles and Goals. Harney County Restoration Collaborative. Available at: http://highdesertpartnership.org/what-we-do/harney-county-restoration-collaborative/common-ground-principles-and-goals.html.

Notes

1. This program was called the Federal Forest Health Program in 2013–2015 and changed to the Federal Forest Restoration Program in 2015–2017.
2. Per Hessburg and Agee (2003), the Inland Northwest is the catchment area of the Interior Columbia River Basin in the coterminous USA, which includes "all of Washington and much of Oregon east of the crest of the Cascade Mountain Range, excluding the Northern Great Basin and Owyhee River Uplands; nearly all of Idaho north of the Owyhee Uplands and Snake River Plains and portions of northwestern and southwestern Montana extending to the Continental Divide."
3. Co-facilitated Ochoco Forest Restoration Collaborative from 2011 to 2013 and Umatilla Forest Collaborative Group from 2013 to 2015.
4. Developed socioeconomic monitoring indicators and participated in monitoring of Lakeview Stewardship Group's Collaborative Forest Landscape Restoration Program work (White et al. 2015).
5. Conducted focus groups with Wallowa-Whitman Forest Collaborative, Umatilla Forest Collaborative Group, Blue Mountains Forest Partners, Harney County Restoration Collaborative, and Ochoco Forest Restoration Collaborative in 2015 to assess their self-reported progress

toward accelerated restoration. Reviewed state grants and collaborative activities performed using them (Davis et al. 2015a; White et al. 2015, 2016).
6. Conducted case studies of the specific design of collaborative processes for building shared agreement on the Blue Mountains Forest Partners and Deschutes Forest Collaborative Project in 2015 (Nuss and Davis 2015; Davis et al. 2015a, b).

References

Abrams, J., Davis, E. J., & Moseley, C. (2015). Community-based organizations and institutional work in the remote rural west. *Review of Policy Research, 32*(6), 675–698.

Abrams, J., Kelly, E., Shindler, B., & Wilton, J. (2005). Value orientation and forest management: The forest health debate. *Environmental Management, 36*(4), 495–505.

Ansell, C., & Gash, A. (2007). Collaborative governance in theory and practice. *Journal of Public Administration Research and Theory, 18*(4), 543–571.

Baker, M., & Kusel, J. (2003). *Community forestry in the United States: Learning from the past, crafting the future.* Washington, DC: Island Press.

Beebe, J. (2001). *Rapid assessment process: An introduction.* 224 pp. Walnut Creek, CA: AltaMira Press.

Brown, S. J. M. (2012). The Soda Bear Project and the Blue Mountains Forest Partners/USDA Forest Service Collaboration. *Journal of Forestry, 110*(8), 446–447.

Butler, W. H., Monroe, A., & McCaffrey, S. (2015). Collaborative implementation for ecological restoration on US public lands: Implications for legal context, accountability, and adaptive management. *Environmental Management, 55*(3), 564–577.

Cheng, A. S., & Sturtevant, V. E. (2012). A framework for assessing collaborative capacity in community-based public forest management. *Environment Management, 49*(3), 675–689.

Coulter, K., Boggs, D., Macfarlane, G., St. Clair, J., Garrity, M., Marderosian, A., et al. (2015). *Collective statement on collaborative group trends.* Blue Mountains Biodiversity Project. Available at: https://bluemountainsbiodiversityproject.org/collective-statement-on-collaborative-group-trends/ (Last accessed August 28, 2017).

Davis, E. J., Nuss, M. L., & Hughes, J. R. (2015a). *Science and collaborative decision-making: A case study of the Kew Study.* Case Study Research Brief #3, Forest Research Laboratory, Oregon State University. Available at: http://hdl.handle.net/1957/56559.

Davis, E. J., Cerveny, L., Nuss, M. L., & Seesholtz, D. (2015b). *Oregon's forest collaborative groups: A rapid assessment.* Research Contribution Summary #1, Forest Research Laboratory, Oregon State University. Available at: http://hdl.handle.net/1957/55791.

Davis, E. J., White, E. M., Cerveny, L. K., Seesholtz, D., Nuss, M. L., & Ulrich, D. R. (2017). Comparison of USDA Forest Service and stakeholder motivations and experiences in collaborative federal forest governance in the western United States. *Environmental Management, 60*(5), 908–921.

DellaSala, D. A., Olson, D. M., Barth, S. E., Crane, S. L., & Primm, S. A. (1995). Forest health: Moving beyond rhetoric to restore healthy landscapes in the inland Northwest. *Wildlife Society Bulletin (1973–2006), 23*(3), 346–356.

Franklin, J. F., Hagmann, R. K., & Urgenson, L. S. (2014). Interactions between societal goals and restoration of dry forest landscapes in western north America. *Landscape Ecology, 29*(10), 1645–1655.

Fuller, L., Marzano, M., Peace, A., Quine, C. P., & Dandy, N. (2016). Public acceptance of tree health management: Results of a national survey in the UK. *Environmental Science & Policy, 59,* 18–25.

Goldstein, B. E., & Butler, W. H. (2010). Expanding the scope and impact of collaborative planning: Combining multi-stakeholder collaboration and communities of practice in a learning network. *Journal of the American Planning Association, 76*(2), 238–249.

Hessburg, P. F., & Agee, J. K. (2003). An environmental narrative of inland northwest United States forests, 1800–2000. *Forest Ecology and Management, 178*(1), 23–59.

Hessburg, P. F., Agee, J. K., & Franklin, J. F. (2005). Dry forests and wildland fires of the inland northwest USA: Contrasting the landscape ecology of the pre-settlement and modern eras. *Forest Ecology and Management, 211*(1), 117–139.

Hessburg, P. F., Churchill, D. J., Larson, A. J., Haugo, R. D., Miller, C., Spies, T. A., et al. (2015). Restoring fire-prone inland Pacific landscapes: Seven core principles. *Landscape Ecology, 30*(10), 1805–1835.

Heyerdahl, E. K., Loehman, R. A., & Falk, D. A. (2014). Mixed-severity fire in lodgepole pine dominated forests: Are historical regimes sustainable on

Oregon's Pumice Plateau, USA? *Canadian Journal of Forest Research, 44*(6), 593–603.

Jenkins, A. F. (1997). Forest health: A crisis of human proportions. *Journal of Forestry, 95*(9), 11–14.

Keane, R. E., Hessburg, P. F., Landres, P. B., & Swanson, F. J. (2009). The use of historical range and variability (HRV) in landscape management. *Forest Ecology and Management, 258*(7), 1025–1037.

Langston, N. (1995). *Forest dreams, forest nightmares: The paradox of old growth in the Inland West.* Seattle: University of Washington Press.

Margerum, R. D. (2011). *Beyond consensus: Improving collaborative planning and management.* Cambridge, MA: MIT Press.

Monroe, A. S., & Butler, W. H. (2016). Responding to a policy mandate to collaborate: Structuring collaboration in the collaborative forest landscape restoration program. *Journal of Environmental Planning and Management, 59*(6), 1054–1072.

Mutch, R. W., Arno, S. F., Brown, J. K., Carlson, C. E., Ottmar, R. D., & Peterson, J. L. (1993). Forest health in the Blue Mountains: A management strategy for fire-adapted ecosystems. Gen. Tech. Rep. PNW-GTR-310. Portland, OR: U.S. Department of Agriculture, Forest Service, Pacific Northwest Research Station. 14 pp.

Nuss, M., & Davis, E. J. (2015). *Formalizing decisions: A case study on collaborative zones of agreement.* Case Study Brief #2, College of Forestry, Oregon State University. Available at: http://ir.library.oregonstate.edu/xmlui/handle/1957/56364.

Oregon Board of Forestry. (2009). *Achieving Oregon's vision for federal forestlands.* Oregon: Oregon Board of Forestry. 59 pp.

Oregon Wild. (2012). *Restoring eastern Oregon's dry forests: A practical guide for ecological restoration.* Available at: https://www.oregonwild.org/sites/default/files/pdf-files/Eastside_Restoration_Handbook.pdf.

Patton, M. Q. (2005). *Qualitative research.* Hoboken, NJ: Wiley.

Proctor, J. D. (1998). Environmental values and popular conflict over environmental management: A comparative analysis of public comments on the Clinton forest plan. *Environmental Management, 22*(3), 347–358.

Schultz, C. A., Jedd, T., & Beam, R. D. (2012). The collaborative forest landscape restoration program: A history and overview of the first projects. *Journal of Forestry, 110*(7), 381–391.

Seager, S. T., Ediger, V., & Davis, E. J. (2015). *Aspen restoration and social agreements: An introductory guide for forest collaborative groups in central and eastern Oregon.* Portland, OR: The Nature Conservancy.

Shindler, B. A., Brunson, M. W., & Stankey, G. H. (2002). *Social acceptability of forest conditions and management practices: A problem analysis* (p. 68). Portland, OR: US Department of Agriculture, Forest Service, Pacific Northwest Research Station.

Smith, W. B., Miles, P. D., Perry, C. H., & Pugh, S. A. (2009). *Forest resources of the United States, 2007: A technical document supporting the forest service 2010 RPA assessment.* General Technical Report-USDA Forest Service, (WO-78).

Stine, P., Hessburg, P., Spies, T., Kramer, M., Fettig, C. J., Hansen, A., et al. (2014). *The ecology and management of moist mixed-conifer forests in eastern Oregon and Washington: A synthesis of the relevant biophysical science and implications for future land management.* Gen. Tech. Rep. PNW-GTR-897. Portland, OR: U.S. Department of Agriculture, Forest Service, Pacific Northwest Research Station. 254 pp.

Summers, B. M. (2014). *The effectiveness of forest collaborative groups at reducing the likelihood of project appeals and objections in eastern Oregon.* Master's thesis, Portland State University, Portland.

Thorson, T. D. (2003). *Ecoregions of Oregon.* Reston, VA: US Department of the Interior, US Geological Survey.

Tidwell, T. (2012). *Statement to the committee on agriculture, subcommittee on conservation, energy, and forestry, United States House of Representatives.* US Forest Service land management: Challenges and opportunities for achieving healthier national forests. Available at: http://www.fs.fed.us/sites/default/files/media/types/testimony/HAgC_03-27-2012_Testimony.pdf (Last accessed January 14, 2015).

United States Department of Agriculture Forest Service (USDA Forest Service). (n.d.). *Eastside restoration.* http://www.fs.usda.gov/detail/r6/land-management/resourcemanagement/?cid=stelprdb5423597 (Last accessed January 14, 2015).

Vaughn, J., & Cortner, H. (2005). *George W. Bush's healthy forests: Reframing the environmental debate.* Boulder: University Press of Colorado.

White, E. M., Bennett, D. E., Davis, E. J., & Moseley, C. (2016). *Economic outcomes from the U.S. forest service eastside strategy. Ecosystem workforce program* (Working Paper No. 64). University of Oregon.

White, E. M, Davis, E. J., Bennett., D. E., & Moseley, C. (2015). *Monitoring of outcomes from Oregon's federal forest health program*. Ecosystem workforce program (Working Paper No. 57). University of Oregon.

Williams, M. A., & Baker, W. L. (2012). Spatially extensive reconstructions show variable-severity fire and heterogeneous structure in historical western United States dry forests. *Global Ecology and Biogeography, 21*(10), 1042–1052.

Wondolleck, J. M., & Yaffee, S. L. (2000). *Making collaboration work: Lessons from innovation in natural resource management*. Washington, DC: Island Press.

16

Environmental Ethics of Forest Health: Alternative Stories of Asian Longhorn Beetle Management in the UK

Norman Dandy, Emily Porth and Ros Hague

1 Introduction

Humans respond to invasive pathogens and invertebrates by taking actions that have significant consequences for humans, non-humans, and the wider environment. Although the public generally expresses strong support for managing forest health problems by whatever means are deemed necessary (Fuller et al. 2016), these same people are also significantly concerned about the impacts on non-humans as a result of the management methods being used. Such questions are at the

---

N. Dandy (✉)
Plunkett Foundation, The Quadrangle, Woodstock, UK

E. Porth
Independent Scholar

R. Hague
School of Social Sciences, Nottingham Trent University
Nottingham, UK

core of environmental ethics, which explores the relationships between humans and nature, the intrinsic value of nature, and the consequences of anthropocentrism. Key topics in this area of study include climate change, biodiversity loss, the treatment of non-human species, and environmental aesthetics. In this chapter, we explore the implicitly anthropocentric ethical positions which form the foundation of forest health management decisions. We seek to generate insights into the ethical framings of forest health and 'invasive' species management, which remains a much-neglected debate in both forestry and environmental ethics. Our aim is to demonstrate that extant framings and practices of forest management are not the only options, but also rather one approach amongst a number of alternatives. Many of these frameworks go beyond the anthropocentrism that lies at the core of much environmental degradation.

We have generated three novel narrative accounts, or stories, of the 2012 'outbreak' of Asian longhorn beetle (*Anoplophora glabripennis*, ALB) in Kent, UK, by using three distinct perspectives rooted in environmental and non-human ethics. The 'emergency modality' management response (Collier and Lakoff 2008) to ALB, as with forest health management and biosecurity more generally, was founded on broad utilitarian claims. These claims relate primarily to environmental protection, the conservation of 'native' species, and the prevention of economic damage to commercial forestry. We provide a description of how the management of this outbreak unfolded in practice, followed by three possible alternative accounts of the outbreak. Each account critically reflects on the event through different ethical frameworks focused on the moral status of, empathy for, and the flourishing of non-humans.

The three ethical frameworks that we have chosen represent something of a cross section of perspectives within environmental ethics. 'Biocentrism', developed by Paul Taylor, is a relatively mainstream approach that argues for well-established ethical concepts, including the expansion of rights and associated moral consideration beyond humans and other sentient beings. Lori Gruen's arguments for 'entangled empathy' are, although applied to non-humans, a well-known source of moral obligation. Finally, 'flourishing', and in particular the work of Angela Kallhoff, has a long legacy within ethics, but has only rarely been utilised in relation to non-humans. These three frameworks offer distinct opportunities to reflect on forest health, and whether or how they would be accepted into contemporary approaches to environmental management. Consequently, we position each framework in relation to forest health in the sections below.

The stories we construct through our interactions with these frameworks pay particular attention to decision-making and outbreak management processes. Asking who is included in those processes, and how, has significant implications for both humans and non-humans. Our juxtaposition of these three different approaches to the ALB outbreak will demonstrate how competing ethical claims are framed through and by particular interests. Through this analysis, we will demonstrate how different environmental ethical frameworks may or may not demand varied approaches to managing forest health and result in different outcomes.

Narrative construction has an established place in critical environmental ethics (Clayton 1998; King 1999) and ecofeminism (i.e. Vance 1995). At its root, a narrative is a story or an account of events. Paying attention to the frameworks in which stories are constructed allows us to be open to other voices. In the context of this chapter, this means being attentive to the organisms who are central characters in the story, but whose interests were not given consideration in the environmental management approach that was employed in the ALB 'outbreak'. In this way, narrative offers us a means of imagining possibilities for the non-human beings who were originally ignored or silenced, and we are able to explore how things might have been different had those voices in some way been heard. Our goal with this chapter is to add forest health to the narrative analytic tradition. Such stories have often proved an effective way of revealing the detrimental impact of dominant anthropocentric approaches to environmental management. In her discussion on narrative, animals, and ethics, Vance (1995, 165) argues that most human narrative is written with the intention of explaining or giving meaning to human experience; this means that narratives about nature are inherently anthropocentric. By declaring that storytelling is 'an act done in community' (176), Vance identifies that stories are ethical discourse which model the storyteller's beliefs about human–non-human animal relationships, as well as shape the beliefs of others about these relationships. The alternative narratives in this chapter were generated through the recursive process of connecting the practice of ALB outbreak management and associated policy responses, as they were observed and experienced by the authors through research and documentary sources, with each of the three ethical frameworks. Insights were also generated by comparing the narratives to each other as they were developed.

Following a conventional narrative about how the ALB outbreak and its management unfolded (Section 2), each subsequent section outlines an ethical perspective and is followed immediately by the accompanying alternative narrative. The discussion (Section 6) looks across these three alternative narratives to consider the demands that might be placed on outbreak management if forest managers were to adopt an approach that was more attentive to non-human species.

2 Contemporary Management of the Asian Longhorn Beetle in the UK

In 2009, a lone adult ALB was found in the garden of a homeowner on the outskirts of Paddock Wood, Kent, UK. ALB is a wood boring insect that can cause widespread tree mortality and is able to live on many hardwood tree species (Macleod et al. 2002; CABI 2017). The resident reported the beetle to the Food and Environment Research Agency (Fera) and a local officer then carried out a site survey. The survey revealed no evidence of an infestation, but annual monitoring visits by entomologists from Forest Research (FR) were scheduled for the next four years. This decision was made in line with European level guidance to monitor high-risk sites for at least 3 or 4 years following a beetle discovery. During one of these routine site surveys in early March 2012, evidence was found indicating a possible wild population of ALB. Although *Anoplophora* species had occasionally been intercepted whilst entering the UK since the 1990s, this was the first time a breeding population had been found established in British woodland. Following further investigation, the presence of ALB was confirmed by scientists at FR on 15 March 2012.

The site at the centre of the outbreak had previously been a stonemasonry business. As a regular importer of stone from China, it is likely the beetle was introduced from their 'native' range in East Asia to Paddock Wood through the wood pallets in which the stone was shipped (referred to in phytosanitary terms as the 'wood packaging pathway'). Although ALB had historically been regarded as a benign native species in China, they are now widely regarded as a 'pest'; this has primarily been attributed to a significant increase in monoculture tree plantations in China from the 1960s, which has allowed them to reproduce beyond their usual numbers and led to regular outbreaks beginning in the 1980s and continuing into the present (Haack et al. 2010, 527). In the light of this, it could be said that both the proliferation of this species of beetle in China and its increased accidental transport to new environments all over the world is a direct result of human practice. The outbreak site was semi-rural and relatively densely wooded, although it had traditionally been a hop-growing area. As ALB can be hosted on multiple tree species, the variety of trees in the area provided an ideal environment for the insect to spread once adult beetles had emerged from the wood pallets.

An Outbreak Management Board (OMB) was established on 24 March 2012, comprised of experts from the Forestry Commission and Fera, with the goal of developing an eradication programme. The environmental management options presented were based on the biology of ALB, particularly in relation to its large size and relatively sedentary nature, both of which facilitate its containment. Unlike many pest insect species, it is possible to eradicate ALB once it has become established in an area, as has been demonstrated in cases in Europe and North America (Haack et al. 2010). Based on this evidence, the Board decided upon a programme of 'sanitation felling' (removing and incinerating possible host trees) aimed at the eradication of this 'pest'.

Ground surveys were used to identify host trees. This was difficult from a practical perspective because ALB lives in the tree crown and can be difficult to detect from the ground. Once identified, these trees were felled and inspected for the presence of ALB at all life stages; if found to be infested, other tree species within 100 metres known to be hosts for ALB were felled. Infested branch and stem material was cut and packaged before being sent to FR at their Alice Holt Research Station for processing; unaffected woody material was burned to ash on site. The sanitation felling work was carried out by forestry contractors and finished in mid-August 2012. The analysis of infested material at FR was complete by September that year.

A number of established and novel survey techniques were employed during this outbreak. For example, dog handlers with sniffer dogs trained to detect invasive insect species were brought from Austria and worked in the area in late August. The dogs did not, however, detect any ALB or infested trees. The ground and weather conditions caused difficulty for the dogs, as well as for the handlers who were not always able to conclusively identify which tree the dogs had indicated, due to the dense woodland they encountered. As a whole, evidence of infestation (i.e. exit holes and larvae) was very unevenly distributed with, for example, one large sycamore tree accounting for 88% of the exit holes and 40% of the live larvae and pupae discovered (Straw and Tilbury 2012).

When the live beetle was found in 2009, neither Fera nor the Forestry Commission communicated with local residents about its discovery. However, when the wild population was discovered in 2012, staff from the Forestry Commission, along with the local Fera officer, liaised with forestry contractors and local residents, as well as managed the operational aspects of the eradication programme. Immediately after the outbreak had been identified in 2012, the Fera officer distributed information leaflets to Paddock Wood residents about Citrus longhorn beetle (*Anoplophora chinensis*), which looks very similar to ALB, but lives in the roots of trees, rather than in the crown. The CLB leaflet was distributed because, following from the 'emergency modality' management mindset, there was a strong sense of urgency to notify local people about the outbreak. It was ready to distribute when the ALB were discovered, whereas there was no prepared information regarding ALB. This did cause some confusion for residents. Two consultative meetings were held with landowners and residents within the first ten days of identifying the outbreak. The first meeting, held on site, attracted around 90 people, as well as arboriculturalists from other parts of England. The second meeting focused on a discussion of the pre-selected management response (sanitation felling) with residents and local councillors, which brought together about 30 people. The media response was greatest at the beginning of the outbreak management in March and April, with some follow-up at later stages. Social scientists from FR carried out a postal survey with a selection of residents in Paddock Wood nearly a year after the eradication programme concluded, as well as follow-up qualitative interviews with the residents who were most directly affected by the management. These interviews indicated mixed feelings about the programme, and some concern about the impact of the chosen eradication method on native tree and wildlife species.

In total, 2166 trees were removed from the 14-hectare management site and, of these, 66 trees (3%) were found to be infested. Additional to this, 354 live larvae, 34 live pupae, and 2 live eclosed adult beetles were found during the management effort (the live adult beetles emerged from wood material in the laboratory, not in the field). 46 dead adult beetles were discovered in their tunnels within the wood (Straw and Tilbury 2012). The ALB population in Paddock Wood was considered eradicated in early September 2012. However, the site will continue to be monitored until 2018, more than 2 life-cycle periods of the beetle, until eradication can be officially declared.

For local residents affected by the management effort, there were mixed feelings about its conduct and outcome. A number of communication problems emerged, with issues relating to who was responsible for specific elements of the management. Residents expressed frustrations with the actions of some of those involved in management, and in relation to the seeming disparity between the severity of management and the limited evidence of beetle infestation. Perhaps the overriding sense, however, was sadness at the loss of trees and concern regarding the impact of this on the landscape and its resident wildlife (Porth et al. 2015).

3   Biocentrism

3.1   Biocentrism as an Ethical Perspective

In constructing a biocentric account of forest beetle management, our starting point is Paul Taylor's seminal biocentric environmental ethic centred on 'respect for nature' (Taylor 1981, 2011). Situated firmly within the rationalist tradition of ethics, Taylor's biocentrism focuses on individual organisms as 'teleological centers of life' and is highly structured by a number of stated rules and beliefs. Taylor posits duties towards all non-human (plant and animal) life based on the fact their well-being can be promoted or hindered. An attitude of 'respect for nature' is underpinned by a set of beliefs and basic rules of conduct. The beliefs are that (i) humans and non-humans alike are all members of Earth's 'community of life', (ii) the natural world is interdependent to the extent that the survival of individual organisms is interlinked, (iii) each organism has an individual existence and pursues its own way of life in response to its environment, and (iv) humans are not inherently superior to non-humans.

The four rules of conduct are to (i) avoid harm (*nonmaleficence*), which is the most fundamental duty towards nature, (ii) avoid restricting the freedom of individual organisms to act and develop in their own way (*noninterference*), (iii) avoid deception of any organism capable of being deceived (*fidelity*), and (iv) restore the balance of justice between agent and subject in the case of wrongdoing (*restitutive justice*). A critical implication of the rule of non-interference in the context of an ALB outbreak is the implication of species-impartiality. Taylor highlights this as particularly key to redressing the tendency for people to favour and sympathise with certain species over others, for example prey over predator.

> People get disturbed by a great tree being "strangled" by a vine. And when it comes to instances of bacteria-caused diseases, almost everyone has a tendency to be on the side of the organism which has the disease rather than viewing the situation from the standpoint of the living bacteria inside the organism. If we accept the biocentric outlook and have genuine respect for nature, however, we remain strictly neutral between predator and prey, parasite and host, the disease-causing and the diseased. (Taylor 2013, 156–157)

Detailed consideration of what a 'respect for nature' attitude would entail for invasive species management is entirely lacking from academic literature. However, the stated beliefs and rules do have some profound consequences for beetle management in forests. For example, it places individual trees and individual beetles on an equal footing in their interaction with humans as 'centers of life' (Taylor 1981, 210) and, perhaps much more significantly, it requires us not to harm individual trees, beetles, or other organisms in our response.

However, one other key feature of Taylor's biocentrism is enormously relevant to beetle management: for an organism to fall within the remit and moral purview of 'respect for nature', it must be encountered in its 'wild' state within its 'natural' ecosystem. Concepts of natural or ecosystemic 'balance', 'integrity', and/or 'equilibrium' feature clearly in the construction of this ethic as being the appropriate context in which individual organisms can best pursue their own good. This duty extends only to human (non)interference in natural systems. Taylor explicitly precludes human interference to redress *naturally* occurring changes in ecological relationships and structures.

Taylor's specific biocentric perspective relies on some strong conceptual boundaries between the human and non-human worlds, which many call into question as valid bases for scientific or ethical judgements. They are, however, concepts that are regularly mobilised to justify outbreak management. Furthermore, a 'respect for nature' entails moral duties which generate questions about how humans respond to beetle outbreaks. One of these focuses on the means by which a beetle came to be found outside its 'natural' habitat.

3.2 Outbreak Story 1: A Biocentric Account

In 2009, an adult ALB was found in the garden of a homeowner in Kent, UK. Considering the beetle to be unfamiliar and out of place, the resident reported it to Fera. Local officers subsequently carried out a detailed site survey, which revealed no evidence of an infestation. At this stage of the risk assessment, there was considered to be limited threat to the local environment, but annual monitoring visits by entomologists from FR were scheduled. Three years later, during one of these routine surveys, evidence was found of a wild population of ALB. Although *Anoplophora* species had occasionally been intercepted whilst being transported to the UK, this was the first time a breeding population had been found established in British woodland. Had the stonemasonry business at the centre of the outbreak still been operating, government would have sought financial compensation due to the breach of existing environmental non-interference regulations through the import and introduction of ALB from its native environment.

At this stage, an OMB was established, comprised of experts from relevant governmental agencies. It had the goal of developing a management programme to address the occurrence of ALB within the framework of the overarching Respect for Nature Code of Practice for government bodies. Key principles of this code were the prevention of harm to organisms and protection of the integrity of the natural environment. The management options presented were thus based on the biology and ecology of ALB and the ethical commitments to avoid harm to members of native populations, which was perceived to be intrinsic to an attitude of respect for nature. ALB was swiftly designated as non-native, and given its large size and relatively sedentary nature, both of which facilitate its containment, the OMB recommended a measured and targeted response focused on removal of the beetle population. Fera undertook a rapid environmental assessment pertaining to the nativeness and vulnerability of local tree species. The potentially affected outbreak zone was heavily wooded and featured a number of native (e.g. field maple, willow, and black poplar) and non-native tree (e.g. sycamore and horse chestnut) species. Based on this information, the board decided upon a programme of intensive surveying, safe removal of live non-native beetles, and targeted sanitation felling (removing and incinerating clearly identified infested trees along with non-native possible host trees). This was aimed at the containment and gradual removal of ALB.

Detailed ground and aerial surveys were used to positively identify infested trees. ALB survey is difficult from a practical perspective because the beetle lives in the tree crown and can be difficult to detect from the ground. Therefore, substantial resources had to be allocated to the survey effort. Pheromone trapping was used to monitor and capture any beetles potentially in the local landscape. Once identified, infested trees were felled and inspected in detail for the presence of ALB at all life stages. Other potential host tree species within 100 metres were subject to close survey, and the same process was followed if infested.

Host species that were not directly observed as infested were kept under close observation. Infested branch and stem material was sent to FR for further analysis and to support the development of ALB management methods that were not lethal to host trees. The sanitation felling work was carried out by specialist forestry contractors.

As part of the detailed survey work, sniffer dogs trained to detect specific insect species were brought to the area. The dogs did not detect any ALB or infested trees, although the ground and weather conditions caused difficulty for the dogs and their handlers, making the inspection process challenging. No live ALB were encountered in the field during the management programme, although two emerged in laboratory conditions at FR.

Staff from Fera liaised with forestry contractors and local residents and managed the operational aspects of the removal programme. Shortly after the outbreak had been identified, the Fera officer distributed detailed information leaflets to residents about ALB. Local community members proved to be well-informed about the issue of non-native species. Consultative meetings were held with landowners and residents within the first days of the outbreak. In parallel with this, there was in-depth consultation with non-governmental organisations dedicated to nature protection and to the protection of living beings as part of efforts to take account of non-human stakeholder perspectives. Other meetings focused on the selected management with around 30 residents and local councillors contributing actively to the debate. FR conducted social scientific research with a selection of local residents approximately a year after the removal programme concluded, as well as follow-up work considering the impact on those non-human stakeholders most directly affected by the management. These analyses indicated mixed feelings about the programme, although there was a level of satisfaction amongst residents regarding the limited impact of the management on native species.

In total, 66 trees were found to be infested and were felled. A small number of other trees were felled after initial detection of infestation proved false. Nearly 400 live larvae and pupae were removed from the site. The ALB population took several months to be removed, and the location will continue to be monitored for a number of years before the local environment can be officially declared 'safe'.

4 Entangled Empathy

4.1 Entangled Empathy as an Ethical Perspective

'Entangled empathy' is an ethical framework developed by Lori Gruen (2015) as an alternative ethic for human–animal relationships. This framework is part of an 'ethics of care', which focuses on 'the particularity of caring relationships' (Gruen 2015, 32); it falls within the Feminist Care Tradition in Animal Ethics, as characterised by Carol Adams and Josephine Donovan (Gruen 2015, 35). In contrast to traditional ethical approaches, the care tradition is attentive to context rather than abstraction; relationality instead of individualism; connection over impartiality; and responsiveness to move towards solutions, rather than focusing on conflict (2015, 33–34).

Gruen describes empathy as a 'particular type of attention' which can be considered to be a kind of moral perception (2015, 39).

> Entangled Empathy is a type of caring perception focused on attending to another's experience of well-being. An experiential process involving a blend of emotion and cognition in which we recognize we are in relationships with others and are called upon to be responsive and responsible in these relationships by attending to another's needs, interests, desires, vulnerabilities, hopes, and sensitivities. (Gruen 2015, 3)

By using the word 'entangled', Gruen highlights the multiple ways that we exist in active relationships with human and non-human beings, which 'co-constitute who we are and how we configure our identities and agency'. Entanglement asserts that we cannot disentangle ourselves from these relationships because our lives would no longer make sense (2015, 63). The challenge is to recognise how deeply entangled we are in these relationships and to find ways to be more perceptive and responsive to them.

At its core, to enact entangled empathy requires reflection and correction through a blend of cognition (knowledge) and affect (emotional reaction) about these relationships. 'The empathizer is always attentive to both similarities and differences between herself and her situation and that of the fellow creature with whom she is empathizing' (Gruen 2015, 66).

Entangled empathy encourages one to pay attention to well-being, and to meaningfully consider how one's actions interact within privilege and intersectional oppression (2015, 94). Ultimately, it is hoped that deeper understanding will motivate the empathiser to take action in ways that improve communal well-being.

Gruen insists that entangled empathy is limited in its application to beings who are 'sentient' and 'have experiences' (2015, 67). She does not, however, explicitly define what either of these concepts means to her within the context of this work. Gruen conflates ecosystems (including rivers, meadows, glaciers, etc.) with microbes, insects, and trees in her list of non-human beings who lack sentience and the ability to have experiences with which we, as humans, can either empathise or understand (2015, 70–71). Although this part of her ethical framework could pose problems in its application to the field of forest health, there are good reasons to question her assumptions. First, she makes arbitrary distinctions between, for instance, insects 'inhabiting' a tree and the birds 'who make their homes' there (2015, 70–71). These are superficial, semantic descriptors that create unnecessary emotional distance between species. Second, her suggestion that we can know 'what it is like to be like' (2015, 71) sentient beings, such as a cow or a dolphin, to a greater extent than we can with presumably non-sentient beings, such as an ALB or a tree, can be rejected: any of these experiences would feel particularly alien to a human.

There are many different ways to acquire knowledge about species-typical behaviour and the individual personality of a being. Multispecies ethnography, for instance, is an emerging research method which experiments with different ways of knowing and understanding the experiences of a wide variety of non-human beings, including insects, microbes, and trees. Pioneering research is being done with trees to understand how individuals support one another via, for example, nutrient and water transfer (Simard and Durall 2004) and perhaps 'communicate' with each other (Simard 2016; Wohlleben 2016), and research has long been carried out on the sociality of some insect species. Wagler and Wagler (2011) found that teachers exposed to hissing cockroaches in the classroom only developed more positive attitudes towards those insects over time (although not to other arthropods), and insect zoos have also evidenced attitude change in humans towards insects (Pitt and Shockley 2014). These studies demonstrate it is possible for anyone with curiosity and openness to cultivate experiences that will help them to understand the behaviour and individual personalities of other beings over time, even when those species might initially seem very alien to us.

Finally, the degree of empathy responsible for Gruen (or anyone else) refusing to harm insects and choosing to move them to safety (2015, 70) is irrelevant. This is particularly true in a management-centred narrative where we are focusing on situation outcomes. It may also be that in a situation where there are so many human and non-human beings with whom one can enact entangled empathy, the chosen outcome will still result in the harm of some of these beings, whether they are 'sentient' or not. For all of these reasons, we apply the entangled empathy ethical framework to construct a novel account of the ALB outbreak.

4.2 Outbreak Story 2: An Account of Entangled Empathy

An established population of ALB was discovered in Kent during a regular annual inspection by entomologists from FR in 2012. The semi-rural outbreak area was a mix of widely spaced residential and business properties within relatively dense woodland. Given the ability of ALB to inhabit many hardwood tree species, the variety of trees in the area provided an ideal environment for ALB to spread once one or more adult beetles had emerged unseen from the wood pallets. However, these woods are also home to many other animal species, including colonies of woodpeckers who are a source of pride and a symbol of ecosystem health to some residents.

Once the wild population of beetles had been identified through the annual survey undertaken by FR, they contacted Fera and organised a joint OMB. This board was comprised of experts in entomology and environmental management; social scientists from FR who were able to provide advice about how to manage the social impacts of the outbreak and carry out research to understand local responses to the beetles; and public engagement experts who were able to develop and enact a strategy to communicate with and involve affected residents in the management process. In close consultation with the local community, the OMB was responsible for deciding on a course of action in response to the discovery of the beetle population.

The public engagement experts began by printing ALB fliers and visiting people whose properties could be directly affected by the beetle population and/or a type of eradication programme. These people were invited to participate in a series of consultation meetings, each held only a week apart, which were also attended by a representative from each of the local town council, county council, and parish council, in addition to representatives from relevant environmental third sector organisations.

At the first of these meetings with the OMB, local people and the various representatives were provided with background information about the biology of ALB, and what was currently known about the population of beetles in their community. During this meeting, local people and representatives were invited to speak about their concerns. Although some were preoccupied with potential damage to their properties through any sort of eradication effort, others spoke about their apprehension around how the local landscape might change if many trees were felled. There was also concern about the well-being of the trees themselves and the animals who depended on them for their homes and for food. Although some third sector organisations were primarily preoccupied with the long-term damage that ALB could cause to Britain's landscape and were in agreement with OMB environmental managers that the beetle needed to be fully eradicated, other charities and individuals were concerned about the welfare of ALB. Hearing about how the beetle had—purely through human activity—become invasive in its native habitat in China and then introduced to radically different environments where it was hunted down as an alien, some people felt moved to protect ALB, even though they simultaneously wanted to protect the landscapes that they loved and called home. This instigated some emotional discussions amongst those present about whose lives should be prioritised in this situation.

The first meeting provided the OMB with a wide range of material with which to contemplate possible management options. They also received letters from local children who had been learning about empathy and 'compassion for all species' at school. The children advocated the protection of ALB, whom they likened to refugees who were persecuted in their homeland and then smuggled overseas by traffickers. All members of the OMB worked together over the course of a week to synthesise these various perspectives and, in combination with their expert knowledge, narrow down acceptable management options.

At the second meeting, OMB experts gave a presentation about what had happed in other countries where ALB had been discovered, and how it had been managed. Attendees were then presented with several management options in both presentations and dissemination materials, which they were encouraged to take home and contemplate. The first option included sanitation felling and incineration of all trees which were possible hosts of the beetle in order to eradicate it and protect Britain's trees from the potential future spread of ALB. It would, in time, be possible to replant the area with hybrid trees resistant to the beetle. The second option, seeking to minimise collateral harm to wildlife, involved the use of insecticides and pheromone traps to kill and trap as many beetles as possible in the outbreak area, an attempt to protect as many trees and other animals in the local ecosystem who were dependent on them. Third was the option of using a biological control to eliminate the beetle as had been used elsewhere (Liu et al. 1992; cited in CABI 2017), although there was little sense of whether there could be further environmental impacts from releasing nematodes into the local environment. Finally, given the concerns about the beetles themselves, which the OMB had not expected, and using Bavaria's response to spruce bark beetles in Bavarian Forest National Park as a precedent (Müller and Job 2009), the OMB presented the option of allowing the beetle to continue to exist in its adopted environment. At a third meeting one week later, everyone met again to discuss their thoughts and feelings about these management options and how they would impact the local community.

Based on the discussion and final vote on the four presented options, the OMB made the somewhat surprising recommendation that the beetle be left to its own devices. Given the emerging evidence that the majority of beetles were not surviving in their new environment, managers concluded it would be difficult for ALB to spread outside the local area. However, they also recommended a long-term woodland management plan to ensure the area did not lose species diversity. This would include prioritising the planting of hybrid tree species with resistance to ALB infestation; planting a range of oak and beech species which have not been found susceptible to ALB (CABI 2017); and interspersing conifers throughout the landscape, which are resistant to ALB. The OMB decision also referenced the aforementioned research in which visitors to Bavarian Forest National Park revealed a preference for granting the spruce bark beetle (*Ips typographus*) a right to exist in the park and were disinclined to support outbreak management there (Müller and Job 2009). It was made clear that this management option was an informed decision made by local people about what type of changing landscape they were willing to accept. This decision was then communicated in an official press release by the OMB to media outlets for dissemination.

There was some initial opposition to this policy by stakeholders who continued to be concerned about the long-term implications of this decision for the UK as a whole. However, in general the local community expressed pride about the decision they had made in conjunction with a group of wide-ranging expert stakeholders. Through this process, there was a greater awareness of habitat and local species (including ALB) conservation across the immediate area. A small group of local people also formed an advocacy organisation dedicated to campaigning for the flourishing and well-being of non-humans in environmental decision-making.

At this point, the ALB population remains unobservable to the human community, and trees in the area are still visibly unaffected. A monitoring schedule has been instigated by FR as part of the management plan to ensure the ALB do not begin to unduly disturb the local ecosystem, or unexpectedly spread far beyond local boundaries. The OMB was clear that any subsequently identified ALB outbreaks in other areas of the UK would be subject to the same decision-making process and this case does not necessarily set a precedent for their management.

5 Flourishing

5.1 Flourishing as an Ethical Perspective

Our final account of forest beetle management makes use of the concept of 'flourishing', presented here as an ethical framework which sees plants (including trees) as holding moral status due to their ability to flourish. This narrative is based on the work of Angela Kallhoff (2014). However, the concept of flourishing has also been discussed by Martha Nussbaum as a means of extending her 'capabilities approach' to non-human animals. Human capabilities are 'what people are actually able to do and to be' and are necessary in terms of respect for human dignity (2006, 70). In extending her capabilities approach to non-human animals, Nussbaum argues that it 'offers a model that does justice to the complexity of animal lives and their strivings for flourishing' (2006, 407). As Kallhoff notes, 'flourishing is explored as a basic concept of the good life and one in line with concepts such as "happiness", "well-being" and the like' (2014, 689); a flourishing life is one in which we realise our capabilities.

As a framework for considering the moral status of plants, flourishing contains two key features. The first is that any deliberate act which limits the possibility of flourishing is harmful to a plant: 'if the flourishing of a plant suffers negative effects from human actions, this should be part of the process of an ethical assessment of that action … harm to plants should be part of a moral calculation' (2014, 693). Flourishing, as Kallhoff presents it, requires potential harm to plants to be taken into consideration, but 'there is no moral imperative which says that persons should protect the flourishing of each single plant' (2014, 693). However, she draws on further arguments in favour of protecting plants that derive from the particular value that humans gain from them, such as 'aesthetic experiences and feelings of being "at home" in a specific area' (Kallhoff 2014, 693). The second argument is that flourishing is a means of giving plants an ethical status without anthropomorphising (resorting to human moral theory). This begins from the premise of the plant, rather than adding plants to existing human reasons for granting moral consideration (2014, 694). Unlike ethical frameworks which rely on species-based features such as sentience or suffering, flourishing can be applied across species boundaries and gives humans a means 'to interpret non-human nature' (2014, 694).

Kallhoff presents three conditions sufficient for flourishing. First, that a plant remains viable (can react to external stress and maintain its performance, thus sustaining life). Second, a plant is able to accomplish a typical life cycle. Third, its characteristics remain those 'both of a plant which has a specific life-form and of a more specific organism, generally fitting its species description' (2014, 687). Therefore, any discussion of felling trees as part of an eradication programme would be required to give full moral consideration to the trees felled. That the trees in question may also be of worth because of the aesthetic or recreational value they offer to humans would also be considered, but such a discussion should be mindful that any threat to a tree's capacity to flourish is in and of itself a significant moral harm.

5.2 Outbreak Story 3: An Account of Flourishing Plant Life

The presence of ALB was confirmed by scientists at FR on 15 March 2012, after carrying out regular non-invasive monitoring in the area. Given the regularity of stone imports by the business previously sited at the centre of the outbreak area, the UK government would have been justified in pursuing the company as they had breached biosecurity and thus endangered the flourishing of trees in this area. However, the business was no longer operational.

An OMB was established on 24 March 2012, comprised of experts from Fera and the Forestry Commission, which included a Plant Ethics officer. Their goal was to develop an ethically sensitive eradication programme. The discussion centred around the need to contain and then eradicate the outbreak in order to facilitate the ongoing flourishing of as many trees in the area as possible. Consideration was given to the avoidance of harm to trees which were not infested and also to those trees which contained only grubs that may or may not develop into ALB. The grubs could be harmless and, even if they did prove to be ALB, it is unlikely they would survive to develop into adult beetles. Trees already hosting ALB were no longer viable and not capable of achieving a full life cycle and so could not be considered to be flourishing; these trees would be felled. The ethical motivation for felling in this instance was to ensure the flourishing of as many other trees as possible by removing infested trees (all remaining trees were viable and could maintain a full life cycle and develop characteristics of their specific type of tree). An additional and significant consideration was to protect as many trees as possible, given the value that local residents derived from them.

The initial management programme focussed on the area of the existing infestation. Arboriculturalists climbed trees within a 100 metre radius of the discovered ALB population to inspect the crowns for more beetles. Branches of trees were removed and taken to the Alice Holt Research Station to be examined for any grubs and to test their DNA. Branches were given a numerical code corresponding to the tree they came from (the tree had a temporary label placed on it to make identification possible), so if any grubs were found it would be possible to establish which tree they came from. If adult ALB or grubs with ALB DNA were detected, then that particular tree was felled and other possible host trees within 100 metres were closely monitored. Out of respect for the flourishing of the tree, sanitation felling was only carried out on trees that were confirmed to contain ALB.

The operational aspect of the eradication programme was coordinated by the local Fera officer who liaised with the Forestry Commission, contractors (including arboriculturalists), and representatives from woodland advocates such as The Woodland Trust. The local Fera officer had more knowledge of the trees in the area and of the community, so was best placed to coordinate on the ground. The Fera officer also contacted scientists at FR who were able to provide detailed descriptions and pictures of ALB. These images and descriptions were carefully displayed on flyers which were distributed to residents. This information was also given to the media for dissemination and the Forestry Commission launched a social media campaign to raise awareness of this particular beetle. Residents and anyone else who derived value from the trees, as well as landowners, were invited to an initial meeting where the local Fera officer explained the planned initiative to tackle ALB. Residents were assured that no trees would be felled unless it was clear they contained ALB. A scientist from FR also attended the meeting to offer clear guidance on how to detect the beetle, to provide information on the life cycle of the beetle, and to explain the types of trees which were at risk. The flourishing of as many trees as possible would be given priority at all times, and this was particularly in line with the views of residents who had no desire to see their trees felled unless absolutely necessary.

Sixty-six infested trees were removed. Monitoring will continue until 2018 (beyond 2 life cycles of the beetle), with special attention paid to the 'eradication zone', but all trees will be regularly inspected for the presence of ALB.

6 Discussion

This chapter has reflected on the ethical framing of forest health management with an emphasis on non-human and environmental ethical perspectives: biocentrism, entangled empathy, and flourishing. The three alternative stories of 'outbreak' management have shed light on the relationships between some specific management actions, their impacts on non-humans, and their often implicit ethical underpinnings. We do not claim to have provided exhaustive, comprehensive, or unchallengeable applications of these chosen ethical frameworks. Instead, our goal has been to open up and critically reflect on the dominant anthropocentric framings of forest health management.

One of the most striking outcomes of this analysis is that stronger commitments to non-humans would not necessarily result in radically different outbreak management approaches to those currently followed. None of the ethical positions we have presented—respect for nature, a commitment to empathetic engagement with non-humans, and attributing a higher moral standing to plants on account of their ability to flourish—would result in an outright rejection of management methods, such as felling, that are lethal to non-humans. The use of these methods in relation to particular non-humans could be justified within each framework. However, looking across our narratives and at their founding ethical frameworks, two interrelated recommendations for outbreak management emerge.

First, these perspectives suggest the need to vastly increase the surveying and analysis efforts that precede the implementation of management on the ground. This is required to minimise harm to non-humans. All three of our alternative stories told of substantial inspection and investigation work at early stages that subsequently underpinned a more precise and targeted set of management actions. This included both technical and biological assessments of individual trees, along with ecological and epidemiological analyses of the wider environment to assess its vulnerability and the extent of the outbreak. Our narratives also suggest the need for a much greater understanding of the management context, as each story described the in-depth consideration of alternative stakeholder perspectives, whether through strong consultative processes with local human residents and community members or through 'noticing' non-humans such as by proxy representation or empathetic engagement. Indeed, this echoes Anna Tsing's advocacy of 'arts of noticing' as methods for building our appreciation of multispecies assemblages (Tsing 2015, 22–25). However, as has been noted before (MacKenzie and Larson 2010; Porth et al. 2015), the dominant 'emergency modality' of outbreak management often crowds-out the participation of many relevant human stakeholders and the expression of their perspectives. This same modality more or less bulldozes (perhaps literally in some cases!) opportunities for 'noticing' (Tsing 2015) non-humans.

The second, very much interrelated, recommendation made by our alternative stories would be for a substantive shift and increase in the allocation of resources to outbreak management. The above-mentioned processes of investigation and taking notice would require significant investment of personnel, skills, and technology, particularly in their initial development. A primary driver of current approaches to outbreak management is the minimisation of economic costs (both of the management scheme itself, and any consequent environmental or resource damage). Therefore, careful consideration of the costs of these activities relative to one another is important. Having said this, evidence suggests that public support for forest health management is strong, forests are very highly valued as places for wildlife, and there is clear support for particular management methods which minimise potential impacts on 'non-pest' wildlife (Fuller et al. 2016). This may indicate a widespread, yet unacknowledged, acceptance amongst stakeholders of higher costs associated with outbreak management. The government agencies responsible for managing environmental outbreaks could leverage this support to access additional funds to better 'notice' or otherwise account for non-humans. Furthermore, there have been significant recent steps forward to account for non-humans in public policy. For instance, in 2017 legal frameworks in New Zealand and India were extended to include non-human elements of natural systems—most specifically, rivers—as having clear moral rights worthy of consideration (Safi 2017). These precedent-setting decisions could underpin policy to develop more effective processes to 'notice' non-humans in environmental management.

The ethical frameworks we employ here require different levels of stakeholder engagement and afford distinct reflections on forest health. For example, whilst biocentrism and flourishing are examples of environmental ethics which are relatively easy to translate to the case of forest health, entangled empathy demands greater interpretation and justification. This does not detract from the value of entangled empathy as a perspective for understanding, in this instance, forest health management. It is common for environmental ethics to be adapted from earlier frameworks in order to fit environmental debates. Virtue ethics, for example, is now used in environmental debates to better understand human relationships to nature, but was previously concerned with broader political questions about how we should live.

Notably, flourishing is the only perspective that gives exclusive attention to plants (trees in this instance). Both entangled empathy and biocentrism involve consideration of living beings more broadly—the beetles as well as the trees. This illustrates important questions about what and who counts when it comes to moral consideration, and which non-human living organisms should be afforded a moral status. Finally, biocentrism and flourishing specifically enjoin us to avoid harm. With these two perspectives, it is considered both logically and morally inappropriate to cause harm and this should consequently be part of our consideration as human beings. Whilst entangled empathy would also not advocate harm, it provides a more positive approach to our relation with the more-than-human world. Rather than emphasising our capacity to harm, Gruen's framework stresses our capacity to engage:

to actually entangle ourselves with these 'others'. Consequently, of the three frameworks discussed, it could be said that entangled empathy encourages us the most to be open to the lives of all others.

7   Conclusion

In this chapter, we have used three ethical frameworks to open discussion of the status of non-humans within forest health management. Our aim is to highlight the impact that outbreak management has on non-humans and to challenge deeply entrenched justifications for management intervention. We have considered one specific 'outbreak', which had a particular epidemiology and constituted a particular set of threats to human values, to the environment, and to non-human beings. ALB are by no means the most potentially damaging 'pest' to threaten British forests, and it is important to note that the application of these three ethical frameworks to other outbreaks may well have resulted in different stories.

More work is required in forestry and environmental ethics to unpack the issues that this chapter has begun to explore. However, our analysis leads us to advocate the allocation of greater resources to outbreak management. Most notably, this requires forest managers to undertake improved investigation and stakeholder consultation prior to deciding on a management programme, both of which must be context-specific. These measures have the potential to underpin a substantial reduction in harm to non-human stakeholders.

References

CABI. (2017). *Data sheet: Anoplophora glabripennis (Asian longhorned beetle)*. Accessed March 1, 2017. http://www.cabi.org/isc/datasheet/5557.

Clayton, P. (1998). *Connection on the ice: Environmental ethics in theory and practice*. Philadelphia: Temple University Press.

Collier, S. J., & Lakoff, A. (2008). The problem of securing health. In A. Lakoff & S. J. Collier (Eds.), *Biosecurity interventions: Global health and security in question* (pp. 7–32). New York: Columbia University Press.

Davidson, I. (2017, March 15). Whanganui River given legal status of a person under unique Treaty of Waitangi settlement. *New Zealand Herald*.

Fuller, L., Marzano, M., Peace, A., Quine, C., & Dandy, N. (2016). Public acceptance of tree health management: Results of a national survey in the UK. *Environmental Science & Policy, 59,* 18–25.

Gruen, L. (2015). *Entangled empathy: An alternative ethic for our relationships with animals*. Brooklyn, NY: Lantern Books.

Haack, R. A., Hérard, F., Sun, J., & Turgeon, J. J. (2010). Managing invasive populations of Asian longhorned beetle and citrus longhorned beetle: A worldwide perspective. *Annual Review of Entomology, 55,* 521–546.

Kallhoff, A. (2014). Plants in ethics: Why flourishing deserves moral respect. *Environmental Values, 23,* 685–700.

King, R. J. H. (1999). Narrative, imagination, and the search for intelligibility in environmental ethics. *Ethics and the Environment, 4,* 23–38.

Mackenzie, B. F., & Larson, B. M. H. (2010). Participation under time constraints: Landowner perceptions of rapid response to the Emerald ash borer. *Society and Natural Resources, 23,* 1013–1022.

Macleod, A., Evans, H. F., & Baker, R. H. A. (2002). An analysis of pest risk from an Asian longhorn beetle (*Anoplophora glabripennis*) in the European Community. *Crop Protection, 21,* 635–645.

Müller, M., & Job, H. (2009). Managing natural disturbance in protected areas: Tourists' attitude towards the bark beetle in a German national park. *Biological Conservation, 142,* 375–383.

Nussbaum, M. (2006). *Frontiers of justice: Disability nationality, species membership*. Cambridge, MA and London: Belknap Press of Harvard University Press.

Pitt, D. B., & Shockley, M. (2014). Don't fear the creeper: Do entomology outreach events influence how the public perceives and values insects and arachnids? *American Entomologist, 60*(2), 97–100.

Porth, E., Marzano, M., & Dandy, N. (2015). 'My garden is the one with no trees': Residential lived experiences of the 2012 Asian longhorn beetle eradication programme in Kent, England. *Human Ecology, 43,* 669–679.

Safi, M. (2017, March 21). Ganges and Yamuna rivers granted same legal rights as human beings. *The Guardian*.

Simard, S. (2016). *How trees talk to each other. TED Talk*. Accessed April 1, 2017. https://www.ted.com/talks/suzanne_simard_how_trees_talk_to_each_other?language=en.

Simard, S. W., & Durall, D. M. (2004). Mycorrhizal networks: A review of their extent, function, and importance. *Canadian Journal of Botany, 82*, 1140–1165.

Straw, N., & Tilbury, C. (2012). *Report on the outbreak of Asian longhorn beetle (Anoplophora glabripennis) at Paddock Wood, Kent in 2012*. Farnham: Forest Research.

Taylor, P. W. (1981). The ethics of respect for nature. *Environmental Ethics, 3*, 197–218.

Taylor, P. W. (2011). *Respect for nature: A theory of environmental ethics* (25th Anniversary Edition). Princeton: Princeton University Press (Originally published 1986).

Taylor, P. W. (2013). Respect for nature: A theory of environmental ethics. In M. Boylan (Ed.), *Environmental ethics* (2nd ed.). Malden, MA: Wiley Blackwell.

Tsing, A. (2015). *The mushroom at the end of the world: On the possibility of life in capitalist ruins*. Princeton: Princeton University Press.

Vance, L. (1995). Beyond just-so stories: Narrative, animals, and ethics. In C. J. Adams & J. Donovan (Eds.), *Animals and women* (pp. 163–191). Durham: Duke University Press.

Wagler, R., & Wagler, A. (2011). Arthropods: Attitude and incorporation in preservice elementary teachers. *International Journal of Environmental and Science Education, 6*(3), 229–250.

Wohlleben, P. (2016). *The hidden life of trees: What they feel, how they communicate: Discoveries from a secret world*. Vancouver: Greystone Books.

17

Towards a More-Than-Human Approach to Tree Health

Alison Dyke, Hilary Geoghegan and Annemarieke de Bruin

> …*humans are not the only ones caring for the Earth and its beings—we are in relations of mutual care*
> (Puig de la Bellacasa 2010, p. 164)

> *I am the Lorax, I speak for the trees, I speak for the trees, for the trees have no tongues.*
> (Dr. Seuss 1971)

1   Introduction

The climate is changing, and the number of pests and diseases affecting trees is increasing. For Anderson and Bows (2012, p. 640) writing in *Nature Climate Change*: 'The world is moving on and we need to have

A. Dyke (✉) · A. de Bruin
Stockholm Environment Institute, University of York, York, UK

H. Geoghegan
Department of Geography and Environmental Science,
University of Reading, Reading, UK

© The Author(s) 2018
J. Urquhart et al. (eds.), *The Human Dimensions of Forest and Tree Health*,
https://doi.org/10.1007/978-3-319-76956-1_17

the audacity to think differently and conceive of alternative futures'. Nowhere is this more pressing than in the context of tree health and the ways that we respond to and manage plant health risks (plant biosecurity). As Atchison and Head (2013, p. 951) argue: 'We cannot appeal to a past or stable Nature, separable from human activity, as the basis of decision-making'. We must pay closer attention to the entanglements between humans and non-humans in order to 'energise our thinking about new ways of living in the world' (Atchison and Head 2013, p. 965). In the light of the complexities of disease and biosecurity, human and non-human relations are being re-theorised. In this chapter, we examine how we might reimagine tree health by starting with the trees themselves and our research engagement with them.

This chapter begins by exploring the roles that trees play within human society, considering how humans define trees and the values people associate with trees, what contradictions surround tree health management, and what social science research exists in this area. Section 2 then discusses current theorisations of human and non-human relations, with Section 3 describing a two-day event in Epping Forest, North London, entitled 'In conversation with oak trees'. Section 4 draws together the theoretical lessons from the event in Epping Forest, and finally, Section 5 highlights some of the applications of this approach for tree health and plant biosecurity research.

1.1 Trees and Human Values

Let's begin with 'what is a tree?' There are many scientific definitions of trees, commonly referring to trees as perennial plants with woody stems, supporting branches and leaves. When more details are added to the description, problems and exceptions arise. For instance, many trees don't have a single stem, either by accident of growth, by grazing or by active human management. Most of these scientific descriptions regard trees as individual entities, but when we consider how closely a tree lives with other organisms, it is hard to draw boundaries between what is part of the tree and what is not. Similarly, the relationships between trees of the same species are close and entangled. When a tree reproduces vegetatively, by suckers growing off its roots, are its offspring still a part of the parent tree or are they separate individuals? The challenge of untangling 'what a tree is' is further complicated when scientific definitions are met with very different social science understandings of trees. As Jones (2014, p. 112) describes: 'there are always multidirectional flows of actions, meanings and *feelings* as communities and agencies respond to trees and act with and upon them'. Trees, as in the wider case of plants, '*emerge* as an assemblage of shared differences from other beings, where common capacities manifest in different material form' (Atchison and Head 2013, p. 955).

Trees pose a particular challenge in that they live very different lives from humans over timescales that can span several generations of human life. The lives of managed trees, for example, are both accelerated and often truncated in order to bring them closer to human timescales of production and investment. Over its lifespan, a tree will host many different organisms that will have some negative or positive effect on its health. Furthermore, as a tree moves into old age (if it isn't taken as a timber harvest) it will have a long period when it is no longer growing vigorously, may lose limbs or can become hollow. This begs the question, is what a human might perceive as decline an undesirable state for a tree? Viewing trees through human eyes, it is difficult not to anthropomorphise and to impose our understandings onto trees. However, recognising the norms associated with anthropomorphism may allow us to move beyond it to a new 'differently human' understanding of trees.

For humans, particularly those living in the UK, there are 'strong values associated with the countryside and rural spaces, and the cultural, affective and symbolic meanings of woods and trees' (Pidgeon and Barnett 2013, p. 6). As a consequence, Western management of tree health is governed from a very human perspective that makes preservation of the current state of the environment a priority. In turn, it puts the lives of trees above the organisms that depend on them. As a result, attitudes to and management of trees are full of contradictions.

1.2 Tree Health Management and Social Science Responses

This chapter responds to Sinden's (1990, p. 9) request to find new ways of 'living happily with trees' by starting with the trees themselves and our research engagement with them in the context of UK plant health policy. Recent documents such as the *Tree Health and Plant Biosecurity Management Plan* (Defra 2014) are focussed on 'threats' and the desire to create biosecure space, specifically to manage and avoid attacks on UK trees from non-native pests and diseases.

Following the spread of *Phytophthora ramorum* to Japanese larch in 2009 in the UK, the discovery of *Chalara* dieback of ash in the autumn of 2012 raised the profile of tree health to a priority issue for the Department for Environment, Food and Rural Affairs (Defra). In this period, social science research supported the agenda of prioritisation, governance and response in the context of biosecuring space. In 2013, when Defra commissioned a social scientist (Hall 2013) to review the role of social science in tree health, the resulting brief focused on a securitised approach to human attitudes and behaviours post-outbreak. Tomlinson et al. (2015) used similar language in their study on the governance of urban tree health issues.

The alternative to these securitised narratives, specifically the potential for co-existence with disease, is seen as failure, whilst tacitly accepted for some tree health issues. Examples include Dutch elm disease, where no solution has been found (Harwood et al. 2011), or knopper gall, where the issue is seen as minor. Porth et al. (2015 and this volume) question the appropriateness of an emergency modality to managing tree health issues. Drawing on lived human experience, they propose a more open approach that would promote trust and foster biosecure citizenship.

With much work focusing on the consequences, rather than the causes of tree health issues, this emphasis is suggestive of a realisation that the causes are so complex that to control all human and non-human factors would be beyond any capacity to act. As the number of 'threats' to tree health increases, the causes and consequences of tree health issues become unmanageable and uncontainable. We need, therefore, to think differently and imagine alternatives (Atchison and Head 2013; Barker 2010; Hinchliffe et al. 2013). This chapter responds to the challenge by introducing theories from human geography, science and technology studies (STS) and the environmental humanities on human and non-human relations that propose a shift away from human exceptionalism (Bastian 2017). Such an approach involves unsettling the existing ways of doing tree health research and enables us to question notions of what constitutes healthy and unhealthy and the potential to research *with* rather than *on* trees, fungi and beetles (Bastian et al. 2017).

1.3 PuRpOsE: Researching Acute Oak Decline

This chapter is based on the work of the PuRpOsE (Protecting Oak Ecosystems) project,[1] which investigates the context of Acute Oak Decline (AOD). AOD is a syndrome that was first described by Denman et al. (2014). A great deal of uncertainty surrounds AOD, though symptoms include: stem bleeds that are associated with two previously unknown bacteria; dieback of branches in the crown of the tree; and, in some trees affected by stem bleeds, the presence of the oak jewel beetle, an *Agrilus* beetle that has long been associated with old oak trees. Other environmental factors such as drought are thought to be involved, but the relationships are not clear. Affected trees appear to go through cycles of symptoms and recovery. Some trees die suddenly, whilst others recover. Our multidisciplinary PuRpOsE project aims to address some of those unknowns by looking more broadly at the context of affected trees to understand the factors that put trees at risk, mapping the risk of oak trees to AOD when taking into account climate and soil data and other factors that may affect oak health in the future. Imagining a future where oak decline will be widespread, researchers are investigating alternative tree species to understand which of these could replace oaks in terms of the ecosystem services they currently provide. These results all feed into engagement with stakeholders to develop adaptive or mitigating tree health management practices.

Underpinning all of this, we recognise AOD as a loose and undefined syndrome—entangling trees, invertebrates, bacteria, water, humans and others—which offers an apt case study through which to begin investigating the more-than-human worlds of tree health. This marks the beginning of a turning point in tree health management by commencing the work of reconnecting humans and non-humans in AOD and revealing the possibility of 'a more relational, less managerial alternative to biosecurity' (Nading 2013, p. 68).

2 Current Theorisations: Towards a More-Than-Human Approach to Tree Health

We have already established that plants are challenging to think with and that ambiguity surrounds what constitutes health or successful management. These issues have been compounded by tight legislation and a desire to biosecure space. Whilst tree health has largely been the preserve of the natural sciences, we use this section to introduce natural scientists, foresters, plant pathologists and others to the social sciences and environmental humanities and explore the implications of the false nature/culture binary.

2.1 Nature–Society Relations

There has been a long-standing social science interest in how non-humans and their lives are understood and valued by humans (Castree 2013). Beginning with Arne Naess's 'Deep Ecology' understanding of ecological interdependence (Luke 2002) and moving through David Abram's (1997) theorisation of embodied and affective engagement with more-than-human worlds in the mid-1990s, there has been an academic movement towards the recognition of the perspectives of non-humans. In human geography, researchers have been examining the interconnections between nature and society and how they differ culturally and spatially. In acknowledging the active role of non-humans in shaping the world, researchers have examined the politics, histories and geographies of nature (Hinchliffe 2007). Here, nature is no longer regarded as passive, static and mute. Instead, the non-human world resists and unsettles. In her book *Hybrid Geographies*, Sarah Whatmore (2002, p. 3) advocates for a more-than-human approach to understanding how the world is made and remade, whereby 'other modes of travelling through the heterogeneous entanglements of social life' are explored. Such approaches, Whatmore suggests, 'attend closely to the rich array of the senses, dispositions, capabilities and potentialities of all manner of social objects and forces assembled through, and involved in, the co-fabrication of socio-material worlds' (2006, p. 604). A key thinker from STS, Bruno Latour has influenced research in this area on 'matter', arguing for researchers to attend to *all* participants that 'are gathered in a thing to make it exist and maintain its existence' (2004, p. 246). This thinking is useful in understanding trees in the context of tree health. Rather than viewing trees merely as natural resources, they 'become again things, mediating, assembling, gathering many more folds' (2004, p. 248). Latour's work resonates with changes in the field of human geography towards more fully understanding the 'material' world, with Hinchliffe (2008) suggesting that topics become much more interesting when we include non-humans. Jones and Cloke expand this further in their book *Tree Cultures* (2002, p. 1), in which they argue that 'nature-society relations are continually unfolding in the contexts of specific places, in which meanings arise from particular interactions between different assemblages of social, cultural and natural elements'. Feminist STS scholar Donna Haraway (2008) has been instrumental in making sense of these encounters between humans and non-humans. She suggests that when people are 'in touch' with things—in her example she talks about dogs—then people begin to care for and develop a sense of 'response-ability' for them. This is echoed in the work of feminist theorists by notions of the ethics of care, particularly developed in the area of human–animal relations in the work of Gruen (2015) and Donovan and Adams (2007). In the next section, we discuss recent research on human and non-human entanglements in the context of disease.

2.2 Disease: Entanglements of Humans and Non-humans

Defining health as 'the combination of practice and epistemology by which people confront *disease*, the manifestation of symptoms associated with biophysical disorder' (Nading 2013, p. 60), we witness the entangled relationship between humans and non-humans. In the biological sciences, a disease triangle shows the combination of pest/pathogen, host and environment in which disease manifests (McNew 1960). There is no mention of human agency. Instead, plant diseases are explained as the result of pathogens that can be biotic, such as fungi, nematodes, bacteria, viruses and/or abiotic factors, such as environmental conditions relating to temperature, moisture, light and chemicals (Agrios 2005; Baudoin 2007). Such pests and pathogens have found a susceptible host and a favourable environment. Plant disease management is offered as the solution to reduce the damage caused by disease, with diagnosis—the identification of the correct pathogen—being key to any management strategy.

The identification of insects and bacteria as pests and pathogens, and the accompanying securitised language of threat, attack and security, has led many people—including scientists—to 'treat [pests and diseases] as Others, objects of cultural scorn and as subjects of detached strategies of technological control' (Nading 2013, p. 61) and human and non-human relations are regarded as pathological rather than normal. This approach in both the UK and elsewhere has led scholars in the humanities and social sciences to question which bodies and lives are fostered, protected, managed, threatened or killed (Haraway 2008; Collard 2012). In the case of tree health, the complexity of control (Atchison 2015) and 'this ambivalent interdependence between life and death, between co-existence and instrumental relations' (Beisel et al. 2013, p. 10) has yet to be fully discussed.

In the face of such a legacy of scientific plant health research and technical management, we challenge the contemporary disease paradigm by reimagining the disease triangle as an entanglement of humans and non-humans that 'live with' rather than manage against 'disease'. This more-than-human approach acknowledges: that humans, non-humans and their disease relations are 'anything but static' (Nading 2013, p. 63); that in order to take this approach seriously researchers and others must 'step away from the modernist dismissal of nature and non-humans as anything but resources' (Bastian et al. 2017, p. 2); that this way of working and thinking means 'we may envisage a different biopolitics of living with these plants' (Atchison and Head 2013, p. 956); and finally, that researchers are not always 'free' to think with plants because of the complicated and contradictory notions of health and management, the desire to biosecure space, the need for impact agendas and to acquire research funding.

A body of work around biosecurity and invasive species is emerging with attention being paid to opening up to include humans and management/biosecurity programmes (Barker 2008). Two studies in particular inform our thinking here. We address each one in turn, before employing them both in our discussion of methodological approaches to work of this nature. First, Hinchliffe et al.'s novel theorisation of 'borderlands' in relation to the entanglements and intensities that constitute animal health issues, whereby 'disease is understood as relational: that is, both integral to, and always part of, an entangled interplay of environments, hosts, pathogens and humans' (2013, p. 532). Here, 'disease and the responses to it are marked more by intense entanglements of hosts, environments and institutions than a simple geometry of fixed objects invading pure, or more or less resilient, spaces' (2013, p. 540). Whilst biosecurity and keeping disease 'out' is not the focus of our chapter, we do want to think differently about disease, specifically because disease does not neatly inhabit spatial and temporal borders.

Second, we are inspired by Atchison and Head's work on plants where they draw on the work of feminist theorists (Bennett 2009; Haraway 2008) to challenge Western colonial thought in terms of conceptualisation of the body and the individual and 'discourses of defence, invasion, and fear' (2013, p. 953). They instead acknowledge 'the planty subjects with whom we cohabit, as well as greater ethical engagement with questions of our mutual living and dying' (Atchison and Head 2013, p. 965).

Writing about invasion, Atchison and Head describe it as 'a relational process in which many different lives – human and non-human – are embedded together' (2013, p. 952). In her work on the control of invasive plants in Australia, Atchison reveals the range of practices involved in eradication and highlights the urgent need to pay closer attention to the humans and non-humans that form 'biocommunities'. Biosecurity thus gives life to new entanglements, with plants resisting control and new collaborations emerging both between humans and with non-humans (Atchison 2015). Research in this area must interrogate this complexity and critique established modes of diagnosis, management and desirable future(s). One way of achieving this, as advocated by Atchison who draws on wider geographical interests in the nature of scientific experiment (Davies 2013; Greenhough 2012), is to consider biosecurity as experimentation, whereby 'scientists, field practitioners, human and nonhuman together enact biosecurity as an experiment in co-existence through embodied learning and adjustment' (2015, p. 1709). We respond to these theoretical challenges by considering how researchers in a multidisciplinary team might make a step change to embrace non-humans in the practice of their work.

3 In Conversation with Oak Trees

3.1 Approaches to More-Than-Human Participation

Whilst ethnographic methods have been favoured by the aforementioned researchers in order to reveal what is taking place (Nading 2013; Atchison 2015; Hinchliffe et al. 2013), we have been inspired by Bastian's (2017) 'speculative experiment' in more-than-human participatory research. Bringing traditional participatory research methods together with more-than-human research, Bastian, her colleagues, and their 'fellow enquirers'—dogs, bees, water and trees—ask:

> What might it mean to invite 'the more-than-human' to be an active participant, and even partner, in research? How are prevailing ways of conceiving research in terms of issues of knowledge, ethics, consent and anonymity challenged and transformed when we think of the more-than-human as a partner in research? How might it be possible to transform

existing frameworks, practices and approaches to research? What would this transformed research look like? (Bastian et al. 2017, p. 1)

In asking what matters to non-humans, Bastian et al. used a series of workshops entitled 'In conversation with…' to cross borders and connect human and non-human worlds. These events focused on being in conversation with non-humans (bees, dogs, water and trees) and those humans who have a close relationship to them, either through work or leisure. To do this effectively, humans and non-humans have to be given equal weight, 'in ways that are situated, embodied and non-homogenising' (Bastian et al. 2017, p. 3). These speculative experiments were about interrogating the issues of power and agency central to traditional participatory research. We employed Bastian et al.'s more-than-human approach in a two-day event 'In conversation with oak trees' in order to attune ourselves to the trees in our research and explore new ways of working and thinking differently about tree health. The event took place in September 2016 at Gilwell Park, a mixed woodland park, and in Epping Forest, North London.

Using diverse ways of knowing to engage with non-humans, our 'In conversation with oak trees' event brought to life Atchison's (2015, p. 1699) observation that 'A framework of experiment in co-existence […] offers an opportunity to move beyond a human interventionist debate and thus has the potential to re-engage concerned publics and scientists alike in the ongoing challenge of living with invasive species'. Through our experiment, we expanded Hinchliffe et al.'s work that 'engage[s] with infected life as part of a *borderlands* within a mutable disease environment' (2013, p. 532) and its focus on the fluid and interactive spatialisation of disease and intense entanglements, in order to consider breaking down boundaries of expertise, power relations, seniority, hierarchies and disciplines associated with particular roles and activities in a research project that can stifle the possibility of researching and living differently. We also extended Atchison and Head's discussion of 'relational intensions' (2013, p. 953), which allow us to access new perspectives on established processes and spaces, acknowledges uncertainty in regulation and risk, attends to the temporalities of invasion, and understands the specificities of each species.

We examined more fully how humans and non-humans mingle in a research context focused on 'infected life', mounting a serious challenge to prevailing notions of tree health management, asking how we might transform our understanding of oak ecosystems and AOD and transforming ourselves as researchers.

3.2 The Participants and New Ways of Working

The event brought together a subset of the researchers involved in the PuRpOsE project. We asked that each project work package (soil analysis, risk mapping, ecosystem services and narratives) was represented, giving participants with backgrounds in microbiology, molecular biology, soil science, ecohydrology, community ecology, woodland ecology, silviculture, cultural geography and political ecology at different levels of seniority thereby making this event truly interdisciplinary. This also served to recognise the different mandates of the organisations involved in the research. In total, there were 14 participants, including the three organisers. In addition, a woodland manager, a conservation manager and a wood turner joined the conversation at different times. Non-human participants were oak trees and their surrounding contexts at Gilwell Park and Epping Forest.

Rather than using conventional scientific knowledge exchange practices through conference-style presentations, the event involved practices of attunement, listening, attention, conversation, encounter and storytelling. Table 1 lists the activities and describes their intended outcomes. The activities enabled humans and non-humans to be in conversation in the borderlands at a range of intensities, such as moments of being in intense relation with trees in a diversity of 'states', encountering infected life beyond the laboratory and in quiet contemplation at dawn or via textual accounts.

3.3 The Start of a Conversation

Before meeting at Gilwell Park, the participants received a very brief outline of the two-day programme via email and were asked to be

17 Towards a More-Than-Human Approach to Tree Health

Table 1 Outline of the 'In conversation with' workshop

	Session	Activity	Intent
Day one	Introduction to the workshop	Introduction of key concepts and agreement on principles of engagement with (non-)humans throughout the event	Setting the scene for a different mode of working
	Tales of trees	Exploration of some preparation questions (e.g. what kind of tree would you be?) and a visit to Epping Forest with two forest managers discussing their approach to tree health management and AOD and to observe oak trees in context, encountering infected life beyond the laboratory	Moving through and beyond anthropomorphism and using storytelling for a location-focussed encounter between (oak) trees and humans
	Handling wood, a story from a local wood turner	A wood turner talked about his journey into wood and presented bowls made of different woods	Understanding human cultural connections with trees through viewing the life of a tree from within
Day two	Good morning with oak trees	Reading of a short extract from the poem by Alice Oswald 'Tithonus: 46 Minutes in the Life of the Dawn', followed by an invitation to spend some time alone with trees and engage senses beyond the visual	Giving an opportunity for intense attention to difference and to develop attunement
	Reflections so far	Facilitated conversation that invited oral reflections on developing more-than-human understandings	Bringing out reflections on more-than-human approaches
	Methodologies of PuRPOsE	5-minute preparation and a short pitch by everyone of the methods (to be) used in their part of the PuRPOsE project	Building understanding across disciplines
	'Silly questions'	Discussion based on post-it notes capturing a silly question or particular interest from each individual	Building a critical understanding informed by more-than-human thinking
	Narratives of Oak tree health and AOD	Narratives of AOD were created in groups using available materials (pictures, natural materials, the site archives) and presented to the group in whichever way seemed appropriate (acting, singing, presenting, walking, etc.)	Moving beyond scientific narratives of disease and gain greater understanding of cultural and historical contexts
	Lessons learned and the project work packages	Meeting in work package groups to bring more-than-human understandings to working with trees into the research going forward	Moving the abstract discussions of earlier in the event into planning for future work
	Revisiting principles of engagement with (non-)humans	Revisiting the principles we developed on day one based on our conversations with oak trees to form a set of principles to be used across the project	Establishing consensus and giving a concrete outcome to the event

willing to engage in an experiment. Each person was asked to do some preparatory work in thinking about their relationships with trees by answering a set of questions. Examples were: How did they see their expertise and role within the project? If they were a tree what kind of tree would they be and why? What did they think about AOD? These responses were used in the event as a way to get to know each other.

At the outset of the workshop participants arrived at Gilwell Park with a sense of apprehension as no one really knew what was going to happen. Participants said they were 'up for joining in with whatever was planned' and expressed their appreciation for being 'out of the office' in the wooded surroundings of Gilwell Park. The organisers (authors of this chapter) then opened the event with a short presentation about two important concepts that had inspired the design of the event: 'borderlands of health and ill-health' (Hinchliffe et al. 2013; Hinchliffe and Bingham 2008) and a 'more-than-human' approach to research (Bastian et al. 2017). Sharing our activities for the next two days (detailed in Table 1), we posed two overarching questions to participants: How might we transform our understanding of oak ecosystems and AOD, and what might it mean to *research with* trees and *live with* disease?

3.4 The Conversations

3.4.1 In Conversation with Epping Forest—Tree Histories

Before visiting the oak trees of Epping Forest, we spent some time sharing our responses to questions designed to facilitate thinking about how we relate to trees, with discussion focused on the how the response of a tree might differ from that of a human. With these thoughts in mind, we went on a walk with the woodland and conservation managers of Epping Forest to give us an insight into the lives of oak trees. The visit gave voice to the historic management practices in Epping Forest. From around 1365 (Dagley and Burman 1996) branches were considered communal and were pollarded (cut back to the trunk) on a 13- to 15-year cycle by local people. However, in 1878 the Epping Forest Act prohibited further pollarding and by the beginning of the twentieth century all such activity ceased (Dagley and Burman 1996). We were introduced to a tree that was a composite of a tree trunk of over 400 years old and branches that were around 200 years old illustrating the tree as a fluid shape-shifter: 'While the trunk of the tree ages, its canopy is rejuvenated and the life of the tree can be extended hugely, often by many centuries' (Dagley 2016). That tree embodied its historic relation with humans, the practice of pollarding illustrating how the different temporal resolutions of trees and humans collide.

Over time, due to the lack of pollarding and environmental stresses (such as compaction), several of the veteran trees in the Forest are suffering ill health. This has had severe implications for current management strategies. These trees now have very large limbs of around 200 years old and stems that have hollowed over time, with increased risk for humans as well as the trees themselves as the limbs are very heavy and not well supported by the trunk. Since the 1980s, the forest management team have reintroduced pollarding activities and carefully monitoring how best to support the trees to keep as many alive as possible. The forest managers take into account the climatic conditions in current and previous years in order to identify the level of stress a tree is likely to be under. These management practices suggest an attunement to the needs of the tree and a relationship between humans and non-humans of interdependent care.

3.4.2 In Conversation with Wooden Bowls

Before dinner on the first night, the voices of different kinds of wood were given life by a wood turner. He had brought with him a large selection of bowls made from different tree species. Each bowl had a different story to tell. He talked about their different colours, textures and their different uses. One bowl showed how the health of the tree was represented in the wood with a fungus affecting the colour and structure in a way that told an additional story line that was interwoven with the fabric of the tree (see Fig. 1). The wood turner emphasised that only through hands-on experience was he able to really get to know the different species of trees and how to manage their wooded 'personalities'.

Fig. 1 A turned bowl showing fungal spalting

3.4.3 In Conversation with Human Experts

We returned to the overarching question of how we might transform our understanding of oak ecosystems and AOD in the 'Silly Questions' activity. Participants of the PuRpOsE research team brought their own expertise and experience of working with tree health into the conversation. The activity offered a space in which people could query their own and other peoples' expertise as well as reflect on how their perspective on AOD had changed.

Participants were given time to reflect individually and write their question or reflection on a piece of paper which was then added to the wall. The questions were then grouped into themes and some were discussed in small groups. This exercise gave rise to a set of questions related to the different roles of organisms in the context of AOD. One researcher posed that 'Pathogens are just a member of the community', a reflection on the value judgement put on pathogens in the story of tree health. Another point was raised when referring to AOD as a 'disease created out of natural ecology - not [an issue of] a non-native invasion/ poor sanitation'. This reflects that disease may not be easily preventable when it is related to the host and environment. Another participant asked 'If the abiotic stress mechanism is the problem, what other symptoms are occurring in oak trees and oak ecosystems?' and 'Time scale of beetle vs bleeding - is it not associated or causal?' These questions point to the current uncertainty in the understanding of the assemblage of AOD and to the complexity and temporality of organisms within the disease triangle giving rise to new questions for investigation.

3.5 Using Different Senses: Being in Intense Relations

The organisers observed how participants used different senses during each activity and through this started to build an intense relation with their surroundings. In the walk in Epping Forest, participants moved through the landscape looking for symptoms of health and ill health. When we came upon a tree with symptoms of the beetle and bleeds, a small group of participants rushed to the tree itself to observe it closely, feel the bark, listen to the cavities underneath it by knocking on the bark and discussing the spatial location and context of this unhealthy tree (see Fig. 2). Others remained at a distance at first, but after the

Fig. 2 The PuRpOsE team inspects a tree with symptoms of AOD

forest manager had stopped talking they also went to look at the symptoms more closely. When later asked why people rushed to the tree that showed symptoms, participants said they were intrigued by these signs of disease 'in the flesh'.

Participants themselves were asked to reflect on using different senses during the 'Reflections so far' activity on the second day. Participants revealed how they had tried to connect with a tree's perspective during the 'Good morning with oak trees' activity, namely what it could be like to be a tree, and how trees sensed noise, sunlight and space. One participant highlighted the sense of feeling and touch of sunlight on the skin and wondered whether a tree would feel that change from night to day as a good thing, as food would be able to be produced through photosynthesis. Another reflected on the sense of space that a tree takes up and decided to experience this by walking the circumference of the canopy and the trunk. Another reflected on the noise of the landscape and wondered whether trees experience noise? They themselves were very aware of the noise of the motorway nearby, although another participant reflected on being a source of noise themselves when they became aware that they were sharing the space with a squirrel.

3.6 Our Principles for Engagement with Oak Ecosystems

Following (Bastian et al. 2017) and feminist traditions in the ethics of care (Gruen 2015; Donovan and Adams 2007), we asked what it might mean to *research with* trees and *live with* disease. To bring this into our ways of working in the PuRpOsE project, we chose to develop a set of principles of engagement with oak ecosystems, relating to both humans and non-humans. We agreed on a set of principles at the start of our event, which were refined at the end based on our conversations with oak trees. Our 'Principles of Engagement' (see Table 2) offer a clear outcome from our experiment and afford other researchers some insight into how we might now work and think differently about tree health.

Table 2 Principles of engagement (underlined text was added at the end of the event)

Engagement with humans
- Respectful listening—Preparedness for different understandings
- Looking after each other—<u>To be inclusive of all abilities and career levels</u>
- Avoiding disturbing others with distractions
- Ensuring that those who want to contribute can speak
- Biosecurity: Taking care to not spread disease and to help non-humans

Engagement with non-humans
- Keep disturbance to a minimum
- To be respectful at all times
- To appreciate all trees, including young saplings that may become veteran trees
- <u>Encourage the use of other senses and attunement to the non-human</u>
- <u>Consider temporality, the past and future as well as the present</u>
- <u>Consider geographic scales in addition to the one in which you work</u>
- <u>Consider the community of non-humans rather than the individual trees studied</u>

4 Becoming Differently Human

In this chapter, we have moved beyond an active, securitised and human-centric management paradigm towards one that is more holistic, where the voice and agency of non-humans are acknowledged and valued and in which human agency in determining the health of non-human entities is more fully accounted for. We now discuss the implications of this work in adopting differently human perspectives.

Being in conversation with oak trees gave us the opportunity to explore new ways of working and thinking in relation to tree health. Whilst the immersive activity of being in the company of trees discussed in this chapter is in its infancy, our conversations in Epping Forest largely related to attuning ourselves to trees, suspending professional scientific identities and drawing other aspects of self forward to become 'differently human'. As we reveal below, time, visceral and embodied experiences, becoming care-ful and reimagining our research subjects are important aspects of any attempt to develop more 'concrete' ways of working with plants or the more-than-human more generally.

4.1 Human Identities

Being in close proximity to trees meant many participants were able to move beyond professional and expert modes of being, creating encounters in the borderlands of academic disciplines and interests. Several participants mentioned that they were aware they had been listening to the forest managers with their 'project hat' on. By the morning of the second day, some participants began to feel that they were participating more directly with the trees and the Forest. Their perspective as professional scientists had morphed into a conversation between humans and non-humans.

4.2 Visceral and Embodied Experiences

Going into the woods at dawn and sharing the experiences of the wood turner and woodland managers afforded participants many opportunities for different human experiences and to draw those into their professional lives. As a result, new ways of being, doing and thinking emerged in relation to our research, and also to our personal and collective engagements with trees. We were aware that some participants found it easier than others to access visceral and embodied experiences. To this end, we sought activities that focussed attention on senses beyond the anthropocentric concentration on the visual, in the half dark, allowing touch and hearing to take over the privilege that daylight gives to the visual. Being in the woods for no purpose other than to experience being *with* trees also allowed for visceral and embodied experiences that are otherwise usually marginalised in the research process.

4.3 Time

Becoming attuned to oak ecosystems and gaining a deeper understanding of the history, management practices and contexts of oak ecosystems are a slow process. This reflects tree time itself as much slower than human time. Our event gave us the time and space to reveal new histories involving politics, economics and attitudes to trees. One of the most obvious illustrations of this were the ancient oaks of Epping Forest, for whom management changes 200 years ago are still playing out and impacting on their vulnerability to ill health in the present.

4.4 Becoming Care-Ful

Through different embodied experiences with trees, matters of fact were transformed into matters of care: 'transforming things into matters of care is a way of relating to them, of inevitably becoming affected by them, and by modifying their potential to affect others' (de la Bellacasa 2011, p. 99). Working in a multidisciplinary group that was (un)comfortable with unfamiliar ways of working had a way of allowing participants to not only step outside their professional boundaries, but to also revitalise their expertise by uncovering care within their disciplines. We were reminded that ecology does not take sides in competition between species or that a long decline may not necessarily be an undesirable state for a tree.

4.5 Reimagining Our Research Subjects

The event inspired some new ways of questioning, specifically giving space for some fundamental reflections on AOD and tree management, and freedom to contemplate new thoughts resulting from the experience with trees. Participants were able to ask new and difficult questions in the 'Silly Questions' and narrative development sessions, which brought to the fore the uncertainty that many of the team were feeling about the actuality of AOD. During the event, our research 'subject', Acute Oak Decline, was identified as a word, an idea, a challenge, a question, a health issue, an imagined syndrome and a physical manifestation. Discussing and being in touch with oak trees with AOD symptoms, and acknowledging this uncertainty, enabled us to situate AOD within oak health issues more generally. This reimagining/repositioning felt like a more comfortable place from which to address oak health, rather than containing it within the conventional boundaries of pests and disease.

5 Non-human Perspectives at Work in Tree Health Research

'In conversation with oak trees' was a first step towards enacting a more-than-human approach in tree heath. Our success in uncovering and accessing attunement to the more-than-human with a group of researchers in a multidisciplinary team illustrates that we have made some significant progress. The challenge now is to find the language and practices to bring this research into policy and practice arenas. Following our newly defined principles of engagement, this will involve putting management and policy actors into conversation with trees at future workshops.

Shifts need to be made not only in attitudes to trees and other organisms, but also to the way that biosecurity practice views scientific knowledge. If scientific knowledge can be regarded as referential, capable of shifting and changing, then biosecurity practice can shift more readily in response. We have identified trees as fluid shape-shifters, where health and ill health are relational rather than distinct and separate states. A very simple recognition of this would be to avoid using individual trees as units of analysis.

In on-the-ground management, direct caring relationships with trees already exist and were clearly evidenced by the Epping Forest managers in their consideration of the location and context of each tree and the stresses that are at play. Drawing this care to the fore means building attunement by considering the borderlands of trees: the fuzzy borderlands of individuals and community; between species; between timescales; across states of being, health and ill health; and ways of relating to the world. In practice, this means stepping out of the laboratory, stepping out of disciplinary boundaries and invite more-than-human perspectives to influence our work and help us think differently.

By taking non-human life and the entanglements of human/non-humans seriously, the complexity and the role of humans in the co-creation of disease can be addressed. The work presented here, and existing research on the co-creation of disease, borderlands and intentional relationalities, make it possible to: first, acknowledge the human-created factors in tree ill health such as human global trade and movement; and second, explore the potential to live *with* ill health, offering a more viable option to address those factors which it is not possible to fix. Indeed, the notion of *living with* disease is not about doing nothing, rather it is about rethinking our human notions of time, challenged by the lower temporal resolution of trees and the higher temporal resolution of bacteria, and the ways in which humans force trees to live truncated and accelerated lives in order to become more resilient. Taking these complexities into account requires a reworking of the neat traditional disease triangle into a web of entanglement. Time and temporality, space and relations between trees add further dimensions. The presence of humans in the co-creation of ill health becomes an under and over-lying layer. By attuning ourselves to these new possibilities and imagining alternative futures, we challenge business as usual in tree health research.

Note

1. The 3-year project (2016–2019) is funded by the Biotechnology and Biological Sciences Research Council's (BBSRC) Tree Health and Plant Biosecurity Initiative (THAPBI) and is a collaboration between the universities of Reading, York, Oxford, the Centre for Ecology and Hydrology, Forest Research, and the James Hutton Institute.

References

Abram, D. (1997). *The spell of the sensuous: Perception and language in a more-than-human world*. New York: Vintage Books.

Agrios, G. N. (2005). *Plant pathology* (5th ed.). Burlington, MA: Elsevier.

Anderson, K., & Bows, A. (2012). A new paradigm for climate change. *Nature Climate Change, 2*(9), 639–640. Available at http://www.nature.com/doifinder/10.1038/nclimate1646 (Accessed February 8, 2017).

Atchison, J. (2015). Experiments in co-existence: The science and practices of biocontrol in invasive species management. *Environment and Planning A, 47*(8), 1697–1712.

Atchison, J., & Head, L. (2013). Eradicating bodies in invasive plant management. *Environment and Planning D: Society and Space, 31*(6), 951–968.

Barker, K. (2008). Flexible boundaries in biosecurity: Accommodating gorse in Aotearoa New Zealand. *Environment and Planning A, 40*(7), 1598–1614.

Barker, K. (2010). Biosecure citizenship: Politicising symbiotic associations and the construction of biological threat. *Transactions of the Institute of British Geographers, 35*(3), 350–363. Available at http://doi.wiley.com/10.1111/j.1475-5661.2010.00386.x (Accessed July 12, 2016).

Bastian, M. (2017). Towards a more than human participatory research. In: M. Bastian, O. Jones, N. Moore, & E. Roe (Eds.), *Participatory research in more-than-human worlds*. London: Routledge.

Bastian, M., Jones, O., Moore, N., & Roe, E. (2017). Introduction. More-than-human participatory research. Contexts, challenges, possibilities. In M. Bastian, O. Jones, N. Moore, & E. Roe (Eds.), *Participatory research in more-than-human worlds*. London: Routledge.

Baudoin, A. B. A. M. (2007). The plant disease doughnut, a simple graphic to explain what is disease and what is a pathogen. *Plant Health Instructor*. Available at http://www.apsnet.org/edcenter/instcomm/TeachingArticles/Pages/PlantDiseaseDoughnut.aspx (Accessed February 16, 2017).

Beisel, U., Kelly, A. H., & Tousignant, N. (2013). Knowing insects: Hosts, vectors and companions of science. *Science as Culture, 22*(1), 1–15. Available at http://www.tandfonline.com/doi/abs/10.1080/09505431.2013.776367 (Accessed February 8, 2017).

Bennett, J. (2009). *Vibrant matter: A political ecology of things*. Durham, NC: Duke University Press.

Castree, N. (2013). *Making sense of nature*. Abingdon: Routledge.

Collard, R.-C. (2012). Cougar—Human entanglements and the biopolitical un/making of safe space. *Environment and Planning D: Society and Space, 30*(1), 23–42. Available at http://epd.sagepub.com/lookup/doi/10.1068/d19110 (Accessed February 8, 2017).

Dagley, J. (2016). What is a Pollard?—Epping Forest News—City of London. *City of London webpages*. Available at https://www.cityoflondon.gov.uk/things-to-do/green-spaces/epping-forest/news/Pages/pollards.aspx (Accessed February 17, 2017).

Dagley, J., & Burman, P. (1996). The management of the Pollards of Epping Forest: Its history and revival. In H. J. Read (Ed.), *Pollard and veteran tree management 2* (pp. 29–41). London: Corporation of London. Available at http://www.ancienttreeforum.co.uk/wp-content/uploads/2016/12/Pollard-Veteran-Tree-Management-2-1993.pdf (Accessed February 17, 2017).

Davies, G. (2013). Mobilizing experimental life: Spaces of becoming with mutant mice. *Theory, Culture and Society, 30*(7–8), 129–301.

de la Bellacasa, M. P. (2011). Matters of care in technoscience: Assembling neglected things. *Social Studies of Science, 41*(1), 85–106.

Defra. (2014). *Tree Health Management Plan*. Available at www.gov.uk/defra (Accessed February 14, 2017).

Denman, S., Brown, N., Kirk, S., Jeger, M., & Webber, J. (2014). A description of the symptoms of acute oak decline in Britain and a comparative review on causes of similar disorders on oak in Europe. *Forestry, 87*(4), 535–551. Available at http://forestry.oxfordjournals.org/cgi/doi/10.1093/forestry/cpu010 (Accessed November 8, 2016).

Donovan, J., & Adams, C. J. (2007). *The feminist care tradition in animal ethics: A reader*. New York: Colombia University Press.

Dr. Seuss. (1971). *The Lorax*. Collins.

Greenhough, B. (2012). Where species meet and mingle: Endemic human-virus relations, embodied communication and more-than-human agency at the Common Cold Unit 1946–90. *Cultural Geographies, 19*(3), 281–301.

Gruen, L. (2015). *Entangled empathy: An alternative ethic for our relationships with animals*. New York: Lantern Books.

Hall, C. (2013). October 2013 (RPC PB 2013/09). *The role of social science in policy making for tree and plant health and biosecurity*. pp. 1–4.

Haraway, D. J. (2008). *When species meet*. Minneapolis: University of Minnesota Press.

Harwood, T. D., Tomlinson, I., Potter, C., & Knight, J. D. (2011). Dutch elm disease revisited: Past, present and future management in Great Britain. *Plant Pathology, 60*(3), 545–555. Available at http://doi.wiley.com/10.1111/j.1365-3059.2010.02391.x (Accessed February 17, 2017).

Hinchliffe, S. (2007). *Geographies of nature: Societies, environments, ecologies*, London: Sage.

Hinchliffe, S., & Bingham, N. (2008). Securing life: The emerging practices of biosecurity. *Environment and Planning A, 40*(7), 1534–1551.

Hinchliffe, S., Allen, J., Lavau, S., Bingham, N., & Carter, S. (2013). Biosecurity and the topologies of infected life: From borderlines to borderlands. *Transactions of the Institute of British Geographers, 38*(4), 531–543.

Jones, O. (2014). Urban places of trees: Affective embodiment, politics, identity, and materiality. In L. A. Sandberg, A. Bardekjian, & S. Butt (Eds.),

Urban forests, trees, and greenspace: A political ecology perspective (p. 331). Abingdon: Routledge.

Jones, O., & Cloke, P. J. (2002). *Tree cultures: The place of trees and trees in their place*. Oxford: Berg.

Latour, B. (2004). Why has critique run out of steam? From matters of fact to matters of concern. *Critical Inquiry, 30*(2), 225–248.

Luke, T. W. (2002). Deep ecology: Living as if nature mattered: Devall and sessions on defending the Earth. *Organization & Environment, 15*(2), 178–186. Available at http://oae.sagepub.com/cgi/doi/10.1177/10826602015002005 (Accessed June 26, 2017).

McNew, G. L. (1960). The nature, origin, and evolution of parasitism. In J. G. Horsfall & A. E. Dimond (Eds.), *Plant pathology: An advanced treatise* (pp. 19–69). New York: Academic Press.

Nading, A. M. (2013). Humans, animals, and health: From ecology to entanglement. *Environment and Society: Advances in Research, 4*(1), 60–78. Available at http://openurl.ingenta.com/content/xref?genre=article&issn=2150-6779&volume=4&issue=1&spage=60.

Pidgeon, N., & Barnett, J. (2013). *Chalara and the social amplification of risk*. Available at www.gov.uk/defra (Accessed February 8, 2017).

Porth, E. F., Dandy, N., & Marzano, M. (2015). "My garden is the one with no trees": Residential lived experiences of the 2012 Asian longhorn beetle eradication programme in Kent, England. *Human Ecology, 43*(5), 669–679.

Puig de la Bellacasa, M. (2010). Ethical doings in naturecultures. *Ethics, Place & Environment, 13*(2), 151–169. Available at http://www.tandfonline.com/doi/abs/10.1080/13668791003778834 (Accessed February 27, 2017).

Sinden, N. (1990). *In a nutshell: Manifesto for trees and a guide to growing and protecting them*. London: Common Ground.

Tomlinson, I., Potter, C., & Bayliss, H. (2015). Managing tree pests and diseases in urban settings: The case of oak processionary moth in London, 2006–2012. *Urban Forestry & Urban Greening, 14*(2), 286–292.

Whatmore, S. (2002). *Hybrid geographies: Natures, cultures, spaces*. London: Sage.

Whatmore, S. (2006). Materialist returns: Practising cultural geography in and for a more than human world. *Cultural Geographies, 13*(4), 600–609.

18

Towards an Agenda for Social Science Contributions on the Human Dimensions of Forest Health

Mariella Marzano and Julie Urquhart

1 Introduction

Although there is still much to learn, we can find a substantial body of work on understanding the ecological dimensions of tree pests and diseases, but until recently the much needed analysis on the human dimensions has largely been missing despite acknowledgement of the significant part that human behaviours and decision-making play in tree health. The chapters in this book and the references that they cite go some way to address the human dimensions gap and to lay the foundations for future social and economic research in tree health. The IUFRO working party (7.03.15—Social dimensions of forest health) strives to bring together social scientists and economists working on tree health

M. Marzano (✉)
Forest Research, Edinburgh, UK

J. Urquhart
Countryside & Community Research Institute,
University of Gloucestershire, Oxstalls Campus,
Gloucester, UK

issues and provides a forum for sharing ideas, knowledge and methodologies from across the globe in recognition that pest and diseases cross many sociocultural, economic and political borders.

There are a broad range of trees and their associated pest and diseases, native and non-native, covered in this book including Acute oak decline, Asian longhorn beetle (ALB), *Phytophthora ramorum*, Ash dieback, Emerald ash borer, Oak processionary moth, Mountain pine beetle, Dutch elm disease, *Xylella fastioda* and more. In this context of outbreaks of tree pest and diseases, management and adaptation or future threats, researchers in this book have worked with many stakeholders including local communities, indigenous peoples, scientists, government agencies, NGOs, businesses, policy and decision-makers in villages, cities and rural forests. Many of the chapters in this book highlight the significance of collaboration, partnership and engagement, which suggests that better biosecurity necessitates inclusion of different knowledges, values, expectations and aspirations. Allen et al. (Chapter 11) underline why it is important to involve stakeholders in tree health highlighting that people must be given the opportunity to have a role in decision-making that affects them, but also greater participation ensures that social, cultural and economic impacts are also considered alongside ecological effects (see also Marzano et al., Chapter 12; Davis et al., Chapter 15). Often there is a focus on the consequences of tree health rather than the causes as these often seem too complex and difficult to control (Dyke et al., Chapter 17), perhaps requiring changes in people's behaviours such as recreationists visiting forests, consumers purchasing plants, producers and traders importing or selling live plants and wood products, those involved in large-scale planting programmes or forest management generally (Marzano et al., Chapter 12). However, Urquhart et al. (Chapter 7) and Price (Chapter 10) warn that there is no simple way to capture the interests, concerns and responses of individuals and groups and several chapters highlight ways in which local narratives of disturbance compete with scientific ones (e.g. Mattor et al., Chapter 14; Lambert et al., Chapter 5; Prentice et al., Chapter 4; Gürsoy, Chapter 3). Culturally embedded conceptions of the natural world often inform the construction of, and responses to, pest and disease outbreak events, and thus, it is important to incorporate an understanding of human–nature interactions as well as different agencies (human and non-human) into pest management deliberations (e.g. Prentice et al., Chapter 4; Fellenor et al., Chapter 6; Dandy et al., Chapter 16; Dyke et al., Chapter 17; Williamson et al., Chapter 2). It is also necessary to place tree health concerns at local scales within a wider global context of market pressures, harvesting practices, formal regulations and governance processes (e.g. Dragoi, Chapter 13; Keskitalo et al., Chapter 8; Jones, Chapter 9).

To understand, analyse and communicate about the complex tree health landscape, the authors in this book have adopted a variety of research methodologies and tools such as literature reviews, social media analysis, historical documents, face-to-face interviews, workshops, questionnaire surveys, Q methodology, rubrics and scenarios or narrative development. Through these different approaches, the chapters make important contributions on the human dimensions of forests and tree health in different geopolitical and sociocultural contexts. This chapter attempts to summarise the contributions, all of which are needed to inform tree biosecurity policy and management planning, and concludes by proposing an agenda for future social science research in this field. The synthesis presented in the following sections incorporates knowledge, values and attitudes, governance processes, risk communication and engagement and different way of investigating and understanding tree health.

2 Values

What people value will have significant implications in terms of their own behaviours and action as well as their acceptance of management responses. Gürsoy (Chapter 3), for example, highlights that villager perspectives on tree health crucially depend on the values they attribute to different trees. In the forest villages of Turkey, fruit trees found in gardens and orchards are particularly significant for their economic value. Gürsoy explores the symbolic spaces that trees inhabit such as the garden (domestic), orchard (domestic), forests (wild) and how this influences villager perspectives on who has responsibility for monitoring of trees and any interventions. The majority of forests are owned by the state, and Gürsoy notes that while forest pests are observed by villagers, unless they appear in the domestic space, pests are felt to be the responsibility of others, even though forest villagers have the right to utilise their local forests.

The difficulties of managing for pests and diseases when there are multiple, competing stakeholder values and interests is exemplified by Prentice et al. (Chapter 4). In the USA, they found that community responses to the catastrophic Mountain pine beetle (MPB) outbreak in Colorado was structured through several lenses—the local economy, policies and the biophysical landscape—but also how they interact with nature (e.g. livelihood versus recreation). Prentice et al. present a number of key stakeholder viewpoints on the reasoning for the MPB epidemic. The forest service suggests that lack of 'aggressive' management has led to a proliferation of mature forest stands that are aesthetically pleasing but vulnerable to MPB attack. Industry stakeholders see the MPB outbreak as a result of a diminished industry presence and point to wider conservation priorities (carried out by the forest service) that limit silvicultural practices to create habitat for designated animal species. The environmental perspective identified native MPB disturbance as important for forest succession and believe that forests will eventually recover. The potential impact of pest management on biodiversity is of greater concern than the pests themselves. There were also divisions between local communities with the more affluent and recreation/amenity-oriented communities supporting minimal intervention. Less affluent communities whose livelihoods were, or had been, linked to the forest industry were more supportive of intensive forest management and felt that these bigger outbreaks of MPB were a result of their disenfranchisement from the forest. Thus, Prentice et al. demonstrate how powerful environmental narratives are constructed within entangled sociocultural, environmental and economic histories that all play a role in how pest threats are perceived and acted upon. These findings chime with Urquhart et al.'s (Chapter 7) Q methodology study of residents in a community in South East England affected by Ash dieback. Here, perceptions about management and concern about the impacts of the outbreak were related to people's fundamental environmental worldviews, such as their beliefs about the vulnerability or resilience of nature, together with their beliefs about whether Ash dieback had arrived in the UK on imported nursery stock or had blown in on the wind.

3 Contested Knowledges

While there are a growing number of studies that highlight low knowledge levels amongst a range of publics on tree pests and diseases (e.g. Marzano et al. 2015; Fuller et al. 2016; Urquhart et al. 2017), Lambert et al. (Chapter 5) and Mattor et al. (Chapter 14) investigate important issues around whose knowledge counts. In New Zealand, Lambert et al. explore how Māori indigenous knowledge is contesting mainstream science perspectives on tree health. Calling for joint approaches to managing tree pest and diseases, they highlight the need to bridge the cultural gap between local indigenous knowledge and western scientific views of forests and their management. This call is not only for Māori representation in decision-making or governance roles but also including Māori methods and priorities for protecting forests. There is very little published evidence on the impacts of pest and diseases on social and cultural values and identity. In this chapter, Lambert et al. present the example of Kauri (a sacred Māori tree species) dieback (*Phytophthora agathidicida*). Local Māori have responsibility for all Kauri on their tribal land, and a failure to protect Kauri reflects on the *mana* (respect, authority, status and spiritual power) of tribal elders and future generations. The urgency of responding to Kauri dieback has led to greater involvement and leadership from Māori and involvement of Mātauranga Māori (knowledge and wisdom) at all levels of management including development of a Kauri cultural health index. Māori approaches to measuring impact include how people feel spiritually when they enter the forest—an assessment that does not sit well with traditional quantitative approaches to risk assessment and can be met with resistance from the western science community. The recent development of the Māori Biosecurity Network is a move to empower participation of local peoples and ensure that indigenous voices are included in wider biosecurity issues.

Gürsoy (Chapter 3) also emphasises that forest villagers will not always have the same understanding or perspectives as tree health scientists and that there should be greater dialogue and respect for local knowledge. At the same time, villagers signalled a desire for scientific knowledge on new and emerging diseases and for mitigation activities in forests for trees they value. As with many of the chapters in the book, Gürsoy highlights a need for greater collaboration amongst key stakeholders such as the forestry administration, villagers and scientists. Like Prentice et al. (Chapter 4), Mattor et al. (Chapter 14) also researched the impacts of MPB, but in this case study the authors were concerned about drinking water resources. The over-arching aim of the authors was to explore differences in knowledge bases between water managers and scientists and to assess the extent to which there was knowledge exchange about impacts in principle (scientific data) and practice (water managers' experience) between the different parties. Underlying the proliferation of the MPB epidemic is climate change with warmer weather creating drought stress in conifers. Tree mortality in large numbers can create problems for watercourses in terms of water quality, yield and flow. The authors maintain that while successful knowledge exchange can lead to changes in attitudes and behaviours, cultural differences between water scientists and managers over what constitutes evidence act as a barrier to interventions. In presenting scientific evidence, there is often a lack of understanding of what informs manager decision-making, which is often not based on scientific research but more on past experiences and traditional approaches. Mattor et al. suggest that scientific research often does not take into account tacit knowledge so research findings can have limited application. When surveying water managers, those who indicated high knowledge levels did read published scientific evidence and were more likely to be involved in collaborative water programmes. Nevertheless, the authors identified low levels of knowledge relating to MPB impacts compounded by the fact that managers had not yet experienced any evidence of detrimental effects linked to the beetles. Thus, stakeholder perceptions of risk and interpretations of actions required were entirely different to the recommendations of scientific research.

4   Understanding Risk

Urquhart et al. (Chapter 7) specifically focus on risk perceptions, emphasising the need for a good understanding of how experts and publics view risks around tree health. How risks associated with tree pests and diseases are perceived at multiple scales will play an important role in attitudes and behavioural responses. Thus, we need to know more about factors influencing risk perceptions including official pest communication, social networks, personal experiences and trust in those who manage outbreaks (see also Porth et al. 2015; Mackenzie and Larson 2010). Using Ash dieback as an example, Urquhart et al. highlight the complex interactions between government bodies managing disease outbreaks, media coverage of outbreak events and the diverse and adaptive risk perceptions of stakeholders and publics. The authors employ the Social Amplification of Risk Framework (SARF) to investigate how people make sense of different risks and the interactions between risk communication from external sources and their own identities, values, beliefs and experiences. They found that experts' view of risk was relatively dynamic and drew on a wide range of evidence, not just technical risk assessments and official information but less tangible forms such as prior experience, social networks, anecdotes and the media (see also Matter et al., Chapter 14). Policy makers were sensitive to reputational risk, and thus, tree health decisions were made that related to perceived social acceptability (and to be seen to be doing something) rather than empirical evidence of real impact or concern. Urquhart et al. note that risk understanding is not a linear process and that policy and expert priorities can be reassessed in light of media and public scrutiny. A key element, they suggest, is trust in the governance process and the institutions responsible for managing and communicating about the risk.

5   Governance and Collaborative Processes

The governance of tree health and analysis of existing governance structures is an important contribution to tree health studies, particularly in scoping the contribution of non-state actors to biosecurity processes and practices (Marzano et al. 2017). Many of the chapters in this book investigate and recommend collaborative processes and partnerships because, as Keskitalo et al. (Chapter 8) point out, while there is range of potential legal instruments and incentives, challenges remain with implementation on the ground. Keskitalo et al. emphasise the complexities within the European plant health system dealing with free trade between member states as well imports from non-EU states (see also MacLeod et al. 2010; Holmes et al. 2017). Health certificates such as plant passports for risky plant material are one way for national authorities to regulate and monitor potential threats, but Keskitalo et al. suggest that plant passports are not standardised across Europe and do not include non-regulated (new and emerging) pests. Crucially, higher-level regulatory systems cannot control the minutia of daily practices across a range of sectors that may threaten biosecurity. The authors present a case study of the nursery sector in Europe characterised by a strong system of inspection. Citing a survey of plant nurseries, the authors describe how nurseries identified a concern about pests and diseases and maintained that they regularly check plants for known pests. However, they also acknowledged 'risky' practices such as reusing storage containers that have been washed rather than disinfected, untreated water sources and failing to check plants that are purchased from another nursery. Keskitalo et al. noted the importance of collaborative processes for raising awareness and building capacity for better biosecurity, but they also highlighted limited integration between agencies and nurseries and between nurseries and research.

In this book, authors are quite right to point out that much existing work on the human dimensions of forest health often focuses on individual values but not so much on collaborative groups or collective action. Increasing threats to forest health have fuelled new ways of collaborating as we have seen with Lambert et al. and the Māori Biosecurity Network (Chapter 5). In the USA, Davis et al. (Chapter 15) map the development of forest collaborative groups (FCGs) as a way for state forests to communicate with a broad range of local stakeholders including environmentalists, forest industry, local communities and others as well as handling differing (sometimes competing) interests over forest management and resources. The areas where FCGs are being trialled is characterised by a declining forest industry and loss of livelihood (see also Chapter 4). As they are voluntary entities, Davis et al. explore how FCGs work in practice and ways in which they could be improved. FCGs were set up to identify social acceptability of forest management interventions, avoid litigation from those disagreeing with the interventions and speed up planning timelines. The authors found that FCGs were generally more successful when there is a collaborative body or group that can organise and sustain itself. However, they did find that forest health issues to be discussed were primarily introduced by the forest service or scientists although some FCGs were starting to lead with their own knowledge and perspectives or funding their own monitoring programmes. Davis et al. warn that not all stakeholders will participate in their FCG, and these groups do not guarantee that there will not be public objection to interventions. Although the focus of FCGs appears to be on wildlife issues, they do provide a useful framework for thinking about pests and diseases and collaborative processes and can provide a snapshot of stakeholder views over forest management issues. Key lessons were that consultation responses from FCGs are dynamic and do not represent an enduring social licence to operate; rather, collaboration is an iterative process. FCGs are not legally organised entities that employ staff and have access to funds so there is a need to manage expectations of what they can achieve. FCGs currently operate in isolation, and Davis et al. felt that there was scope for greater knowledge sharing and learning between the groups.

6 Knowledge Exchange and Research Tools

Developing tools to facilitate knowledge exchange was a key feature of a number of chapters. Jones (Chapter 9) maintains that there is a strong economic argument for public support of plant health policies.

However, Price (Chapter 10) poses the question: how do you value impacts of tree diseases? He then goes on to explore whether contingent valuation is a useful method to assess whether it is worth expending resources to control or mitigate effects of pests and diseases. Price suggests that some services have a market benefit like water regulation and carbon dioxide fixing but other non-market goods such as cultural services (e.g. aesthetics) are more intangible. Thus, economics have looked to provide these values through contingent valuation like Willingness To Pay (WTP) for environmental improvements or to accept compensation for deterioration. Both Price and Jones highlight issues with validity and accuracy of valuations. Price (Chapter 10) calls for careful design of WTP questions and suggests that while it is useful to provide participants with enough information so that they can make informed judgements, there is a danger that too much information will prompt expectations that responses should be based on expertise and judgements regarding the public good rather than simply their own 'self-interested' preferences. Price advises that questions be neutral and refrain from value-laden terms such as 'disease'. Regardless of research responses, there is still an issue of how to translate economic findings into policy-relevant recommendations. Jones (Chapter 9) ponders on which economic methodologies can provide the best information in the shortest time—to fit in with policy decision-making in the context of significant uncertainty—and with limited resources. Jones advocates the use of bio-economic models to help assess the effectiveness of different management options on natural resources. Notwithstanding the need for empirical data, which is often lacking (see also Marzano et al. 2017), integrated bio-economic modelling can help determine the economic efficiency of interventions such as prevalence when found (how established is it), predicted rate of spread, judgements of impacts per host, the value of the host (to ecosystem services including human well-being), efficacy of control options and engaging stakeholder interests and capacity.

Stakeholder engagement was the key theme for Allen et al. (Chapter 11). A number of the chapters in this book have already observed growing recognition of the need for partnership-based approaches to tree health that include multiple stakeholders and their perspectives. Allen et al.

note that it is rare for biosecurity programmes to provide practical guidance or tools on working in multi-stakeholder contexts requiring not only expertise in technical activities but a greater understanding of organisational and social processes. The authors discuss an action-research approach with biosecurity agencies in New Zealand to develop a rubric, signalling a move away from top-down communication to greater engagement and dialogue and trust building. The key aim in New Zealand is to enhance the surveillance system for pests and diseases, and rubrics, the authors suggest, can be a template to instruct and evaluate activities and provide a framework for learning. Agency relationships with communities can be developed during periods of 'quiet' (e.g. surveillance or monitoring) in preparation for crises situations (e.g. eradication of incursions). Allen et al. maintain that rubrics can aid in developing communication and engagement processes but also facilitates thinking through the 'bigger picture' of biosecurity so it is both a long-term learning process and product.

Approaches to learning was a key element for Marzano et al. (Chapter 12) who focus on technology development for early detection of pests and diseases. At present, most countries rely on trained inspectors to detect pests and pathogens, mainly via visual inspections. However, given the volume of inspections required, the finite amount of resource usually available and the huge practical challenges associated with these inspections, this task is extremely difficult and the efficiency of detection is low. Thus, the authors highlight the demand for new and better methods for detecting tree pests and pathogens along trade pathways and in the wider environment. They highlight that technological innovations require close collaboration and interactions between researchers, end users, manufacturers and markets set within the broader context of social norms and the regulatory environment. The authors discuss the use of a learning platform, a concept that builds on learning alliances (Sutherland et al. 2012) which encourages multi-stakeholder knowledge exchange, dialogue and social learning to promote greater engagement and input into outputs and outcomes. In the context of early detection technologies, the aim of the learning platform was to move beyond provision of information or broader consultation towards greater decision-making and active engagement with the process of technology development. The learning platform involved having to engage innovative tools to attract the interest of participating stakeholders and to encourage scientists to present technological ideas and invite feedback in accessible ways. Marzano et al. explored how evolving interactions between individuals and groups can influence the scope and speed with which technologies are developed. However, it is not a linear process and technologies are unlikely to be fully functional in a normal funded project lifecycle of 3–4 years. Rather, technology development is often supported by previous projects and other ongoing projects. In this context, fundamental questions were raised around 'who pays?' as tree health technologies are not merely products for consumer consumption. Will the lack of market potential (because of a narrow user base) limit innovation and what is needed to provide non-market-based stimulus? The authors believe that stakeholder engagement through the learning platform did influence technology development and raised important questions about how products move from concept to production and use. However, they recognised a need to be able to assess by how much stakeholder engagement can improve the socio-technological innovation process to encourage the prioritisation of participatory approaches over other activities.

This book also includes useful tools and research that have wider implications for tree health and biosecurity in the future. For example, Dragoi (Chapter 13) explored the development of a training tool to assist forest managers in selecting which trees should be kept as standing deadwood for biodiversity to meet FSC criteria. The tool is being trialled in post-socialist Romania where tracts of forests, formally under control of a communist government, have been restituted to landowners. However, the transition has been difficult particularly as the private sector is required to follow the same forest code as state-owned forests with little guidance on how to manage their forests. Forest managers and landowners face further difficulties of moving to a new system of certification and environmentally-sensitive logging due to an increasingly fragmented and bureaucratic governance system that has led to overharvesting and rent seeking. The tool (inspired by operant learning theory) encourages social learning and is focussed on training foresters to identify which trees should be harvested (healthy, salvage and sanitation) and which should be left in the forest to grow or as standing deadwood. The training will help foresters to consider multiple issues (e.g. harvesting, biodiversity obligations, disease management) and is likely to be more cost-effective in terms of the time required to visit forests for each single issue. The tool will also allow foresters to continually collect data for monitoring and facilitates reflections on what constitutes a healthy forest.

How people engage with knowledge and information about the world has changed dramatically over recent decades, especially with the growth of digital technologies. This, asserts Fellenor et al. (Chapter 6), has implications for how tree health issues are viewed and understood by stakeholders and publics. The authors undertook a rapid evidence review on User-Generated Content (UGC), which relates to blogs, social networking sites, wikis, social commerce sites and discussion or opinion (e.g. trip advisor) forums. They found little detailed exploration of UGC aside from statements and assumptions that social media is a good thing. Many organisations and individuals will use the internet as one way of communicating with their audiences, and the authors suggest that UGC not only provides information to users and social networks, but users are themselves data sources. They note that UGC creates socio-technical material involving the trees, social media users and technological devices (e.g. smartphone monitoring systems). While UGC is never value-free, online interactive sites such as social media could be beneficial for forest health, providing real-time data. However, the authors warn that we don't understand enough about online communities and their relationships with forests and, consequently, there is a tendency to idealise what can be achieved through this interaction. Like Urquhart et al. (Chapter 7), the authors question how UCG and social media change our perceptions of the world and of ourselves.

7   Differing Approaches to Exploring Human Dimensions of Forest Health

Recently, Marzano et al. (2017) called for the inclusion of other social science perspectives that have been missing so far from explorations of the human dimensions of tree health. Historical analyses and ethics are two disciplinary areas that can provide important insights into current concerns and priorities around pest and disease outbreaks. Both Prentice et al. (Chapter 4) and Williamson et al. (Chapter 2) highlight that more modern conceptions of nature, natural landscapes and how forests should look and feel are potentially contributing to their vulnerability to pests and diseases and that tree health should be viewed within a broader historical context. Williamson et al. take a historical approach to understand the potential contribution of forest management to pest outbreaks by exploring documentary evidence available in the UK from the sixteenth century onwards. The authors found that while there has been no large-scale pest or disease event prior to the twentieth century, tree health issues are not a new phenomenon and that the trade in live plants and trees existed—sometimes on a substantial scale—for centuries. They suggest that earlier generations viewed tree ill health as normal with diseased trees being felled and sold. Interestingly, the prevalence of oak, ash and elm in the British landscape only came into being from the seventeenth century despite there being at least 25 other native species that could grow into reasonably sized trees and were previously linked to specific regions and English counties. It is likely these species were favoured because of their ability to thrive in a wide range of habitats and for the value of their wood. However, an important difference compared with today is that trees were felled at a relatively young age or at least when they reached the size required for whatever commercial or domestic product was needed. The authors attribute the appearance of older, mature trees in the countryside with rapid social change coupled with the rise of conservation-based organisations with their own idealised constructions of nature and natural landscapes. This has led to unrealistic expectations in modern times that trees will stay healthy if left to grow into old age when history suggests that the most rigorously managed treescapes were the most healthy. Williamson et al. warn against continued conservation-based attempts to replicate existing woodlands and retaining large, over-mature and dead wood in the landscape, asking: are tree diseases an artifice of allowing trees to grow too old? Lessons from history indicate that the current 'artificial' landscape presents us with opportunities for the future by identifying a number of minority native species suited to specific regions that could make their comeback.

Social constructions of nature, forests and tree health are introduced in several of the chapters in this book. Fellenor et al. (Chapter 6) comment on how digital technology is increasingly mediating how some 'virtually' engage with forest health, rather than direct exposure or experience of tree pests and disease outbreaks. Dyke et al. (Chapter 17) were particularly interested in exploring ethical approaches to identify how humans impose their own values onto trees with the use of labels and concepts that often dictate who or what is allowed to live or be killed. Dandy et al. (Chapter 16) provide three different ethical framings to reflect on different approaches to management of Asian longhorn beetle (ALB). These framings—biocentrism, entangled empathy, flourishing—demonstrate how different approaches of seeing human–nature relationships can result in very different outcomes for managing forest health. For example, the biocentric approach insists that we remain neutral towards all species without favouring one species over another, which suggests that beetles, trees and human interests are equal and none should be harmed. However, biocentrism refers to a wild state in a natural ecosystem rather than non-native invasions. Entangled empathy, on the other hand, proposes multiple ways in which we have an active 'caring' relationship with humans and non-humans that feeds into discussions about whose lives should be prioritised in outbreak situations and where empathy may lie (e.g. could ALBs be viewed as refugees?). The flourishing framework involves the attribution of human values and perspectives such as decisions over what is healthy and able to flourish or not. Flourishing would dictate the felling of infected trees only (rather than all potential hosts) as these are unlikely to flourish. Dandy et al. stress that the development of narratives and alternative outcomes involving ethics and non-human agency doesn't mean that felling or other forms of pest management would be rejected, but the authors call for a 'noticing' of non-humans. Dyke et al. (Chapter 17) also emphasise the need to include non-human agency in considering tree health management and to move beyond scientific narratives of disease. They question management terminology around security, defence and invasion and advocate a focus on coexistence or living with 'invasive' species, which they stress does not equate to doing nothing but rather places constructions of health and ill health in the broader context of how trees, people, beetles and bacteria coexist in space and time.

8 Framing the Future of Research into the Human Dimensions of Tree Health

Taken as a whole, the chapters in this volume represent the first synthesis of social science approaches to address tree health issues and bring together interdisciplinary researchers from across the world to exemplify the diverse and rich contributions that social scientists can offer in tackling the growing threat from tree pests and diseases. As this field of scholarly interest develops, the book provides a useful applied and theoretical foundation on which to build an agenda for future research activity. In terms of a path forward, we suggest a number of key areas of focus.

Firstly, it is clear that dealing with tree pest and disease outbreaks is complex and involves navigating a broad set of actors at a range of spatial scales. This requires recognition of the diverse values that are implicated in tree health outbreaks. As contributions in this volume have shown, different stakeholders will have diverse, and sometimes conflicting, values about how outbreaks should be managed. Therefore, as in other areas of environmental management, stakeholder participation and co-management are an important strategy for successful outbreak management. This involves dialogue between stakeholders, outbreak managers and policy makers to build better governance mechanisms, and social scientists can provide empirical evidence to support this process.

Secondly, in a domain traditionally dominated by natural science, a key challenge will be to develop closer interdisciplinary engagement between natural and social scientists. Policy makers and research funders have a role to play in this regard, by recognising the value that social science 'evidence' can bring and by ensuring that policy processes support the integration of natural and social research (Marzano et al. 2017). As the economics chapters in this book (Chapters 9 and 10) suggest, economic models are good for estimating impacts on some ecosystem services, such as carbon sequestration, but less so for estimating impacts on cultural values, such as aesthetics, spiritual values or existence values. This is a gap that social scientists can help to fill, but long-term commitments to fund social science are required in order to build and sustain research capacity in this field.

Thirdly, while much of the book focuses on applied contributions, several chapters illustrate the rich and varied way in which social science can contribute conceptually. For instance, Chapters 16 and 17 in particular make important claims about the largely unquestioned anthropogenic approach to tree health management, arguing that environmental ethics can provide alternative lenses that consider the value of non-humans and, as a consequence, may shift management priorities.

Finally, tools, methods and conceptual frameworks are required that recognise the complexity and dynamic nature of the human dimensions of tree health. As such, this calls for drawing on existing and new innovative approaches from across the social sciences that are relevant in practice and address real-world problems. The editors and authors of this volume very much hope that other social scientists, and also arts and humanities scholars, will bring their own disciplinary expertise to this growing area of research that is the human dimensions of forest and tree health.

References

Fuller, L., Marzano, M., Peace, A., Quine, C. P., & Dandy, N. (2016). Public acceptance of tree health management: Results of a national survey in the UK. *Environmental Science & Policy, 59*, 18–25. https://doi.org/10.1016/j.envsci.2016.02.007.

Holmes, T. P., Allen, W., Haight, R. G., Keskitalo, E. C. H., Marzano, M., Pettersson, M., et al. (2017). Fundamental economic irreversibilities influence policies for enhancing international forest phytosanitary security. *Current Forestry Reports, 3*(3), 244–254. https://doi.org/10.1007/s40725-017-0065-0.

Mackenzie, B. F., & Larson, B. M. H. (2010). Participation under time constraints: Landowner perceptions of rapid response to the Emerald ash borer. *Society & Natural Resources, 23*(1), 1013–1022. http://doi.org/10.1080/08941920903339707.

MacLeod, A., Pautasso, M., Jeger, M. J., & Haines-Young, R. (2010). Evolution of the international regulation of plant pests and the challenges for future plant health. *Food Security, 2*(1), 49–70. https://doi.org/10.1007/s12571-010-0054-7.

Marzano, M., Dandy, N., Bayliss, H. R., Porth, E., & Potter, C. (2015). Part of the solution? Stakeholder awareness, information and engagement in tree health issues. *Biological Invasions, 17*(7), 1961–1977. http://doi.org/10.1007/s10530-015-0850-2.

Marzano, M., Allen, W., Dandy, N., Haight, R., Holmes, T., Keskitalo, E. C. H., et al. (2017). The role of the social sciences in understanding and informing tree biosecurity policy and planning: A global synthesis. *Biological Invasions, 19*(11), 3317–3332. https://doi.org/10.1007/s10530-017-1503-4.

Porth, E. F., Dandy, N., & Marzano, M. (2015). "My garden is the one with no trees": Residential lived experiences of the 2012 Asian Longhorn Beetle Eradication Programme in Kent, England. *Human Ecology, 43*(5), 669–679. https://doi.org/10.1007/s10745-015-9788-3.

Sutherland, A., da Silva Wells, C., Darteh, B., & Butterworth, J. (2012). Researchers as actors in urban water governance? Perspectives on learning alliances as an innovative mechanism for change. *International Journal of Water, 6*(3/4), 311–329. https://doi.org/10.1504/IJW.2012.049502.

Urquhart, J., Potter, C., Barnett, J., Fellenor, J., Mumford, J., Quine, C. P., et al. (2017). Awareness, concern and willingness to adopt biosecure behaviours: Public perceptions of invasive tree pests and pathogens in the UK. *Biological Invasions, 19*(9), 2567–2582. https://doi.org/10.1007/s10530-017-1467-4.

Index

A

acute oak decline (AOD) 40, 449, 450, 456–458, 460, 461, 465, 472
adaptive management 146, 147, 158, 289
amenity contexts 86
anthropocentrism 420
AOD. *See* acute oak decline (AOD)
apple-pie [values] 250, 255, 257, 261
ash 9, 15, 22, 25–27, 30–34, 36, 40, 148, 166, 168, 172–178, 180, 183, 184, 187, 194, 237, 238, 317, 332, 338, 424, 448, 472, 475, 484
  as coppice 29, 32, 33
  as timber 24, 26, 29–32, 34, 36, 40, 331
  uses of 28, 30, 32, 34

ash dieback (*Hymenoscyphus fraxineus*) 13, 17, 22, 25, 168, 172–177, 180, 183–185, 187, 215, 237, 249, 472, 475, 477
Asian longhorn beetle (*Anoplophora glabripennis*) 17, 420–429, 431–434, 435, 437–439, 442, 472, 485
Atchison, Jennifer 446, 453–455
Austropuccinia psidii 15, 111, 272, 283

B

bark beetle 81, 82, 92, 93, 358, 361, 368, 435
bark, economic importance of 29, 30, 35, 173, 335, 341, 461
Bastian, Michelle 449, 454, 455

beetle infestation 81, 369, 425
biocentrism 420, 425, 439, 441, 485
bio-economic model 221–223, 227, 480
biosecure behaviours 175, 182
biosecurity 5, 6, 10, 11, 13, 15, 16, 109–112, 116, 121–124, 126–129, 131, 132, 147, 175, 186, 188, 213, 227, 240, 270–276, 278, 280, 281, 284, 286, 287, 289, 292, 300, 302, 306, 316, 322, 325, 326, 420, 446, 450, 453, 454, 466, 467, 472, 476, 478, 481
   system 123, 271, 272, 278, 291, 300, 318
black poplar 31
blame 78, 184
boxwood tree (*Buxaceae*) 68
   boxwood blight (*Cylindrocladium pseudonacivulatum*) 69
   health 70
   V. Buxi 69

C

citizen science 145, 151, 153, 318
citizen [values] 257, 261
climate change 2, 12, 22, 41, 59, 60, 80, 89, 140, 193, 206, 302, 349, 355, 356, 359, 361, 371, 405, 445, 476
   perception of 12, 14, 59, 80, 140
coal 28, 36, 98
Colorado, USA 14, 77–79, 81, 83, 85, 86, 88–94, 98, 99, 101–103, 357, 362, 364, 365, 368, 374, 375, 474

communication 10, 12, 16, 18, 144–148, 153, 155–159, 167, 170, 173, 177, 179, 180, 185, 186, 215, 271, 273–276, 280, 281, 287, 289, 291, 292, 356, 357, 359–361, 363, 364, 369, 370, 373, 375, 376, 407, 425, 473, 477, 481
   theory 16, 17, 148, 170, 276
community contexts 81, 94
community engagement 15
compensation 236, 243, 244, 428
contingent referendum 244, 259
contingent valuation 236, 239, 243–245, 250, 251, 255, 259, 428
   controversies over 49, 67
   clear felling 66
   Finnish style of forest management 66
   forest management techniques 66, 67
   goats entering forests 70
   selection method 66, 67
Cook, Moses 24, 32, 39
coppice 29, 33
   composition of 33
   management of 29, 33
   replanting of 33
   uses of wood from 32
costs 11, 129, 203, 219, 223, 227, 241, 251, 311, 313, 370, 371, 373, 440
crowdsourcing 142, 145, 151, 159, 335
cultural services 212, 239, 480
CVM. *See* contingent valuation method (CVM)

D

deadwood 245, 331, 334, 339, 343, 344, 347, 348, 350, 482
Dendroctonus ponderosae 14, 85, 203
detection technologies 16, 301, 302, 306, 312, 315, 317, 319, 323, 481. *See also* technology
  air spore trapping 301
  citizen science 318. *See also* pheromone traps (beetle traps)
  co-design 319
  co-development 301, 319
  commercialisation 17, 300, 302, 323
  cost 325
  multispectral imaging 301
  non-market stimulus 323
  pheromone traps (beetle traps) 434
  reliability 317
  volatile organic compounds ('sniffer' technologies) 301
  water-borne spore trapping 301
digital forestry 148
  Internet of Things (IoT) 150
disinfectant-by-products (DBP) 359, 367–370
Dutch elm disease (*Ophiostoma ulmi*) 36, 177, 180

E

East kent 176, 184
ecological imaginaries 78, 102, 103
economics 3, 6, 97, 212, 217, 235, 261, 464, 486
ecosystem services 2, 151, 212, 214, 229, 235, 240, 255, 299, 335, 386, 456, 486

elm 22–24, 26, 30–32, 35, 39, 40, 177, 237, 249, 255, 337, 448, 472
  uses of 30–32, 34
embedding 245
emergency modality 420, 440
empathy (engaged empathy) 103, 420, 430–432, 434, 439, 441, 485
engagement 10, 12, 16–18, 111, 120, 121, 126, 127, 129, 142, 145, 147, 148, 153, 167, 270, 271, 273–276, 279–281, 286, 288, 289, 291, 292, 301, 303–306, 316, 319–321, 324, 325, 357, 361, 388, 410, 433, 439–441, 446, 448, 450, 453, 457, 462, 463, 466, 472, 473, 480–482, 486. *See also* involvement
environmental ethics 6, 17, 420, 441, 442, 487
environmental narratives 14, 79, 83, 84, 94, 474
Epping Forest 446, 455, 457, 458, 461, 465
European mistletoe (*Viscum album ssp. Austiacum (Wiesb.) Vollman*) 69
evaluation 48, 197, 217, 251, 257, 261, 276, 280, 282, 289, 290, 292, 341, 350, 372, 410
expert elicitation 363, 375

F

farmland trees in England 36
  management of 25, 27, 33, 37, 42
  numbers of 25, 27, 36, 41

species composition 25
flourishing 420, 435–439, 441, 485
fodder 27
Food and Environment Research Agency (FERA) 422
forest collaborative groups 17, 384, 386, 388, 391, 395, 402, 409, 479
forested watersheds 358, 369
forest nurseries 195–197, 201–205
Forest Research (FR) 5, 8, 173, 176, 422–425, 427, 429, 432, 435, 437–439, 467
forests 2, 14, 47–49, 51–56, 59, 66, 69, 77, 79, 80, 85, 87–91, 93–95, 101, 102, 111, 112, 116, 118, 122, 129, 139, 140, 146, 148, 149, 151, 156–158, 166, 188, 196, 211, 311, 325, 332, 334, 335, 337, 341, 343, 348, 350, 356, 358, 368, 370, 373, 383, 385, 387, 388, 394, 395, 402–405, 426, 442, 472–475, 479, 482–484
Forest Service, U.S. 78, 85–95, 98, 99, 101, 104, 384, 386–389, 393, 395, 397, 399–402, 408, 474, 479
Forest Stewardship Council (FSC®) certification 337
fuel 3, 28, 29, 34, 42, 97, 146

G

gardens 8, 48, 51, 52, 54, 56, 72, 155, 286
Geels, Frank W. 301, 302, 304, 323, 324
global trade 2, 13, 467

governance 3, 10, 15, 16, 111, 120, 129, 147, 323, 387, 448, 473, 475, 478, 486
Gruen, Lori 420, 430

H

Harmful mammals and birds 62
  bears 62
  watch against 62
  wild boar 62
hazard 169, 170, 178, 181
Head, Lesley 35, 39, 446, 453, 455
headlining 249, 261
Hertfordshire 22, 25, 26, 32, 34, 41
heuristic 171, 177
historic range of variability 403, 404, 409
holly 27
hornbeam 31, 34, 41
human/non-human relations 446, 449, 452
  attunement 456
  co-creation (of disease) 466, 467
  differently human 463
  embodied experience 463, 464
  more-than-human 441, 450–452, 463, 466
human wellbeing 3, 480

I

illegal fellings 334
imports, of wood, timber and plant materials 24, 36, 37, 197, 199
indigenous 7, 23, 41, 59, 64, 110, 112, 118, 122, 124, 129, 131, 207, 472, 475
  communities 6, 14, 112, 118, 129

knowledge 15, 63, 475
methodologies 6, 111, 472
information bias 253
information communication technologies 139
SMART phones 150
interior Northwest 387
International Union of Forest Research Organizations (IUFRO) 5, 10, 12, 471
invasive alien species 15, 152, 193
involvement 10, 15, 111, 129, 180, 207, 217, 277, 313, 316–318, 334, 368, 388, 475

K

Kallhoff, Angela 420, 436
Kasperson, Roger E. 168, 170
Kent 23, 33, 173, 182, 184, 422, 432
knowledge exchange 17, 303, 356, 357, 359, 361, 363, 373, 375, 376, 476, 479

L

landscape 3, 7, 13, 25, 30, 37, 39, 41, 42, 66, 78, 82, 85, 90, 91, 97, 102, 103, 173, 187, 214, 226, 237–240, 249, 251, 255, 258, 260, 261, 302, 388, 389, 399, 403–405, 412, 425, 433, 435, 461, 473, 484
learning platform 17, 303, 305, 307, 313, 316–318, 320, 481, 482
learning alliance 303, 481

Socio-technological Learning Laboratories (SLLs) 305, 306, 311, 319, 325
legislation 10, 16, 199, 205, 207, 214, 450
lexicographic [values] 250
lime, small-leafed (*Tlia cordata*) 30, 32, 41
local knowledge 14, 49, 59, 66, 400, 476
and experience 66
lodgepole pine 89, 90, 92, 201, 252, 358. *See also Pinus contorta*

M

Māori 15, 110–113, 116, 118, 120–132, 270, 273, 286, 287, 475, 478
maple 31, 32, 41, 428
market failure 213
media 15, 57, 119, 141–145, 147, 148, 153–155, 158, 165, 168, 170, 172–184, 186, 187, 274, 333, 424, 438, 473, 477, 483
Millennium Ecosystem Assessment 2
moral satisfaction 245, 247, 254
mountain pine beetle 14, 17, 77, 85, 89, 90, 93, 203, 356, 358, 362, 374, 472, 474. *See also Dendroctonus ponderosae*
community response to 77, 474
management of 89
outbreak of 14, 17, 77, 85, 90, 358, 362, 474

Myrtle Rust 15, 111, 123–128, 284–286, 306, 311, 319, 325. *See also Austropuccinia psidii*

N

Nading, Alex 450, 452
Narrative 83, 86, 92, 101–103, 395, 420–422, 432, 465, 473
New Zealand 13, 16, 109–113, 115, 116, 119, 122, 124, 125, 127, 130–132, 269, 270, 272, 273, 275, 278, 283–285, 293, 441, 475, 481
non-humans 17, 419–421, 426, 439–442, 446, 450–456, 462, 464, 485, 487
non-market 217, 225, 323, 482
Northamptonshire 22, 24, 26, 27
Nourse, Timothy 27, 30

O

oak 22–24, 29–33, 35, 39, 40, 49, 67, 155, 332, 435, 449, 454–458, 461–466, 472
  numbers of 25, 31, 34
  uses of 30
oak decline 22, 23, 38, 39, 449
Operant/Conditional Learning Theory 16, 17, 339, 482
Oregon, eastern 17, 384–386, 390–393, 395–399, 402, 405, 408–410
outbreak 81, 87, 93, 94, 96, 97, 99, 173, 237
management of 14, 77, 81, 82, 87–93, 95–97, 102, 125, 212, 215, 218, 227, 420–424, 428, 432, 437, 439–442, 484, 486

P

passive use value 239
perceptions 10–12, 14, 49, 59, 60, 80, 82, 83, 87, 98, 111, 142, 165, 167, 169, 170, 172, 174, 177, 180, 184, 187, 240, 289, 363, 374, 386, 406, 475, 477
  of climate change 12, 13, 59, 80, 140
  of tree diseases and health 49, 60, 249
Phytophthora 15, 111, 113, 116, 130, 180, 195, 196, 199, 201, 203, 204, 221, 238, 448, 475
phytosanitary 6, 116, 214, 423
Pine processionary moth (*Thaumetopeo pityocampa* Schiffemüller) 60
Pinus contorta 79, 207, 240
Pinus ponderosa 388
plant health regulation 15, 206
policy 3, 6, 9, 10, 13–15, 41, 42, 110, 147, 154, 156, 167, 170, 172, 174, 176, 178, 179, 182, 187, 188, 195, 211–213, 216, 218, 219, 221, 222, 225, 227, 229, 252, 270, 301, 302, 304, 308, 312, 314, 323, 332, 337, 385–387, 409, 410, 422, 441, 448, 466, 472, 477, 480, 486

makers 9, 173, 176, 177, 179, 185, 218, 280, 304, 324, 385, 409
making 16, 212, 219
pollards 26–28, 32, 35
   age of 35
   decline of 458
   management of 27
   numbers 34, 36
ponderosa pine 79, 85, 88, 388, 394, 405, 406. *See also Pinus ponderosa*
processes 47, 79, 80, 82, 89, 141, 142, 148, 158, 166, 167, 169, 170, 173, 176, 224, 240, 271, 274, 292, 301, 320, 326, 339, 362, 386, 389, 393, 396, 411, 421, 440, 455, 473, 478, 481, 486
   cultural 82, 166, 169, 170, 185
   institutional 82, 169, 170
   psychological 158
   social 142, 171, 174, 271, 283, 409, 481
public 8, 9, 12, 16, 51, 54, 89, 93, 101, 102, 116, 121, 141, 142, 145–147, 151, 152, 155, 166–170, 172–188, 195, 212–215, 218, 226, 228–230, 237, 240, 241, 245, 250, 251, 253, 278, 290, 300, 322, 325, 332, 334, 335, 337, 350, 357, 364, 369, 371, 373–375, 383, 384, 386–388, 401, 408, 410, 419, 433, 441, 477, 479
   attention 9, 176, 334
   awareness 51, 117, 286
   concern 174, 178, 179
   public good 213, 214, 230, 323, 386

Q

Q methodology 168, 176, 474
qualitative 6, 9, 18, 219, 362, 365, 390, 425
   approaches 6, 17
   methods 6, 9, 362
   research 7, 9, 48, 365
quantitative 6, 7, 17, 219, 252, 363, 390
   approaches 6, 7, 17, 475
   methods 6, 9
   research 7, 9
questionnaires 238, 240, 250, 251, 254, 256, 259

R

recreational forests 140
Renn, Ortwin 168, 171
reputational risk 178, 477
residents 14, 81, 87, 94–102, 176, 424, 425, 429, 433, 438, 440
ripple effects 170, 187
risk 6, 7, 9–11, 13, 15, 16, 40, 81, 82, 87, 93, 97, 100, 125, 155, 156, 159, 165–182, 185–188, 195, 196, 198, 199, 202, 205, 206, 214, 220, 222, 223, 228, 242, 269, 271, 273–275, 280, 285–287, 292, 356, 367, 387, 400, 405, 422, 439, 449, 455, 459, 473, 477

assessment 174, 179, 187, 221, 427, 475
communication 18, 166, 167, 270, 271, 275, 291, 292
engagement management 12, 270–275, 291, 292, 473
perception 80, 165, 167, 170, 177, 179, 186, 187, 477
signals 170, 173
Rocky Mountains 357, 365, 372, 374
Romania 13, 332, 336, 482
rubrics 271, 276, 277, 288–290, 292, 473, 481

S

salvage/sanitation fellings 237, 257, 333–344, 346–349, 482
scientific views 475
   of forest engineers 63, 65
   of the Forestry Administration 66
scoping 245, 247, 478
self 30, 33, 129, 141, 157, 159, 215, 251, 254, 367, 412, 463, 480
   and material entities 141
   ontology of 156
semi-structured interviews 57, 174, 196, 306, 307
Skinner, B. Frederic 339
Slovic, Paul 168, 171, 274
social amplification 15, 167, 169, 170, 175, 180, 477
   amplification 15, 167, 169, 172, 175, 176, 180, 181
   attenuation 169
   intensification 169

Social Amplification of Risk Framework (SARF) 15, 167, 168, 180, 477
social station of amplification 180
social construction 85
social learning 16, 17, 303, 325, 339, 481
social media 15, 57, 141, 142, 145, 147, 148, 150, 154, 155, 158, 168, 172, 175, 181, 438, 473, 483. *See also* Twitter
social science 3, 5, 7–9, 54, 168, 357, 362, 446, 448, 450, 473, 486
socio-cultural 6, 82, 275, 299, 472, 474
socio-technological innovations 300
   end-users 306
   markets 301
   Multi-Level Perspective (MLP) 302
   socio-technical systems 301
   socio-technological change 308, 319
spruce trees 67
stakeholder mapping 308, 319
stakeholders 7, 12, 16, 18, 101, 141, 143, 145, 147–150, 155, 165, 167, 173, 174, 177, 179, 229, 270, 271, 273–276, 280, 281, 283, 287, 289, 290, 292, 300, 303–308, 313, 314, 316–318, 320–322, 324–326, 384, 386–388, 392, 400–403, 409, 429, 435, 441, 450, 472, 474, 477, 479, 482, 486
   definition of 385
   horticulturalists 308
   inspectors 300, 305, 308

landscape architects 6, 11, 229, 396, 401, 435, 474
nurseries 319, 323, 325
stated preference 236, 249, 259, 261
stories (storytelling) 83, 101, 181, 421, 439, 440, 442
strategic bias 243, 258
Suffolk 23, 25, 27
surveillance system 228, 275, 278–280, 282, 283, 285, 290, 291
   general surveillance system 278–281
Sweden 13, 195, 196, 200, 201, 205, 207
symbolic [values] 53, 246, 254, 258

T

Taylor, Paul 420, 425
Technology Readiness Levels (TRLs) 306, 315, 321, 322, 324, 326
timber industry 81, 86, 91–95, 101
timber trees 26, 29, 33, 36
   age of felling 37
   principal species of in early-modern England 29
total organic carbon (TOC) 358, 359, 367–371
Tree Health & Plant Biosecurity Initiative 5
tree pathogens 2
tree pests 2–4, 8, 13, 16, 167, 181–183, 187, 212, 225, 236, 300, 301, 303, 306, 308, 310, 319, 471, 477, 485
trees for biodiversity (TFB) 333, 334, 337–341, 344, 346–349

trihalomethanes (THM) 368, 371
Turkey 13, 14, 48–55, 61, 62, 69, 473
   Aegean Region, TR3 49, 60
   Central Anatolia Region, TR7 58
   Eastern Black Sea Region, TR9 58, 62, 68
   Eastern Marmara Region, TR4 65
   forest ownership 48
   Forestry Administration 53, 60, 68, 70
   Forests under Turkish law 53
   İstanbul Region, TR1 60
   Mediterranean Region, TRA 49
   Western Black Sea Region, TR8 61
   Western Marmara Region, TR2 60, 62
Twitter 142, 143, 147, 154, 175, 181

U

uncertainty 9, 97, 159, 167, 177, 178, 216, 217, 220, 223, 227, 237, 363, 449, 461, 465
United Kingdom (U.K.) 50
United States (U.S.) 13, 14, 17, 50, 82, 358, 361, 362, 383, 384, 387, 388
UNPICK 8, 188
   project 167
user generated content 15, 139, 158, 483
   and communication 139
   data integration 152
   volunteered geographic information 153

V

validity tests 259
values 3, 11, 48, 85, 98, 140, 171, 179, 186, 207, 217, 220, 225, 236, 239–241, 246, 248–250, 252–258, 260, 261, 269, 348, 384, 387, 399, 401, 404, 410, 442, 446, 447, 473, 475, 478, 485, 486

W

warm glow 247, 252–254
water quality 17, 356–359, 367–376, 476
Web 2.0 141, 146, 153, 158
   data mining 144, 145
   social media 141, 142
wellbeing 3, 48, 52, 129, 212, 426, 431, 433, 435, 480. *See also* human wellbeing
wildfire 17, 79, 367, 369, 371, 373, 384, 385, 387, 390, 400, 405
willingness to pay (WTP) 17, 236, 237, 240–247, 249–253, 255, 257, 259, 260, 480
wood-pastures 33, 34
   characteristic trees of 33
World Trade Organization (WTO) 10, 16, 199, 205, 206
WTP. *See* willingness to pay (WTP)

X

Xylella, *Xylella fastidiosa* 180, 215, 472

Y

Yorkshire 22, 28, 29

CPSIA information can be obtained
at www.ICGtesting.com
Printed in the USA
LVHW06*0910280518
578668LV00006B/13/P